# Biomedical Materials for Multi-functional Applications

**Series Editors**

Arnab Chanda, Centre for Biomedical Engineering, Indian Institute of Technology Delhi, New Delhi, Delhi, India

Sarabjeet Sidhu, Department of Mechanical Engineering, Sardar Beant Singh State University, Gurdaspur, Punjab, India

The series 'Biomedical Materials for Multi-functional Applications' is intended for the audience with an interest in biomaterials for various applications, including bio-implants, bio-instrumentation, biomechanics, and tissue engineering. The book titles under this series will bring different areas of biomedical engineering, bioengineering, materials science, mechanical engineering, and biosciences under one umbrella of biomedical materials. The books in this series will provide deep insights into the state-of-art of biomedical materials and deal with diverse and interdisciplinary aspects of biomaterials, ranging from basic science to host responses, as well as real-time applications. Some of the topics include but are not limited to: functional materials, smart materials, biomimetic materials, polymers, composites, ceramics, metals, alloys, and nanomaterials. This series will provide the most recent advances in biomedical materials, including design, synthesis, processing, and characterization. This series will accept both authored and edited volumes, including monographs, textbooks, reference works, and handbooks. This series aims to publish new advances in biomedical materials that have been made in the fabrication and development of biomaterials for different applications and the advancement of principles, theories, and designs. Each book title in the series will give a comprehensive overview of pioneering topics, and the readers will gain new knowledge and insights about the topic. As a collection, the series on 'Biomedical Materials for Multi-functional Applications' is anticipated to benefit a wide audience in academia, the research community, and industry.

Deepa Suhag · Swati Kaushik · Vinod Bala Taxak

# Handbook of Biomaterials for Medical Applications, Volume 1

## Fundamentals

Springer

Deepa Suhag
Amity Centre for Nanotechnology
Amity University
Gurugram, Haryana, India

Swati Kaushik
Arjuna Naturals Pvt. Ltd.
Aluva, Kerala, India

Vinod Bala Taxak
Emeritus Professor
Maharshi Dayanand University
Rohtak, Haryana, India

ISSN 2731-9695 ISSN 2731-9709 (electronic)
Biomedical Materials for Multi-functional Applications
ISBN 978-981-97-4820-4 ISBN 978-981-97-4818-1 (eBook)
https://doi.org/10.1007/978-981-97-4818-1



This Springer imprint is published by the registered company Springer Nature Singapore Pte Ltd.
The registered company address is: 152 Beach Road, #21-01/04 Gateway East, Singapore 189721, Singapore

# Preface

*Handbook on Biomaterials for Medical Applications: Fundamentals* is an essential monograph that encapsulates the integration of cutting-edge technologies within the sphere of biomedical materials science. This volume is crafted with precision to bridge the complex theoretical underpinnings with their pragmatic applications in modern medical practices, underscoring the versatile roles of these materials in enhancing therapeutic efficacy.

The evolving landscape of biomedical materials necessitates a synthesis of diverse disciplinary insights—from chemistry and biology to engineering and clinical medicine. This monograph serves as a comprehensive guide through the complex terrain of novel biomaterials, advanced drug delivery systems, innovative tissue engineering, and the burgeoning field of theranostics. It stitches together each theme to furnish a holistic view, revealing how these materials are pivotal in propelling forward medical treatments.

Aimed at scholars, researchers, and healthcare practitioners, this text is enriched with numerous illustrations, diagrams, and tables that not only augment the comprehension of intricate concepts but also facilitate a practical understanding of applications. The monograph's lucid and educational narrative ensures that both seasoned experts and novices can grasp and apply the insights shared within these pages.

As you navigate through this monograph, we aspire to spark a robust enthusiasm for innovation and foster a profound comprehension of how multifunctional biomedical materials are set to redefine the horizons of healthcare. This journey invites you to delve into the confluence of materials science and therapeutic innovation, setting the stage for future breakthroughs that promise to transform medical paradigms.

Embark on a detailed exploration of the advancements in biomedical materials for multifunctional therapeutics, an endeavor that promises to reshape the future of medical science and therapy.

Gurugram, India Dr. Deepa Suhag

# Acknowledgements

This book would not have been possible without the invaluable contributions of those who dedicated their expertise and time. I am particularly grateful to Ph.D. scholars Ms. Raksha Rathore and Ms. Ankita Yadav, whose tireless efforts have enriched the content of this monograph.

I must also extend my heartfelt thanks to Dr. Arnab Chanda, whose constant handholding through the complex process of synthesizing technical knowledge into accessible content has been indispensable. His guidance has been a beacon throughout this journey.

Additionally, I owe a special note of gratitude to Swati Meherishi for her unwavering support and expert advice. Her contributions not only enhanced the scholarly rigor of this work but also made the publication process remarkably smooth.

Thank you all for your dedication and support, which have been vital in bringing this project to fruition.

Dr. Deepa Suhag

# Contents

# About the Authors

**Dr. Deepa Suhag** is Assistant Professor in the Department of Nanotechnology, Amity University Haryana. She has published 17 research articles in major international journals and has filed six patents of which one has been granted by the Government of India. Patent titled "Method for preparation of highly fluorescent biocompatible Sulphur doped graphene quantum dots from affordable agro-industrial bio-waste cane molasses using hydrothermal synthesis for bioimaging application" was granted on 19.05.2022. Her research area spans from biomaterial engineering to wound healing and tissue regeneration. She currently has a combined research funding of 10 million from various funding Indian government agencies. Dr. Deepa Suhag's major collaborators are from CNCI Kolkata, AIIMS Delhi, IIT Delhi, TBRL Chandigarh, INST Mohali, RGCB Kerala, and Harvard University, USA.

**Dr. Swati Kaushik** Senior Scientist at Arjuna Naturals Pvt. Ltd., Cochin, specializes in biotechnology with expertise in nanotechnology, cancer biology, and drug discovery. She earned her Ph.D. in Biotechnology from Amity University, Noida, and has held prestigious fellowships, including the National Postdoctoral Fellowship from the Science and Engineering Research Board. Dr. Kaushik has an extensive publication record with 18 peer-reviewed articles and five granted patents. Her research focuses on innovative cancer treatments and drug delivery systems. She has received numerous awards, including the Young Scientist Award, and has presented her work at various international conferences.

**Prof. Vinod Bala Taxak** is Emeritus Scientist at M.D. University, Rohtak, and the first woman to hold this position. With more than 34 years of experience, she specializes in analytical chemistry, luminescent nanomaterials, and light-emitting metal complexes for OLEDs. Professor Taxak has published approximately 150 research papers and completed significant projects, including a Korean Government Project and a major UGC project. She has been a visiting scientist in South Korea. She has also participated in numerous conferences and served as a reviewer for high-impact journals. Her work includes diffraction patterns accepted by the International Centre of Diffraction Data (ICDD).

# Abbreviations

| | |
|---|---|
| 68Ga-PSMA | Prostate-specific membrane antigen |
| ADCs | Antibody-drug conjugates |
| AFM | Atomic Force Microscopy |
| AI | Artificial Intelligence |
| AMD | Age-related macular degeneration |
| BBB | Blood-brain barrier |
| BMPs | Bone morphogenetic proteins |
| CAD | Computer-aided design |
| CAM | Computer-aided manufacturing |
| CLSM | Confocal laser scanning microscopy |
| CNNs | Convolutional neural networks |
| CNS | Central nervous system |
| CNTs | Carbon Nanotubes |
| CSF | Cerebrospinal fluid |
| CT | Computed Tomography |
| DLP | Digital light processing |
| DLS | Dynamic light scattering |
| DMLS | Direct metal laser sintering |
| DTI | Diffusion tensor imaging |
| E-beam | Electron beam |
| EBL | Electron beam lithography |
| EBM | Electron beam melting |
| ECM | Extracellular Matrix |
| EDS | Energy-dispersive X-ray Spectroscopy |
| EGDMA | Ethylene glycol dimethacrylate |
| ELISA | Enzyme-linked immunosorbent assays |
| EMA | European Medicines Agency |
| EO | Ethylene oxide |
| EPR | Enhanced Permeability and Retention |
| ESCs | Embryonic stem cells |
| ESI | Electrospray ionization |

| | |
|---|---|
| FBR | Foreign body response |
| FDA | Food and Drug Administration |
| FDM | Fused deposition modeling |
| FIB | Focused ion beam |
| FLIM | Fluorescence lifetime imaging microscopy |
| fMRI | Functional magnetic resonance imaging |
| FTIR | Fourier transform infrared spectroscopy |
| GC | Gas Chromatography |
| GFP | Green fluorescent proteins |
| GLP-1Ras | Glucagon-like peptide-1 receptor agonists |
| GMP | Good Manufacturing Practice |
| GNPs | Gold Nanoparticles |
| GO | Graphene Oxide |
| HA | Hydroxyapatite |
| HPLC | High-performance liquid chromatography |
| IMRT | Intensity-modulated radiation therapy |
| IND | Investigational New Drug |
| IP | Intellectual property |
| iPSCs | Induced pluripotent stem cells |
| IVUS | Intravascular ultrasound |
| LCA | Life cycle Assessment |
| LNPs | Lipid nanoparticles |
| LOC | Lab-on-a-chip |
| MALDI | Matrix-assisted laser desorption/ionization |
| MBAA | N, N'-methylenebisacrylamide |
| MDR | Multi-drug resistance |
| Micro-CT | Microcomputed tomography |
| miRNA | MicroRNA |
| MRI | Magnetic Resonance Imaging |
| mRNA | Messenger RNA |
| MS | Mass spectrometry |
| NIR | Near Infrared Region |
| NLCs | Nanostructured lipid carriers |
| NMR | Nuclear Magnetic Resonance |
| NSAIDs | Non-steroidal anti-inflammatory drugs |
| OCT | Optical coherence tomography |
| P(TMC-CL) | Poly (ε-caprolactone-co-trimethylene carbonate) |
| PAAm | Polyacrylamide |
| PCI | Percutaneous coronary interventions |
| PDGF | Platelet-derived growth factor |
| PEG | Polyethylene Glycol |
| PEGDA | Polyethylene Glycol Diacrylate |
| PET | Positron emission tomography |
| PGA | Poly (glycolic acid) |
| PHA | Polyhydroxyalkanoates |

| | |
|---|---|
| PLA | Poly (lactic acid) |
| PLGA | Poly (lactic-co-glycolic acid) |
| PNIPAAm | Poly (N-isopropylacrylamide) |
| POC | Point-of-care |
| PTMC | Poly (trimethylene carbonate) |
| PTT | Photothermal Therapy |
| PU | Polyurethane |
| PVA | Polyvinyl alcohol |
| PVC | Polyvinyl Chloride |
| PVD | Physical Vapor Deposition |
| RES | Reticuloendothelial system |
| ROS | Reactive oxygen species |
| SELEX | Systemic evolution of ligands by exponential enrichment |
| SEM | Scanning Electron Microscopy |
| SHG | Second harmonic generation |
| siRNA | Small interfering RNA |
| SLNs | Solid lipid nanoparticles |
| SLS | Selective laser sintering |
| SPECT | Single-photon emission computed tomography |
| SPIONs | Superparamagnetic iron oxide nanoparticles |
| SPM | Scanning probe microscopy |
| SRS | Stereotactic radiosurgery |
| STED | Stimulated emission depletion |
| STORM | Stochastic optical reconstruction microscopy |
| TAVR | Transcatheter aortic valve replacement |
| TCP | Tri calcium Phosphate |
| TEM | Transmission Electron Microscopy |
| TGF-β | Transforming growth factor-beta |
| VEGF | Vascular endothelial growth factor |
| XPS | X-ray Photoelectron Spectroscopy |
| XRD | X-ray Diffraction |

# Chapter 1
# Introduction to Biomedical Materials

## 1.1 Introduction

In this chapter, we will embark on a journey to explore the fascinating field of biomaterials and their vital role in revolutionizing modern medicine. Whether you're a curious high school student or a budding scientist in college, this chapter will introduce you to the basics of biomedical materials in a clear and understandable manner. So, what exactly are biomaterials? Well, think of them as the unsung heroes of healthcare—materials specially designed to interact with the human body to diagnose, heal, or replace damaged tissues and organs. From artificial joints and dental implants to drug delivery systems [1] and tissue engineering scaffolds, biomaterials come in various forms and serve a wide range of purposes in improving patient health and quality of life.

But why are these materials so important? Imagine a world where medicines can be delivered precisely to where they are needed in the body, where injured tissues can be regenerated, and where medical devices seamlessly integrate with biological systems. That's the promise of multifunctional therapeutics—a cutting-edge approach that harnesses the power of biomaterials to tackle complex health challenges and enhance medical treatments. In this book, we'll delve into the fascinating realm of multifunctional therapeutics and explore how advances in biomedical materials are driving innovation in healthcare. From targeted drug delivery systems that release medications at the right time and place to bioengineered tissues that promote healing and regeneration, the possibilities are truly endless. We'll also discuss the scope of this book and how it aims to contribute to the exciting field of biomedical materials for multifunctional therapeutics. So, whether you're intrigued by the science behind biomaterials or passionate about making a difference in the world of healthcare, join us as we begin on this enlightening journey into the realm of biomedical materials. Get ready to expand your horizons, ignite your curiosity, and uncover the incredible potential of biomaterials [2] to shape the future of medicine.



D. Suhag et al., *Handbook of Biomaterials for Medical Applications, Volume 1*,
Biomedical Materials for Multi-functional Applications,
https://doi.org/10.1007/978-981-97-4818-1_1

In addition to exploring the fundamentals of biomaterials and their role in multifunctional therapeutics, we'll also deep dive into real-world applications and examples that highlight the impact of these materials on healthcare. From the development of innovative medical devices to the creation of personalized treatments tailored to individual patient needs, you'll gain insights into how biomaterials are transforming the way we approach diagnosis, treatment, and prevention of diseases. So, get ready to be inspired by the incredible potential of biomedical materials and their profound influence on the future of healthcare.

### 1.1.1 Overview of Biomaterials

Biomaterials are an integral part of modern healthcare, encompassing a diverse range of materials engineered to interact with biological systems for therapeutic or diagnostic purposes. At its core, biomaterials refer to substances that are intentionally designed and engineered to interface with biological systems to augment or replace natural functions. These materials are meticulously crafted to exhibit specific physical, chemical, and biological properties that enable them to interact harmoniously with the complex environment of the human body. One of the defining characteristics of biomaterials is their biocompatibility [1, 2], which refers to their ability to integrate seamlessly with living tissues without causing adverse reactions or immune responses. Biocompatible biomaterials are carefully selected or engineered to mimic the properties of natural tissues, ensuring compatibility and minimizing the risk of rejection or toxicity when implanted or used in medical devices. This property is crucial for ensuring the safety and efficacy of biomaterials in various biomedical applications, ranging from implants and prosthetics to drug delivery systems and diagnostic tools. In addition to biocompatibility, biomaterials are also designed to possess specific mechanical properties tailored to the intended application. For instance, orthopedic implants require materials with high strength and durability to withstand the mechanical stresses of everyday activities, while tissue engineering scaffolds may require materials with adjustable stiffness and porosity to support cell growth and tissue regeneration. By carefully tuning the mechanical properties of biomaterials, researchers can optimise their performance and functionality for diverse biomedical applications.

Furthermore, biomaterials often exhibit bioactivity, which refers to their ability to interact with biological molecules or cells to elicit specific biological responses. This property is particularly important for materials used in tissue engineering and regenerative medicine, where biomaterials are engineered to stimulate cell adhesion, proliferation, and differentiation to promote tissue regeneration and repair. By harnessing the bioactivity of biomaterials, researchers can create innovative therapies that accelerate healing and enhance patient outcomes in various medical conditions. Biomaterials also play a crucial role in enabling advanced medical technologies and treatments, such as targeted drug delivery systems and medical implants [3]. By encapsulating drugs within biocompatible carriers, researchers can precisely control the release kinetics and distribution of therapeutic agents, maximizing their efficacy while minimizing side effects. Similarly, medical implants and devices made from biomaterials offer innovative solutions for restoring lost or impaired functions, improving the quality of life for patients suffering from various medical conditions.

Biomaterials represent a diverse and dynamic field at the intersection of materials science, biology, and medicine. Defined by their biocompatibility, mechanical properties, and bioactivity, these materials hold immense promise for advancing healthcare through innovative therapies, diagnostics, and medical devices. As researchers continue to explore new materials and techniques, the potential of biomaterials to revolutionize medicine and improve patient outcomes remains boundless.

**Exercise 1.**

**Fill Ups**

**(Human Body, Material Science, Healing, Bedside)**

1. Biomaterials helps in exploring the interface of medicine and ........................
2. The world of biomaterials is all about innovation from bench to ........................
3. Biomaterials are the building blocks of medical innovation and ........................
4. Biomaterials helps in unravelling the secrets of materials designed for the ..................

## 1.1.2 Historical Background

The fascinating story of biomaterials dates back centuries, with early civilizations using natural materials such as wood, bone, and metals for various medical purposes. Ancient Egyptians, for example, employed animal-derived substances like honey and resin for wound healing, laying the foundation for the use of biomaterials in medicine. Similarly, the ancient Greeks and Romans utilized materials such as gold and silver for dental prosthetics, demonstrating early attempts to restore lost or damaged biological structures. The Renaissance period witnessed significant advancements in anatomical studies and surgical techniques [4], leading to increased interest in developing materials for medical applications. Inventors and scientists experimented with materials like ivory, porcelain, and glass in prosthetics and surgical implants, paving the way for the modern field of biomaterials. However, it wasn't until the twentieth century that biomaterials truly began to flourish, driven by rapid developments in materials science, bioengineering, and medical technology. The pioneering work of researchers such as Alexis Carrel and Charles Lindbergh in the early 1900s laid the groundwork for tissue culture techniques and organ transplantation, sparking interest in developing biocompatible materials for medical implants. During World War II, the demand for materials to treat war-related injuries spurred innovation in prosthetics, leading to the development of lightweight and durable materials like plastics and silicone for artificial limbs and organs.

The latter half of the twentieth century witnessed unprecedented growth in biomaterials research, fueled by advancements in polymer science, metallurgy, and biomedical engineering. Breakthroughs such as the discovery of biodegradable polymers and the development of bioactive ceramics revolutionized the field, opening new possibilities for tissue engineering and regenerative medicine. In the 1970s and 1980s, the introduction of biocompatible metals like titanium and stainless steel revolutionized orthopedic and dental implants, significantly improving patient outcomes and longevity.

The turn of the twenty-first century marked a new era of biomaterials innovation, characterized by the convergence of nanotechnology, biotechnology, and materials science. Researchers began exploring novel biomaterials with tailored properties at the nanoscale, enabling precise control over cellular interactions and therapeutic outcomes. Additionally, the emergence of additive manufacturing technologies, such as 3D printing, revolutionized the fabrication of complex biomaterials and medical devices, paving the way for personalized medicine and patient-specific treatments. Today, biomaterials continue to play a central role in modern healthcare, driving advancements in regenerative medicine, drug delivery [3], and medical device technologies. As researchers continue to push the boundaries of biomaterials science, the future holds immense promise for innovative therapies and interventions that will improve patient outcomes and quality of life around the world.

### *1.1.3 Types of Biomaterials*

Biomaterials are materials engineered to interact with biological systems for medical purposes, such as tissue replacement, drug delivery [3], and diagnostic devices (Fig. 1.1). They are categorized based on their origin, composition, and properties (Fig. 1.2). Natural biomaterials, derived from living organisms, include collagen, silk, and chitosan, offering biocompatibility and often promoting tissue regeneration. Synthetic biomaterials, such as polymers like polyethylene glycol (PEG) and poly (lactic-co-glycolic acid) (PLGA), provide versatility in design and adjustable properties [4]. Hybrid biomaterials combine both natural and synthetic components, leveraging the advantages of each to achieve specific biomedical applications. Each type of biomaterial offers unique advantages and challenges, driving ongoing research and innovation in the field of biomaterials science and engineering.

#### 1.1.3.1 Natural Biomaterials

Natural biomaterials, sourced from living organisms, represent a diverse array of materials that have found extensive use in biomedical applications. Collagen, the most abundant protein in mammals, stands out as a versatile natural biomaterial

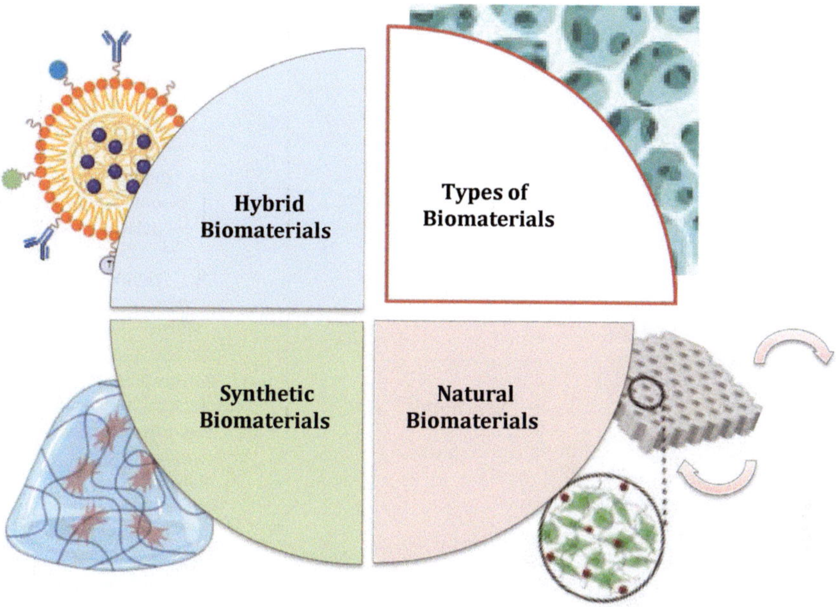


**Fig. 1.1** Illustrating different types of biomaterials

due to its excellent biocompatibility and ability to promote cell adhesion and tissue regeneration. Derived from animal connective tissues, collagen-based biomaterials are widely utilized in wound healing, tissue engineering, and drug delivery systems. Similarly, silk, produced by silkworms and spiders, possesses remarkable mechanical properties, biocompatibility, and biodegradability, making it an attractive natural biomaterial for various biomedical applications [5]. Silk-based biomaterials have been explored for wound dressings, scaffolds for tissue engineering, and controlled drug delivery systems due to their tunable properties and ability to support cell growth and tissue regeneration. Chitosan, derived from chitin found in the exoskeletons of crustaceans, offers biocompatibility, biodegradability, antimicrobial properties, and the ability to promote wound healing and tissue regeneration. Chitosan-based biomaterials [6] have been investigated for applications ranging from wound dressings and surgical sutures to drug delivery systems and tissue engineering scaffolds, owing to their versatility and ability to interact with biological systems.

Other natural biomaterials include alginate, a polysaccharide extracted from seaweed, which forms hydrogels [7] suitable for drug delivery and tissue engineering applications due to its biocompatibility and ability to encapsulate cells and bioactive molecules. Hyaluronic acid, a glycosaminoglycan found in connective tissues, is another natural biomaterial used in biomedical applications for its biocompatibility, viscoelasticity, and ability to promote wound healing and tissue regeneration. Moreover, decellularized extracellular matrices (ECMs) derived from tissues such as

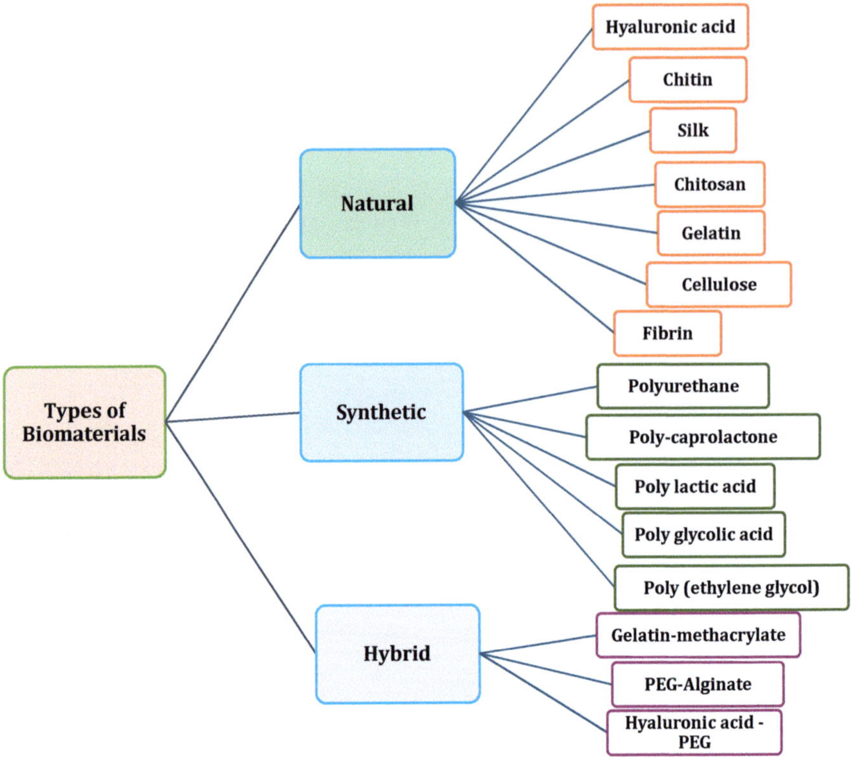


**Fig. 1.2** Classification of biomaterials

heart, liver, and bone offer a complex network of proteins, growth factors, and other bioactive molecules that mimic the native tissue microenvironment. ECM-based biomaterials have been extensively studied for tissue engineering and regenerative medicine applications, providing cues for cell adhesion, migration, proliferation, and differentiation to facilitate tissue repair and regeneration. Overall, natural biomaterials offer a rich source of materials for biomedical applications, providing biocompatibility, bioactivity, and tunable properties that make them invaluable for tissue engineering, drug delivery [3], wound healing, and regenerative medicine. Ongoing research continues to explore new sources, fabrication methods, and applications to harness the full potential of natural biomaterials in addressing biomedical challenges.

**Exercise 2.**

**Fill Ups**

**(Cellulose, Biocompatibility, Tissue Regeneration, Organic Sources, Drug Delivery)**

1. Natural biomaterials are derived from ........................ such as plants, animals, and microorganisms.
2. Examples of natural biomaterials include ............., collagen, silk, and chitosan.
3. Natural biomaterials offer unique advantages such as inherent ...................... and bioactivity.
4. The use of natural biomaterials in medical applications has gained traction due to their ability to promote natural healing and ........................................
5. Research in natural biomaterials continues to expand, exploring novel extraction methods and innovative applications in regenerative medicine and ............................

#### 1.1.3.2 Synthetic Biomaterials

Synthetic biomaterials are engineered materials designed to interact with biological systems for various medical applications. These biomaterials [8] offer precise control over properties such as mechanical strength, degradation rate, and surface characteristics, making them highly versatile and adaptable for specific biomedical needs. Polyethylene glycol (PEG), a synthetic polymer, is widely used in biomedical applications due to its biocompatibility, hydrophilicity, and resistance to protein adsorption. PEG-based biomaterials are utilized in drug delivery systems, tissue engineering scaffolds, and medical coatings to enhance biocompatibility and control drug release kinetics.

Poly (lactic-co-glycolic acid) (PLGA), a biodegradable copolymer synthesized from lactic acid and glycolic acid, is another prominent synthetic biomaterial extensively used in biomedical research and clinical practice. PLGA offers tunable degradation kinetics, mechanical properties, and biocompatibility, making it suitable for drug delivery systems [3, 4], tissue engineering scaffolds, and medical implants. Its ability to degrade into biocompatible byproducts in vivo without causing toxicity makes it particularly attractive for controlled drug delivery applications. Polycaprolactone (PCL), a biodegradable polyester, exhibits slow degradation kinetics and excellent mechanical properties, making it suitable for long-term implantable devices and tissue engineering scaffolds. PCL-based biomaterials have been investigated for applications such as bone regeneration, nerve guidance conduits, and controlled drug delivery systems due to their biocompatibility and ease of processing. Polyurethane (PU) is a versatile synthetic polymer with a wide range of biomedical applications, including medical implants, vascular grafts, wound dressings, and drug delivery systems. PU-based biomaterials offer customizable mechanical properties, biocompatibility, and durability, making them suitable for various medical devices and implants. Furthermore, synthetic hydrogels, such as polyacrylamide (PAAm) and polyethylene glycol diacrylate (PEGDA), offer tunable mechanical properties, swelling behavior, and degradation kinetics, making them valuable for tissue engineering, drug delivery, and regenerative medicine applications [8]. These hydrogels can mimic the native extracellular matrix (ECM) environment and provide a 3D scaffold for cell growth, proliferation, and differentiation.

Synthetic biomaterials play a crucial role in modern biomedical research and clinical practice, offering precise control over material properties and functionalities to address diverse medical challenges. Ongoing advancements in material design, fabrication techniques, and biomedical applications continue to expand the capabilities and potential of synthetic biomaterials in improving healthcare outcomes and patient quality of life.

#### 1.1.3.3 Hybrid Biomaterials

Hybrid biomaterials combine the advantageous properties of both natural and synthetic materials to create versatile platforms for biomedical applications. By integrating components from different sources, hybrid biomaterials can overcome the limitations of individual materials while synergistically enhancing their functionalities. One common approach involves incorporating natural polymers, such as collagen or hyaluronic acid, into synthetic matrices like polyethylene glycol (PEG) or poly (lactic-co-glycolic acid) (PLGA) to improve biocompatibility, bioactivity, and mechanical strength. These hybrid biomaterials offer a unique combination of biocompatibility, mechanical integrity, and tunable properties, making them suitable for a wide range of biomedical applications. For instance, hybrid hydrogels composed of natural and synthetic polymers have been investigated for tissue engineering scaffolds, wound dressings, and drug delivery systems [3], offering enhanced mechanical properties and biological functionality compared to pure natural or synthetic materials alone. Furthermore, hybrid biomaterials can be tailored to mimic the complexity and biochemical cues of the native extracellular matrix (ECM), providing an optimal microenvironment for cell adhesion, proliferation, and differentiation. By incorporating bioactive molecules such as growth factors, peptides, or nanoparticles into hybrid matrices, researchers can create multifunctional biomaterials capable of promoting tissue regeneration, modulating cellular behavior, and delivering therapeutic agents with spatiotemporal control. Moreover, hybrid biomaterials enable the integration of multiple functionalities within a single platform, offering versatility and adaptability for specific biomedical applications. For example, hybrid nanoparticles composed of natural polymers coated with synthetic shells can combine the targeting specificity [7, 8] of natural ligands with the controlled release capabilities of synthetic materials, making them promising candidates for targeted drug delivery and imaging applications.

In addition, hybrid biomaterials hold great promise for personalized medicine and regenerative therapies, as they can be customized to match the unique requirements of individual patients and tissue types. By incorporating patient-derived cells or tissue-specific cues into hybrid scaffolds, researchers aim to develop biomimetic constructs that closely recapitulate the native tissue architecture and function, facilitating better tissue integration and regeneration in vivo. Overall, hybrid biomaterials represent a cutting-edge approach in biomaterials science and engineering, offering tailored solutions for complex biomedical challenges. Through strategic combinations of natural and synthetic components, these materials pave the way for innovative therapies, diagnostic tools, and biomedical devices with enhanced performance and functionality, ultimately improving patient outcomes and quality of life.

**Exercise 3.**

**Choose The Correct Option**

1. Which of the following is NOT a characteristic of synthetic biomaterials?
   a) Biodegradability
   b) Customizability
   c) Mechanical strength
   d) Derived from natural sources
2. What type of biomaterial combines properties of both natural and synthetic materials?
   a) Natural biomaterials
   b) Synthetic biomaterials
   c) Hybrid biomaterials
   d) Polymer biomaterials
3. Which of the following is a commonly used synthetic polymer in biomaterials engineering?
   a) Collagen
   b) Chitosan
   c) Polyethylene glycol (PEG)
   d) Silk fibroin
4. What is the primary advantage of using hybrid biomaterials over pure synthetic or natural biomaterials?
   a) Lower cost
   b) Increased biocompatibility
   c) Enhanced mechanical properties
   d) Faster degradation
5. Which technique is commonly used for the synthesis of hydrogels in biomaterials engineering?
   a) Electrospinning
   b) Freeze-drying
   c) Photolithography
   d) Crosslinking

6. What is the main challenge associated with the use of synthetic biomaterials in tissue engineering?
   a) Lack of mechanical strength
   b) High cost of production
   c) Biocompatibility issues
   d) Limited customizability
7. Which of the following is an example of a synthetic ceramic biomaterial?
   a) Hydroxyapatite
   b) Collagen
   c) Chitosan
   d) Alginate
8. What property of synthetic biomaterials allows for controlled drug delivery?
   a) Porosity
   b) Surface charge
   c) Biodegradability
   d) Mechanical strength
9. Which of the following is NOT a potential application of hybrid biomaterials?
   a) Bone regeneration
   b) Drug delivery
   c) Tissue engineering
   d) Optical fiber manufacturing
10. What is a major advantage of using synthetic biomaterials over natural biomaterials?
    a) Lower cost
    b) Higher biocompatibility
    c) Environmental friendliness
    d) Better control over properties

### *1.1.4 Applications of Biomaterials*

Biomaterials find a wide range of applications in modern healthcare, contributing to advancements in medical devices, drug delivery systems, tissue engineering, and regenerative medicine (Fig. 1.3). One of the most prominent applications of biomaterials is in medical implants, where materials such as titanium, stainless steel, and polymers are used to fabricate orthopedic implants, dental prosthetics, and cardiovascular devices [9]. These implants are designed to replace or augment damaged or diseased tissues, restoring function and improving quality of life for patients suffering from conditions such as osteoarthritis, dental caries, and heart disease. Another critical application of biomaterials is in drug delivery systems, where materials such as liposomes, hydrogels, and nanoparticles are used to encapsulate and deliver therapeutic agents to specific targets in the body. By controlling the release kinetics and distribution of drugs, biomaterial-based drug delivery systems enhance the efficacy of medications while minimizing side effects and improving patient compliance. This technology has revolutionized the treatment of various diseases, including cancer, diabetes, and cardiovascular disorders, by enabling targeted and personalized therapies.

Biomaterials also play a pivotal role in tissue engineering and regenerative medicine, where they are used to design scaffolds, matrices, and substrates that support cell growth and tissue regeneration. Materials such as biodegradable polymers [8, 9], natural hydrogels, and bioactive ceramics provide the necessary structural support and biochemical cues to promote cell adhesion, proliferation, and differentiation, leading to the formation of functional tissues and organs. Tissue-engineered

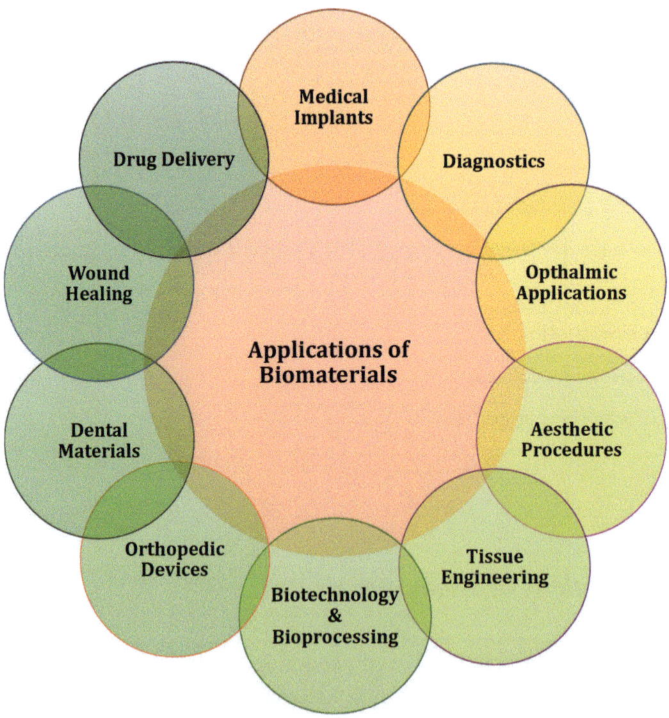


**Fig. 1.3** Various applications of biomaterials in modern healthcare

constructs hold promise for applications ranging from skin grafts and bone substitutes to organ transplantation and wound healing.

In addition to medical applications, biomaterials are also utilized in diagnostic and imaging technologies, where materials such as contrast agents, nanoparticles, and biosensors enable non-invasive detection and monitoring of diseases. These biomaterial-based diagnostics offer advantages such as high sensitivity, specificity, and real-time imaging capabilities, facilitating early disease detection and personalized treatment strategies. Biomaterials also play a crucial role in regenerative dentistry, where they are used in restorative materials, dental implants, and tissue-engineered constructs to repair and replace damaged oral tissues. Biomaterials are increasingly being employed in the field of neural engineering, where they interface with the nervous system to develop implantable electrodes, neural prosthetics, and drug delivery devices for treating neurological disorders. Materials such as conductive polymers, carbon nanotubes, and graphene-based materials exhibit excellent biocompatibility and electrical conductivity, making them ideal candidates for neural interfaces and neurostimulation therapies. These biomaterial-based technologies hold promise for restoring sensory and motor functions in patients with conditions such as spinal cord injury, Parkinson's disease, and epilepsy. Moreover, biomaterials are essential components of regenerative therapies such as stem cell therapy, gene therapy, and tissue engineering, where they provide the scaffold and microenvironment necessary for cell survival, proliferation, and differentiation. By harnessing the regenerative potential of biomaterials, researchers aim to develop innovative treatments for conditions ranging from neurodegenerative diseases and heart failure to spinal cord injuries and limb amputations [10]. These therapies offer hope for restoring lost or impaired functions and improving the quality of life for millions of patients worldwide.

Biomaterials are versatile materials that play a pivotal role in various applications across the field of healthcare. From medical implants and drug delivery systems to tissue engineering and regenerative medicine, biomaterials continue to drive innovation and advancements in medical technologies and therapies. As researchers continue to explore new materials and techniques, the potential of biomaterials to address unmet medical needs and improve patient outcomes remains limitless.

### *1.1.5 Importance of Multifunctional Therapeutics*

The idea of multifunctional therapeutics is super important in healthcare today. Think of it like this: when someone has a complicated illness, it's not always easy to treat with just one medicine or method. That's where multifunctional therapeutics come in. They're like superheroes of medicine because they can do lots of different things at once to help fight diseases. Imagine you have a cold, but also a headache and a fever. Instead of taking three different medicines for each symptom, wouldn't it be awesome if you could take just one that targets all three? That's the kind of magic multifunctional therapeutics can do. They combine different treatments into one, making it easier for patients and often more effective too. These kinds of treatments aren't just for colds though. They're being developed for all sorts of diseases, like cancer and heart problems [11]. And they're not just about making treatment simpler—they can also make it safer and more powerful. By putting different treatments together, doctors can often use lower doses of each, which means fewer side effects for patients. Plus, they can target the disease more precisely, making the treatment more effective.

So, why is this so important? Well, imagine if someone you know has a serious illness. You'd want them to have the best treatment possible, right? Multifunctional therapeutics have the potential to provide just that—better, more personalized treatment options that can make a real difference in people's lives. That's why scientists and doctors are so excited about the possibilities of multifunctional therapeutics and why they're working hard to develop them further.

#### 1.1.5.1 Definition of Multifunctional Therapeutics

Multifunctional therapeutics represent a cutting-edge approach in healthcare, encompassing treatments that integrate multiple therapeutic functions within a single platform. Unlike traditional therapies that typically target a single aspect of a disease, multifunctional therapeutics aim to address multiple disease pathways or symptoms simultaneously. This innovative approach holds great promise for improving patient outcomes by offering more comprehensive and effective treatments for complex medical conditions. At the heart of multifunctional therapeutics is the concept of synergy, where different therapeutic agents work together to enhance the overall treatment efficacy [12]. By combining complementary functionalities, such as targeting specific molecular pathways, delivering therapeutic agents to precise locations in the body, and modulating the immune response, multifunctional therapeutics can achieve synergistic effects that surpass the capabilities of individual treatments. This synergistic approach not only maximizes therapeutic efficacy but also minimizes potential side effects, leading to safer and more tolerable treatment options for patients.

Multifunctional therapeutics have the potential to revolutionize the treatment landscape across various medical domains, including oncology, cardiology, infectious diseases, and neurology. By harnessing the power of multifunctionality, researchers aim to develop innovative treatment strategies that can address the complexities of diseases more effectively while minimizing the burden on patients. As advancements in biomedical engineering, materials science, and drug delivery technologies continue to accelerate, the future of multifunctional therapeutics holds immense promise for improving patient care and advancing personalized medicine.

#### 1.1.5.2 Significance in Modern Healthcare

The significance of multifunctional therapeutics in modern healthcare cannot be overstated. In today's rapidly evolving medical landscape, the burden of complex diseases and conditions continues to grow, necessitating innovative approaches to diagnosis and treatment. Multifunctional therapeutics offer a promising solution to this challenge by providing a holistic approach to managing disease, targeting multiple aspects of pathology simultaneously. This comprehensive approach is particularly beneficial for treating multifactorial diseases, such as cancer, cardiovascular disease, and autoimmune disorders [13], where traditional single-target therapies may fall short in addressing the diverse underlying mechanisms. One of the key advantages of multifunctional therapeutics is their ability to improve treatment efficacy while minimizing side effects. By combining multiple therapeutic agents or modalities into a single platform, multifunctional therapeutics can achieve synergistic effects that enhance treatment outcomes. This means that patients may require lower doses of each therapeutic component, reducing the risk of adverse reactions and improving treatment tolerability. Additionally, the targeted nature of multifunctional therapeutics allows for more precise delivery of therapeutic agents to diseased tissues while sparing healthy tissues, further minimizing side effects, and improving patient quality of life. Moreover, multifunctional therapeutics hold promise for addressing the challenges of drug resistance and treatment failure that often plague conventional therapies. By targeting multiple pathways or mechanisms of disease, multifunctional therapeutics can overcome resistance mechanisms that arise from single-target therapies, providing more durable and effective treatment options for patients. This is particularly relevant in the context of cancer treatment, where tumor heterogeneity and evolution can lead to resistance to individual drugs. Multifunctional therapeutics offer a multifaceted approach to circumventing resistance mechanisms and improving patient outcomes in cancer and other challenging diseases. Multifunctional therapeutics have the potential to streamline treatment regimens and improve patient adherence. Rather than requiring patients to take multiple medications or undergo various treatment modalities separately, multifunctional therapeutics consolidate treatment into a single regimen, simplifying the treatment process [12, 13] and reducing the treatment burden. This can lead to better patient compliance and adherence, ultimately resulting in improved treatment outcomes and reduced healthcare costs. Additionally, the personalized nature of multifunctional therapeutics allows for tailored treatment approaches that address the unique needs and preferences of individual patients, further enhancing treatment efficacy and patient satisfaction. In addition to their therapeutic benefits, multifunctional therapeutics also hold promise for advancing diagnostic and imaging technologies in modern healthcare. By incorporating diagnostic agents or imaging probes into multifunctional platforms, researchers can develop innovative tools for disease detection, monitoring, and imaging. These multifunctional diagnostic platforms offer advantages such as improved sensitivity, specificity, and spatial resolution, enabling early detection of diseases and more accurate assessment of treatment responses. This integration of diagnostic and therapeutic functionalities represents a paradigm shift in healthcare, offering opportunities for earlier diagnosis, personalized treatment planning, and improved patient outcomes across a wide range of medical conditions.

Multifunctional therapeutics represent a transformative approach to healthcare that holds immense promise for improving patient outcomes, reducing treatment burden, and advancing personalized medicine. By targeting multiple aspects of disease pathology [14], overcoming drug resistance, and enhancing treatment efficacy, multifunctional therapeutics offer new possibilities for addressing the complexities of modern healthcare challenges. As research in this field continues to evolve and technologies continue to advance, the future of multifunctional therapeutics looks increasingly bright, with the potential to revolutionize the way we diagnose, treat, and manage diseases in the years to come.

## 1.1.6 Examples of Multifunctional Biomedical Materials

Let's dive into the fascinating world of multifunctional biomedical materials. These innovative materials are like Swiss army knives of healthcare, offering a myriad of capabilities to address diverse medical needs. From targeted drug delivery systems that release medications at just the right time and place to tissue engineering scaffolds [15] that promote healing and regeneration, multifunctional biomedical materials are at the forefront of revolutionizing modern medicine. With their ability to integrate multiple therapeutic functionalities into a single platform, these materials offer exciting possibilities for improving patient outcomes and advancing personalized treatments. So, let's explore some captivating examples of multifunctional biomedical materials and discover how they're shaping the future of healthcare.

### 1.1.6.1 Nanoparticle-Based Drug Delivery Systems

Nanoparticle-based drug delivery systems represent a groundbreaking approach to targeted drug delivery, offering precise control over the release kinetics and distribution of therapeutic agents. These tiny particles, typically ranging in size from 1 to 100 nm, can encapsulate drugs within their structure and deliver them to specific tissues or cells in the body. By functionalizing the surface of nanoparticles with targeting ligands or antibodies, researchers can ensure selective delivery to diseased tissues while minimizing off-target effects. One example of nanoparticle-based drug delivery systems is liposomes, which are spherical vesicles composed of lipid bilayers. Liposomes can encapsulate hydrophobic and hydrophilic drugs within their core and lipid bilayers, allowing for the simultaneous delivery of multiple therapeutic agents. Additionally, liposomes can be modified with targeting ligands or stimuli-responsive moieties to achieve site-specific drug release in response to biological cues or external stimuli.

Another example is polymeric nanoparticles, which are composed of biocompatible polymers such as poly (lactic-co-glycolic acid) (PLGA) or polyethylene glycol (PEG). These nanoparticles can encapsulate drugs through physical entrapment or chemical conjugation and offer advantages such as sustained release, improved stability [15, 16], and reduced systemic toxicity. Furthermore, polymeric nanoparticles can be engineered to respond to specific stimuli, such as pH, temperature, or enzymatic activity, enabling triggered drug release at the desired site of action. In addition to their drug delivery capabilities, nanoparticle-based systems can incorporate diagnostic functionalities, such as imaging agents or biosensors, allowing for real-time monitoring of drug distribution and therapeutic efficacy. Overall, nanoparticle-based drug delivery systems hold immense promise for improving the treatment of various diseases, including cancer, infectious diseases, and inflammatory disorders, by enhancing drug bioavailability, reducing side effects, and overcoming drug resistance.

#### 1.1.6.2 Bioactive Scaffolds for Tissue Engineering

Bioactive scaffolds are another exciting example of multifunctional biomedical materials that play a crucial role in tissue engineering and regenerative medicine. These scaffolds provide a 3D framework that supports cell adhesion, proliferation, and differentiation, facilitating tissue regeneration and repair. By mimicking the extracellular matrix (ECM) of native tissues, bioactive scaffolds create an optimal microenvironment for cellular activities and tissue growth. One type of bioactive scaffold is hydrogels, which are 3D networks of hydrophilic polymers that can absorb and retain large amounts of water. Hydrogels offer advantages such as biocompatibility, tunable mechanical properties, and high water content, making them suitable for a wide range of tissue engineering applications. Researchers can incorporate bioactive molecules, growth factors [17], or cell-laden microspheres into hydrogel scaffolds to enhance cellular responses and promote tissue regeneration. Another example is decellularized extracellular matrix (dECM)-based scaffolds, which are derived from natural tissues through a process that removes cellular components while preserving the ECM architecture and composition. dECM scaffolds retain native biochemical cues and mechanical properties, providing an ideal microenvironment for cell attachment, migration, and differentiation. These scaffolds can be further functionalized with bioactive molecules or growth factors to enhance tissue-specific regeneration and repair. Moreover, bioactive scaffolds can incorporate controlled release systems for delivering growth factors, cytokines, or small molecules that promote tissue regeneration and angiogenesis. By incorporating drug delivery functionalities into the scaffold matrix, researchers can achieve spatial and temporal control over the release of bioactive molecules, ensuring optimal tissue regeneration and integration with the host tissue. Overall, bioactive scaffolds offer versatile platforms for engineering complex tissues and organs, with applications in wound healing, bone regeneration, cartilage repair, and organ transplantation.

#### 1.1.6.3 Stimuli-Responsive Hydrogels for Controlled Drug Delivery

Stimuli-responsive hydrogels are a versatile class of biomaterials that undergo reversible changes in their physical or chemical properties in response to specific stimuli, such as temperature, pH, light, or enzymes. These hydrogels can be engineered to release therapeutic agents in a controlled manner in response to external or internal triggers, offering precise spatiotemporal control over drug delivery. One example is thermo-responsive hydrogels, which undergo phase transition from a liquid to a gel state in response to changes in temperature. By incorporating thermosensitive polymers, such as poly(N-isopropylacrylamide) (PNIPAAm), researchers can design hydrogels that release drugs upon heating, enabling localized and on-demand drug delivery for applications such as hyperthermia therapy [17, 18] or wound healing. Another example is pH-responsive hydrogels, which exhibit changes in swelling behavior or degradation kinetics in response to variations in pH levels. These hydrogels can be designed to release drugs in acidic environments, such as those found in tumors or inflamed tissues, offering targeted delivery and enhanced therapeutic efficacy. By incorporating pH-sensitive moieties, such as acidic or basic functional groups, into the polymer matrix, researchers can tailor the responsiveness of hydrogels to specific pH ranges, enabling precise control over drug release kinetics and distribution within the body.

In addition to their drug delivery capabilities, stimuli-responsive hydrogels can incorporate other functionalities, such as cell encapsulation or tissue engineering. By modulating the cross-linking density or composition of the hydrogel matrix, researchers can create microenvironments that support cell viability, proliferation, and differentiation, facilitating tissue regeneration and repair. Furthermore, stimuli-responsive hydrogels can be combined with imaging agents or biosensors for real-time monitoring of drug release kinetics [12–14] and therapeutic responses, providing valuable insights into treatment efficacy and patient outcomes.

Overall, stimuli-responsive hydrogels offer a promising platform for developing multifunctional biomedical materials that integrate controlled drug delivery, tissue engineering, and diagnostic functionalities. With their ability to respond to specific stimuli and adapt to dynamic physiological conditions, these hydrogels hold great potential for personalized medicine applications, where tailored treatment strategies are essential for optimizing therapeutic outcomes and minimizing side effects.

#### 1.1.6.4 Magnetic Nanoparticles for Theranostic Applications

Imagine tiny particles that are so smart they can both fight cancer and show doctors what's going on inside the body. That's what magnetic nanoparticles are all about. These little guys are made of special materials like iron oxide and cobalt, and they have amazing magnetic powers. Doctors can use magnets to control where these nanoparticles go in the body. One of the important applications of these nanoparticles is that they can help doctors see inside the body using a machine called an MRI. When the nanoparticles are inside, they make certain areas show up clearly on the

MRI scan. This helps doctors find tumors or other problem areas and see how they're doing during treatment. But that's not all—these nanoparticles can also help treat cancer directly [18]. Doctors can use them to deliver heat to tumors using magnets. This heat can make the cancer cells more sensitive to other treatments like radiation or chemotherapy, helping them work better and shrinking the tumor faster. What's even fascinating is that these nanoparticles can carry drugs right to the tumor, like a tiny delivery service. Doctors can attach special molecules to the nanoparticles that stick to cancer cells, so the drugs go straight to where they're needed. This makes the treatment more effective and reduces side effects for patients. Plus, these nanoparticles can also act like spies inside the body, reporting back on how well the treatment is working. Doctors can attach special sensors to them that light up when they find cancer cells. This helps doctors track the progress of the treatment and make any changes needed to keep fighting the cancer. In short, these magnetic nanoparticles are like tiny superheroes in the fight against cancer. They help doctors see inside the body, deliver treatments right to the tumor, and even report back on how things are going. With their help, doctors can give patients better treatments and more hope for beating cancer.

#### 1.1.6.5 Nanoparticles that Act like Cells to Fight Cancer

Some super impressive tiny particles called biomimetic nanoparticles are being used to fight cancer in a smart way. Imagine these nanoparticles as tiny warriors that mimic the behavior of cells to sneak into the enemy territory, which in this case are cancer cells. They're like undercover agents with a mission to deliver powerful drugs right where they're needed the most—the tumor. These nanoparticles are special because they're coated with a disguise made from the outer layer of cells, like those from our immune system [19]. This disguise helps them avoid detection by the body's defences and reach the tumor safely. Once they're there, they use their special targeting abilities to latch onto the cancer cells and deliver their payload of drugs. This targeted attack means less damage to healthy cells and fewer side effects for patients. But these nanoparticles aren't just good at delivering drugs—they're also really clever. They can sense the environment around the tumor and release their drugs only when they're needed the most. It's like they have a built-in alarm that tells them when to attack. This helps make the treatment more effective and reduces the chance of the cancer coming back. Plus, these nanoparticles can also help doctors see how well the treatment is working. They can be loaded with special dyes or imaging agents that light up when they find cancer cells. This means doctors can track the progress of the treatment in real time and adjust if needed. It's like having a spy that reports back on what's happening inside the body. In short, these biomimetic nanoparticles are like tiny superheroes fighting cancer from the inside. They're smart, stealthy, and super effective at targeting tumors while sparing healthy tissues. With their help, doctors can deliver powerful treatments with fewer side effects, giving patients a better chance for beating cancer.

### 1.1.7 Scope of the Book

In this book, we will discover the incredible capabilities and applications of these innovative materials in healthcare. From drug delivery systems to tissue engineering scaffolds, multifunctional biomedical materials are revolutionizing the way we diagnose, treat, and manage diseases.

- **Understanding the Basics**: We'll start by laying the groundwork with a comprehensive overview of the fundamental principles behind multifunctional biomedical materials. From their composition and properties to their fabrication [20] and characterization, we'll explore the key concepts that form the basis of these groundbreaking materials.
- **Exploring Drug Delivery Systems**: One of the most exciting applications of multifunctional biomedical materials is in drug delivery systems. We'll delve into the various types of drug delivery platforms, from nanoparticles to hydrogels, and examine how they can be engineered to precisely target and deliver therapeutic agents to specific sites in the body.
- **Unveiling Tissue Engineering**: Tissue engineering holds immense promise for regenerating damaged tissues and organs, and multifunctional biomedical materials play a crucial role in this field. We'll explore the design and development of bioactive scaffolds, cell-laden hydrogels, and other tissue engineering constructs that mimic the native microenvironment to promote tissue regeneration and repair.
- **Harnessing Theranostic Capabilities**: Theranostics, the integration of therapy and diagnostics into a single platform, is a rapidly growing area of research in healthcare. We'll investigate how multifunctional biomedical materials [21] can be used for both treating diseases and monitoring treatment responses, offering personalized and precise medical interventions.
- **Advancing Imaging Technologies**: Imaging plays a pivotal role in disease diagnosis, treatment planning, and monitoring, and multifunctional biomedical materials are driving advancements in this field [22]. We'll examine how these materials are used as contrast agents, imaging probes, and biosensors to enhance the sensitivity, specificity, and resolution of diagnostic imaging modalities.
- **Addressing Clinical Challenges**: Throughout the book, we'll highlight real-world clinical applications and case studies where multifunctional biomedical materials have made a significant impact on patient care [23]. From targeted cancer therapies to regenerative medicine approaches, we'll explore how these materials are transforming the landscape of modern healthcare.
- **Embracing Future Innovations**: As the field of multifunctional biomedical materials continues to evolve, so too does the scope of possibilities. We'll speculate on future trends and emerging technologies, from personalized medicine and nanomedicine to bioinspired materials and smart implants [24], that hold the potential to revolutionize healthcare in the years to come.
- **Engaging the Reader**: Throughout the book, we'll engage the reader with interactive learning activities, case studies, and thought-provoking questions to deepen understanding and foster critical thinking. Whether you're a student, researcher,

or healthcare professional, this book is designed to inspire curiosity and spark creativity in the fascinating world of multifunctional biomedical materials.

### *1.1.8 Summary*

"Introduction to Biomedical Nanomaterials" is like opening the door to a world of tiny wonders in medicine. In this exciting chapter, we dive into the realm of nanotechnology, where scientists work with materials so small, they're thousands of times thinner than a human hair. These materials, called nanomaterials [2, 5, 6], have big potential in healthcare because they can do incredible things at the tiniest scale. First off, we learn about the building blocks of nanomaterials. They're made up of atoms and molecules, just like everything else, but at this scale, they behave differently. Scientists can engineer these materials to have special properties that make them perfect for tackling medical challenges. Imagine tiny particles that can deliver medicine directly to a sick cell or sensors so small they can detect diseases early on. Next, we explore how nanomaterials are changing the game in diagnosis and treatment. With their ultra-small size, they can travel through the body with ease, reaching places traditional medicines can't. This means better targeting of diseases and fewer side effects for patients [9]. Plus, nanomaterials are like tiny detectives, helping doctors spot problems early and intervene before they become serious. But it's not just about treating diseases—nanomaterials are also paving the way for regenerative medicine. Scientists are using them to create scaffolds that help rebuild damaged tissues and organs from the ground up [22, 24]. It's like giving our bodies a blueprint for healing themselves, leading to faster recovery times and better outcomes for patients.

Of course, with great power comes great responsibility, and we will discuss the importance of safety and ethics in working with nanomaterials. Since they're so tiny, they can interact with our bodies in unexpected ways, so scientists must tread carefully to ensure they're safe for medical use. Ethical considerations also come into play, as we consider the implications of manipulating matter at such a small scale.

Finally, we peer into the future and imagine the possibilities that lie ahead with biomedical nanomaterials. From personalized medicine tailored to each individual's unique biology to revolutionary treatments for previously incurable diseases, the potential is limitless. As scientists continue to unlock the secrets of the nanoworld, we can only imagine the incredible breakthroughs that await us on the horizon.

## Points to Remember

- **Some biomaterials have the remarkable ability to self-heal, mimicking the repair mechanisms found in living organisms. This property holds potential for creating durable and long-lasting materials for various applications.**
- **Many biomaterials are inspired by nature, taking cues from biological structures and processes. For example, the lotus leaf's water-repellent surface has inspired the design of self-cleaning and water-resistant materials.**
- **Nanotechnology is increasingly used in biomaterials to replicate biological structures at the nanoscale, enabling precise control over material properties and interactions with biological systems.**
- **Advances in biomaterials have led to the development of "smart" materials that can respond to external stimuli, such as changes in temperature, pH, or mechanical stress. These materials have applications in drug delivery, tissue engineering, and diagnostics.**
- **Some biomaterials are edible and biodegradable, offering sustainable alternatives to traditional packaging materials and single-use plastics. These materials can help reduce environmental pollution and waste.**
- **Emerging bio fabrication techniques, such as 3D bioprinting and electrospinning, enable the precise engineering of complex tissue structures and organs for regenerative medicine and transplantation.**
- **Biomaterials play a crucial role in space exploration, where lightweight and durable materials are needed for spacecraft components, spacesuits, and life support systems.**
- **Biomaterials are not only limited to scientific and medical applications but also find use in art and design. Artists and designers are incorporating biomaterials into their creations, exploring themes of sustainability, nature, and technology.**

## References


1. Hench LL (1980) Biomaterials. Science (New York, N.Y.) 208(4446):826–831. https://doi.org/10.1126/science.6246576
2. Hubbell JA (1995) Biomaterials in tissue engineering. Bio/Technology (Nature Publishing Company) 13(6):565–576. https://doi.org/10.1038/nbt0695-565
3. Suhag D, Bhatia R, Das S, Shakeel A, Ghosh AB, Singh A et al (2015). Physically cross-linked pH-responsive hydrogels with tunable formulations for controlled drug delivery. RSC Adv 5(66):53963–53972. https://doi.org/10.1039/c5ra07424j
4. Rahmati M, Pennisi CP, Budd E, Mobasheri A, Mozafari M (2018) Biomaterials for regenerative medicine: historical perspectives and current trends. Adv Exp Med Biol 1119:1–19. https://doi.org/10.1007/5584_2018_278
5. Hench LL (1998) Biomaterials: a forecast for the future. Biomaterials 19(16):1419–1423. https://doi.org/10.1016/s0142-9612(98)00133-1

6. Stegemann JP, Wagner WR, Göbel U, Perets E, Leong KW, Leach JK, Gilbert J (2023) The year 2022 in biomaterials research: a perspective from the editors of six leading journals. J Biomed Mater Res Part A 111(9):1298–1308. https://doi.org/10.1002/jbm.a.37529
7. Shakeel A, Singh A, Das S et al (2017) Synthesis and morphological insight of new biocompatible smart hydrogels. J Polym Res 24:113. https://doi.org/10.1007/s10965-017-1267-7
8. Rathore R, Suhag D, Wan F, Verma R, Thakur A, Thakur P (2023) Functionalization of biomaterials. In: Suhag D, Thakur A, Thakur P (eds) Integrated nanomaterials and their applications. Springer, Singapore. https://doi.org/10.1007/978-981-99-6105-4_11
9. Williams DF (1976) Biomaterials and biocompatibility. Med Prog Technol 4(1–2):31–42
10. Suh H (1998) Recent advances in biomaterials. Yonsei Med J 39(2):87–96. https://doi.org/10.3349/ymj.1998.39.2.87
11. Nii T, Katayama Y (2021) Biomaterial-assisted regenerative medicine. Int J Mol Sci 22(16):8657. https://doi.org/10.3390/ijms22168657
12. Liu S, Yu JM, Gan YC, Qiu XZ, Gao ZC, Wang H, Chen SX, Xiong Y, Liu GH, Lin SE, McCarthy A, John JV, Wei DX, Hou HH (2023) Biomimetic natural biomaterials for tissue engineering and regenerative medicine: new biosynthesis methods, recent advances, and emerging applications. Mil Med Res 10(1):16. https://doi.org/10.1186/s40779-023-00448-w
13. Binyamin G, Shafi BM, Mery CM (2006) Biomaterials: a primer for surgeons. Semin Pediatr Surg 15(4):276–283. https://doi.org/10.1053/j.sempedsurg.2006.07.007
14. Cecen B, Hassan S, Li X, Zhang YS (2024) Smart biomaterials in biomedical applications: current advances and possible future directions. Macromol Biosci 24(3):e2200550. https://doi.org/10.1002/mabi.202200550
15. Marin E, Boschetto F, Pezzotti G (2020) Biomaterials and biocompatibility: an historical overview. J Biomed Mater Res Part A 108(8):1617–1633. https://doi.org/10.1002/jbm.a.36930
16. Langer R, Cima LG, Tamada JA, Wintermantel E (1990) Future directions in biomaterials. Biomaterials 11(9):738–745. https://doi.org/10.1016/0142-9612(90)90038-r
17. Lee EJ, Kasper FK, Mikos AG (2014) Biomaterials for tissue engineering. Ann Biomed Eng 42(2):323–337. https://doi.org/10.1007/s10439-013-0859-6
18. Zhang K, Ma B, Hu K, Yuan B, Sun X, Song X, Tang Z, Lin H, Zhu X, Zheng Y, García AJ, Mikos AG, Anderson JM, Zhang X (2022) Evidence-based biomaterials research. Bioact Mater 15:495–503. https://doi.org/10.1016/j.bioactmat.2022.04.014
19. Singh AK, Sundram S, Malviya R (2023) Human-derived biomaterials for biomedical and tissue engineering applications. Curr Pharm Des 29(8):584–603. https://doi.org/10.2174/1381612829666230320103412
20. Williams DF (2009) On the nature of biomaterials. Biomaterials 30(30):5897–5909. https://doi.org/10.1016/j.biomaterials.2009.07.027
21. Troy E, Tilbury MA, Power AM, Wall JG (2021) Nature-based biomaterials and their application in biomedicine. Polymers 13(19):3321. https://doi.org/10.3390/polym13193321
22. Huebsch N, Mooney DJ (2009) Inspiration and application in the evolution of biomaterials. Nature 462(7272):426–432. https://doi.org/10.1038/nature08601
23. Kim BS, Baez CE, Atala A (2000) Biomaterials for tissue engineering. World J Urol 18(1):2–9. https://doi.org/10.1007/s003450050002
24. Bhat S, Kumar A (2013) Biomaterials and bioengineering tomorrow's healthcare. Biomaterial 3(3):e24717. https://doi.org/10.4161/biom.24717

# Chapter 2
# Fundamentals of Biomaterials

## 2.1 Introduction

Biomaterials represent a fascinating intersection of medicine and engineering, offering innovative solutions to diverse challenges in healthcare. At its core, biomaterials refer to materials engineered to interact with biological systems, whether within the human body or in laboratory settings. These materials are designed with specific properties to perform various functions, ranging from providing structural support to facilitating drug delivery and tissue regeneration. In essence, they act as bridges between the living and the synthetic, harnessing the principles of both disciplines to improve medical treatments and outcomes. The importance of biomaterials [1] in medicine and engineering cannot be overstated. In medicine, biomaterials play a pivotal role in revolutionizing patient care by enabling the development of advanced medical devices, implants, and drug delivery systems [2]. From artificial joints that restore mobility to patients with arthritis to bioresorbable scaffolds that aid in tissue regeneration after surgery, biomaterials [3] contribute significantly to enhancing quality of life and extending lifespan. Moreover, they facilitate minimally invasive procedures and personalized treatments, paving the way for tailored healthcare solutions that cater to individual patient needs. In engineering, biomaterials serve as the building blocks for creating next-generation medical technologies and therapies. Engineers leverage their expertise in materials science to design biomaterials with precisely controlled properties, such as biocompatibility, mechanical strength, and degradation kinetics. By understanding the complex interactions between materials and biological systems, engineers can develop implants that seamlessly integrate into the body, medical devices that deliver therapeutic agents with precision, and diagnostic tools that provide insights into disease progression. Furthermore, biomaterials hold immense promise for addressing pressing global health challenges, such as organ transplantation shortages, chronic diseases, and age-related degeneration. Through interdisciplinary collaboration between medical professionals, engineers, biologists, and materials scientists, innovative biomaterial solutions are continually

 
D. Suhag et al., *Handbook of Biomaterials for Medical Applications, Volume 1*,
Biomedical Materials for Multi-functional Applications,
https://doi.org/10.1007/978-981-97-4818-1_2

being developed to tackle these complex issues. Whether it's engineering artificial organs using biocompatible polymers or designing nanoparticles for targeted drug delivery, biomaterials [1, 3] offer endless possibilities for advancing healthcare and improving patient outcomes.

Biomaterials represent a dynamic field at the intersection of medicine and engineering, driven by the quest to enhance human health and well-being. With their unique ability to interface with biological systems, biomaterials hold the key to unlocking groundbreaking medical treatments, diagnostic tools, and therapeutic interventions.

## *2.1.1 Properties of Biomaterials*

Understanding the properties of biomaterials is crucial for their successful application in various biomedical and engineering contexts (Fig. 2.1). These properties encompass a wide range of characteristics, including mechanical strength, biocompatibility, degradation behavior, and surface interactions with biological tissues. Mechanical properties determine the structural integrity and performance of biomaterials, ensuring they can withstand physiological forces without failure. Biocompatibility is essential to prevent adverse reactions when biomaterials come into contact with living organisms, ensuring they integrate seamlessly with biological systems [4]. Degradation behavior is important for bioresorbable materials, as they must degrade at a controlled rate to support tissue regeneration. Surface interactions play a critical role in determining the bio interface between biomaterials [1, 2] and host tissues, influencing cellular responses and ultimately the success of medical interventions. By understanding and tailoring these properties, researchers can design biomaterials with optimized performance for specific applications, ranging from orthopedic implants to drug delivery systems, thereby advancing the field of biomaterials and improving patient outcomes.

### 2.1.1.1 Mechanical Properties

The mechanical properties of biomaterials play a fundamental role in determining their suitability for various biomedical applications. These properties encompass a range of characteristics, including stiffness, strength, toughness, and elasticity, which collectively influence how biomaterials respond to mechanical forces within the body. Stiffness, often quantified by the modulus of elasticity, describes the resistance of a material to deformation under load. Biomaterials with appropriate stiffness are crucial for mimicking the mechanical properties of natural tissues [4], ensuring compatibility and functionality in implants and prosthetics. Strength refers to the maximum stress a biomaterial can withstand before failure, making it a critical factor in designing load-bearing implants and structural components. Biomaterials must exhibit sufficient strength to withstand physiological forces and maintain structural integrity over

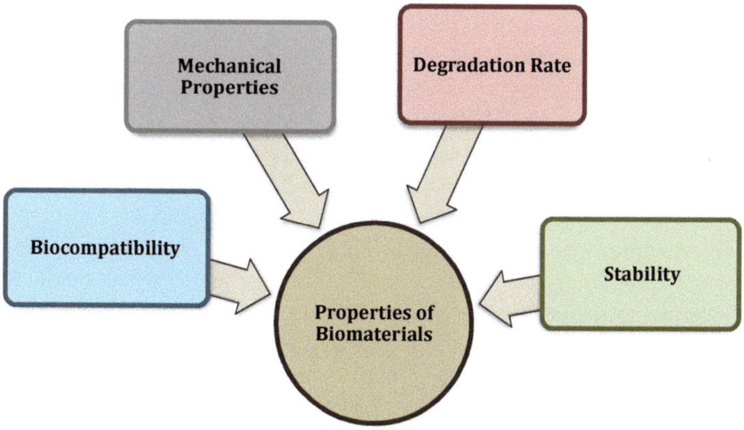


**Fig. 2.1** Major properties of biomaterials in biomedical applications

time, minimizing the risk of mechanical failure or implant loosening. Toughness, on the other hand, measures the ability of a biomaterial to absorb energy and deform plastically before fracturing. Tough biomaterials [1] are essential for withstanding impact and cyclic loading conditions, particularly in applications subject to dynamic forces such as orthopedic implants and cardiovascular devices.

Elasticity, characterized by the material's ability to return to its original shape after deformation, is another important mechanical property of biomaterials. Elastic biomaterials can deform reversibly under load, allowing them to accommodate physiological movements and stresses without permanent damage. This property is particularly crucial in soft tissue engineering and flexible implants, where biomaterials must mimic the elastic behavior of native tissues to ensure proper function and compatibility [5]. Overall, a comprehensive understanding and control of the mechanical properties of biomaterials are essential for designing safe, effective, and long-lasting biomedical devices and implants that meet the mechanical demands of the human body while promoting tissue integration and patient well-being.

#### 2.1.1.2 Biocompatibility

Biocompatibility is an important aspect of biomaterials [1–3] design, ensuring that materials used in medical applications interact safely and effectively with biological systems. It encompasses the ability of a biomaterial to perform its intended function without eliciting harmful effects on living tissues or triggering adverse immune responses. Achieving biocompatibility requires careful consideration of various factors, including the chemical composition, surface properties, and degradation behavior of the biomaterial. By ensuring biocompatibility, biomaterials can integrate seamlessly with the body, promoting tissue regeneration, minimizing inflammation, and ultimately improving patient outcomes. One key aspect of biocompatibility is the surface properties of biomaterials, which dictate their interactions with surrounding tissues and cells. Biomaterial surfaces can be modified through techniques such as coating, functionalization [4, 5], or surface patterning to enhance biocompatibility and promote specific cellular responses. For example, surface modifications can facilitate cell adhesion, proliferation, and differentiation, leading to improved tissue integration and regeneration. Surface modifications can reduce the risk of immune recognition and inflammatory responses, further enhancing the biocompatibility of biomaterials. Another critical consideration in biocompatibility is the degradation behavior of biomaterials, particularly for bioresorbable implants and scaffolds. Degradable biomaterials are designed to gradually break down and be absorbed by the body over time, allowing for tissue regeneration and remodeling. The rate and mechanism of degradation must be carefully controlled to ensure that degradation byproducts are non-toxic and do not induce adverse reactions. By designing biomaterials with tunable degradation kinetics, researchers can tailor the biocompatibility of implants to match the healing process of the surrounding tissues, promoting seamless integration and minimizing inflammatory responses. Moreover, biocompatibility testing is essential for evaluating the safety and efficacy of biomaterials prior to clinical use. In vitro and in vivo studies are conducted to assess the biological response of cells and tissues to biomaterials, including cytotoxicity [5], inflammation, and tissue integration. These tests provide valuable insights into the biocompatibility profile of biomaterials, guiding the design and optimization of future biomedical devices and implants. Through rigorous testing and validation, researchers can ensure that biomaterials meet stringent regulatory requirements and demonstrate superior biocompatibility in clinical settings.

Biocompatibility is a critical consideration in the development of biomaterials for medical applications, ensuring their safe and effective use in the human body. By understanding and optimizing surface properties, degradation behavior, and biological interactions, researchers can design biomaterials that promote tissue regeneration, minimize immune responses, and improve patient outcomes. With continued advancements in biomaterials science and biocompatibility testing, the potential for innovative biomedical solutions to address unmet medical needs continues to grow, ushering in a new era of personalized and regenerative medicine.

#### 2.1.1.3 Degradation and Stability

The degradation and stability of biomaterials are crucial considerations in their design and application, particularly for implants and devices intended for long-term use within the human body. Degradation refers to the process by which biomaterials break down over time, either through chemical, physical, or biological mechanisms. The stability of biomaterials, on the other hand, refers to their ability to maintain their structural integrity and functional properties over time in the physiological environment. Achieving the right balance between degradation and stability is essential for ensuring the safety, efficacy, and longevity of biomaterial-based medical devices and implants [5, 6]. Various factors influence the degradation behavior of biomaterials, including their chemical composition, structural architecture, and environmental conditions. For example, bioresorbable polymers are designed to degrade gradually in the body, allowing for tissue regeneration and remodeling. The rate of degradation can be controlled by adjusting the molecular weight, cross-linking density, and chemical structure of the polymer, ensuring that degradation occurs at a pace compatible with tissue healing processes. In contrast, non-degradable biomaterials such as metals and ceramics are engineered to resist degradation and maintain their mechanical properties over extended periods. The stability of biomaterials is closely linked to their degradation resistance and ability to withstand physiological stresses without undergoing significant structural changes. Stability can be influenced by factors such as temperature, pH, mechanical loading, and exposure to biological fluids. For example, biomaterials implanted in the body may be subjected to enzymatic degradation, hydrolysis, or oxidative processes, which can compromise their stability and performance over time. Therefore, it is essential to select biomaterials with inherent stability and resistance to degradation for applications requiring long-term implantation. Moreover, the degradation and stability of biomaterials can be influenced by surface modifications, coatings, and biomimetic strategies aimed at enhancing their biocompatibility and performance [2, 4]. Surface modifications can improve the resistance of biomaterials to degradation, reduce friction and wear, and promote tissue integration and healing. For example, bioactive coatings can stimulate the formation of a stable interface between the biomaterial and surrounding tissues, enhancing stability and reducing the risk of implant failure.

Biodegradable biomaterials offer the advantage of promoting tissue regeneration and remodeling while gradually resorbing in the body, eliminating the need for surgical removal. However, careful consideration must be given to the degradation kinetics, degradation byproducts, and tissue response to ensure optimal outcomes. By understanding the complex interplay between degradation and stability, researchers can design biomaterials with tailored properties to meet the specific needs of various biomedical applications, from temporary implants to permanent medical devices. Through continued research and innovation, the development of biomaterials with improved degradation profiles and enhanced stability holds promise for advancing regenerative medicine, tissue engineering, and personalized healthcare.

**Exercise 1.**

**Fill Ups**

**(Biocompatibility, Strength, Stress, Structural, Elastic modulus, Cellular response, Toughness)**

1. The mechanical properties of biomaterials include parameters such as................, ................, and ..................., which are crucial for their performance in biomedical applications.
2. Tensile strength measures the maximum ............... a biomaterial can withstand before breaking under tension, providing insight into its structural integrity.
3. The .................. of biomaterials refers to their ability to interact with biological systems without causing harm.
4. Surface chemistry of biomaterials plays a significant role in determining their biocompatibility and .........................
5. The degradation and stability of biomaterials are influenced by factors such as environmental conditions, chemical composition, and ............... properties, ultimately impacting their performance in biomedical applications.

### *2.1.2 Classification of Biomaterials*

Classification of biomaterials is a pivotal aspect of understanding their diverse nature and applications within the fields of medicine and engineering. By categorizing biomaterials based on various criteria such as composition, origin, functionality, degradation behavior, and application, researchers and practitioners can systematically organize and analyze the vast array of materials available for biomedical use (Fig. 2.2). This classification framework provides a structured approach for selecting, designing, and optimizing biomaterials tailored to specific medical needs and requirements. Classification facilitates communication and collaboration among interdisciplinary teams working on biomaterials research and development [2, 5, 6], enabling the exchange of knowledge and expertise across different domains. Overall, a comprehensive classification system enhances our understanding of biomaterials' properties, capabilities, and limitations, paving the way for innovative solutions that address complex healthcare challenges and improve patient outcomes.

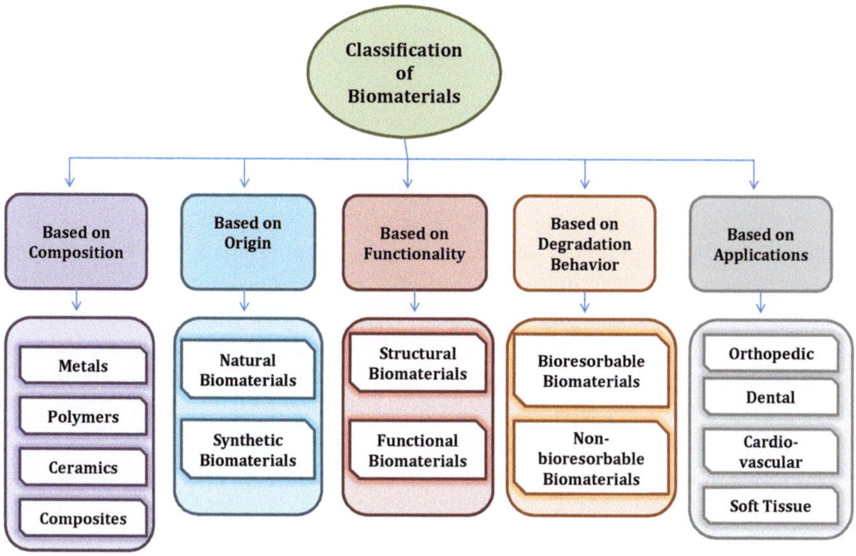


**Fig. 2.2** Classification of biomaterials based on composition, origin, functionality, degradation behavior, and applications

#### 2.1.2.1 Based on Composition

Based on composition, biomaterials are classified into distinct categories, each characterized by their predominant material makeup. This classification scheme encompasses metals, polymers, ceramics, and composites. Metals, such as titanium and stainless steel, offer exceptional strength and durability, making them ideal for load-bearing implants like hip replacements. Polymers, including polyethylene and silicone rubber, provide flexibility and versatility, enabling their use in a wide range of medical devices and drug delivery [7] systems. Ceramics, such as hydroxyapatite and alumina, exhibit excellent biocompatibility and bioactivity, making them suitable for dental implants and bone grafts. Composites combine two or more materials to synergize their properties, offering customizable solutions for complex biomedical applications. Understanding this classification framework is essential for selecting the most appropriate biomaterial for specific medical needs, ensuring optimal performance and patient outcomes.

Metals

Metals are a prominent category of biomaterials widely used in medical applications due to their exceptional mechanical properties, biocompatibility, and corrosion resistance. Titanium and its alloys, such as Ti–6Al–4V, are among the most commonly employed metals in orthopedic and dental implants due to their high strength-to-weight ratio and excellent biocompatibility. Stainless steel, another widely used metal, offers superior corrosion resistance and affordability, making it suitable for various surgical instruments, orthopedic implants, and cardiovascular devices. The robust mechanical properties of metals enable them to withstand physiological loads and provide long-term stability within the body, ensuring the durability and functionality of medical implants. Furthermore, metals can be surface-modified or coated to enhance their biocompatibility and promote osseointegration, the process by which bone integrates with the implant surface [8]. Surface treatments such as plasma spraying, acid etching, and physical vapor deposition (PVD) improve the adhesion of bone cells to metal implants, facilitating faster healing and reducing the risk of implant failure. Bioactive coatings containing hydroxyapatite or growth factors can stimulate bone formation and enhance the integration of metal implants with surrounding tissues, further improving their clinical performance and longevity. Despite their numerous advantages, metals may pose challenges related to their stiffness and potential for stress shielding, wherein the implant absorbs excessive loads, leading to bone resorption and implant loosening. To address these concerns, researchers are developing advanced metal alloys with tailored mechanical properties that closely match those of natural bone, reducing stress shielding and enhancing long-term implant stability. Furthermore, the advent of additive manufacturing techniques, such as 3D printing, allows for the fabrication of complex metal implants with customized geometries and internal structures, offering personalized solutions for patients with unique anatomical requirements. As research in metallurgy and biomaterials engineering continues to advance, metals remain integral to the development of innovative medical devices and implants, contributing to improved patient care and quality of life.

### Polymers

Polymers represent a diverse class of biomaterials valued for their flexibility, versatility, and tunable properties, making them suitable for a wide range of medical applications. Synthetic polymers, such as polyethylene, polyvinyl chloride (PVC), and polyurethane, offer excellent biocompatibility and can be engineered to mimic the mechanical properties of natural tissues. Their ability to be synthesized with precise control over molecular structure and properties allows for the customization of polymers to meet specific biomedical needs [9]. For example, biodegradable polymers like poly (lactic acid) (PLA) and poly (lactic-co-glycolic acid) (PLGA) degrade into non-toxic byproducts within the body, making them ideal for drug delivery systems and tissue engineering scaffolds. Moreover, polymers can be modified through various techniques to enhance their functionality and biocompatibility for specific medical applications. Surface modifications, such as plasma treatment and chemical grafting, can improve cell adhesion, proliferation, and differentiation on polymer surfaces, promoting tissue integration and wound healing. Additionally, polymers can be combined with bioactive agents, nanoparticles, or growth factors to create composite materials with enhanced properties, such as controlled drug release or antibacterial activity. These advances in polymer science have led to the development of novel medical devices, including catheters, sutures, and tissue-engineered constructs, that offer improved biocompatibility and performance. However, challenges associated with polymers include their mechanical strength, degradation kinetics, and potential for leaching of toxic compounds. To address these issues, researchers are exploring advanced polymer processing techniques, such as electrospinning and additive manufacturing, to fabricate biomimetic scaffolds with tailored mechanical properties and hierarchical structures. The integration of smart polymers capable of responding to external stimuli, such as pH, temperature, or light, holds promise for the development of stimuli-responsive drug delivery [7] systems and implantable devices. As polymer science continues to evolve, polymers remain at the forefront of biomaterials research, driving innovation and improving healthcare outcomes through their unique combination of versatility, biocompatibility, and tunability.

Ceramics

Ceramics are a class of biomaterials known for their excellent mechanical properties, biocompatibility, and bioactivity, making them valuable for a variety of medical applications. Hydroxyapatite (HA) and tricalcium phosphate (TCP) are widely used ceramic materials due to their chemical similarity to the mineral component of bone, facilitating bone integration and regeneration. These bioactive ceramics promote osteoconduction, the process by which new bone forms on the surface of the implant, leading to enhanced osseointegration and long-term stability. Ceramics exhibit high compressive strength and hardness [9, 10], making them suitable for load-bearing orthopedic implants, dental restorations, and bone graft substitutes. Furthermore, ceramics offer advantages such as corrosion resistance, radiopacity, and inertness in biological environments, minimizing adverse reactions and inflammatory responses. Zirconia-based ceramics, for example, are highly resistant to chemical degradation and wear, making them ideal for dental crowns, bridges, and orthopedic implants. Ceramic coatings can be applied to metallic implants to improve their biocompatibility, reduce friction, and enhance wear resistance, extending their lifespan and performance. These characteristics make ceramics a preferred choice for critical medical devices and implants where reliability and longevity are paramount.

Challenges associated with ceramics include their brittleness, difficulty in processing complex shapes, and potential for fracture under tensile loads. To address these limitations, researchers are exploring strategies such as ceramic-polymer composites and additive manufacturing techniques to improve the mechanical properties and manufacturability of ceramic-based implants. By incorporating polymers or reinforcing phases into ceramic matrices, composite materials with enhanced toughness and fracture resistance can be achieved, expanding the range of applications for ceramic biomaterials. Moreover, advances in additive manufacturing technologies, such as ceramic 3D printing, enable the fabrication of patient-specific implants with complex geometries and internal structures, offering personalized solutions for improved clinical outcomes. As research in ceramic biomaterials continues to advance, ceramics remain indispensable for a wide range of medical applications, contributing to advancements in regenerative medicine, implantology, and tissue engineering.

### Composites

Composites represent a versatile class of biomaterials that combine two or more distinct materials to synergize their properties and functionalities, offering tailored solutions for diverse biomedical applications. By blending different materials, composites can overcome individual material limitations while enhancing desirable characteristics such as mechanical strength, biocompatibility, and degradation behavior [10]. For example, composite materials composed of polymers reinforced with ceramic or metallic particles exhibit improved mechanical properties, making them suitable for load-bearing implants and tissue-engineered scaffolds. Composites can incorporate bioactive agents or growth factors to promote tissue regeneration and healing, further expanding their utility in regenerative medicine and tissue engineering. Composites offer flexibility in design and fabrication, allowing for the customization of material properties to meet specific application requirements. Composite biomaterials can be engineered with controlled porosity, surface roughness, and degradation kinetics to optimize tissue integration, vascularization, and host response. By tailoring the composition and microstructure of composites, researchers can design materials with tunable mechanical properties and degradation rates, offering personalized solutions for patient-specific needs. These customizable properties make composites attractive for a wide range of medical devices, implants, and drug delivery systems. Furthermore, composites can address challenges associated with individual material classes, such as the brittleness of ceramics or the degradation kinetics of polymers, by combining them into hybrid structures with complementary properties. For instance, bioactive glass-ceramic composites combine the mechanical strength of ceramics with the bioactivity of glass, offering enhanced bone regeneration and integration capabilities. Polymer-ceramic composites can be tailored to match the mechanical properties of native tissues, minimizing stress shielding and improving long-term implant stability [11]. As research in composite biomaterials continues to advance, novel combinations and fabrication techniques hold promise for developing next-generation medical devices with improved performance, biocompatibility, and clinical outcomes.

However, challenges remain in optimizing the processing techniques, interface bonding, and long-term stability of composite biomaterials for clinical translation. Researchers are exploring advanced manufacturing methods, such as additive manufacturing and electrospinning, to overcome these challenges and fabricate complex composite structures with precise control over composition and architecture. Moreover, interdisciplinary collaboration between materials scientists, engineers, and biomedical researchers is essential for advancing composite biomaterials and unlocking their full potential in addressing unmet medical needs and improving patient care.

#### 2.1.2.2 Based on Origin

Based on origin, biomaterials are classified into two main categories: natural and synthetic biomaterials. Natural biomaterials are derived from biological sources such as plants, animals, or microorganisms and include substances like collagen, chitosan, and silk. These materials often possess inherent biocompatibility and bioactivity, making them suitable for various medical applications such as tissue engineering, wound healing, and drug delivery [7]. In contrast, synthetic biomaterials are man-made materials designed specifically for biomedical use, offering advantages such as customizable properties [11], controlled degradation kinetics, and scalability. Understanding this classification based on origin provides insights into the unique properties and applications of biomaterials derived from natural and synthetic sources, guiding their selection and development for specific biomedical needs.

##### Natural Biomaterials

Natural biomaterials, derived from biological sources such as plants, animals, or microorganisms, offer unique advantages for various biomedical applications due to their inherent biocompatibility, bioactivity, and resemblance to native tissues (Fig. 2.3). Collagen, the most abundant protein in the human body, is a commonly used natural biomaterial known for its biocompatibility and ability to promote cell adhesion and tissue regeneration. It serves as a scaffold for cell growth and tissue repair in applications such as wound healing, skin grafts, and cartilage regeneration. Another natural biomaterial, chitosan, derived from chitin found in the exoskeletons of crustaceans, possesses antimicrobial properties and biodegradability, making it suitable for wound dressings, drug delivery systems, and tissue engineering scaffolds. Furthermore, natural biomaterials offer the advantage of being inherently biodegradable and biocompatible, minimizing adverse reactions and immune responses when implanted in the body. Silk, produced by silkworms and spiders, is a versatile natural biomaterial valued for its strength, flexibility, and biocompatibility. It has been utilized in various medical applications, including sutures, tissue scaffolds, and drug delivery [7] vehicles, owing to its biodegradability and low immunogenicity. Natural biomaterials often exhibit unique bioactive properties that can stimulate cellular responses and tissue regeneration, further enhancing their utility in regenerative medicine and tissue engineering. Despite their numerous advantages, natural biomaterials may present challenges such as batch-to-batch variability, limited mechanical strength, and potential immunogenicity. To address these limitations, researchers are exploring strategies to modify and enhance the properties of natural biomaterials [1, 5] through chemical modifications, cross-linking, and composite formulations. By combining natural biomaterials with synthetic polymers or ceramics, researchers can create hybrid materials with improved mechanical properties and tailored functionalities, expanding their range of applications in medical devices and implants. Overall, natural biomaterials represent a valuable resource in biomedical research and

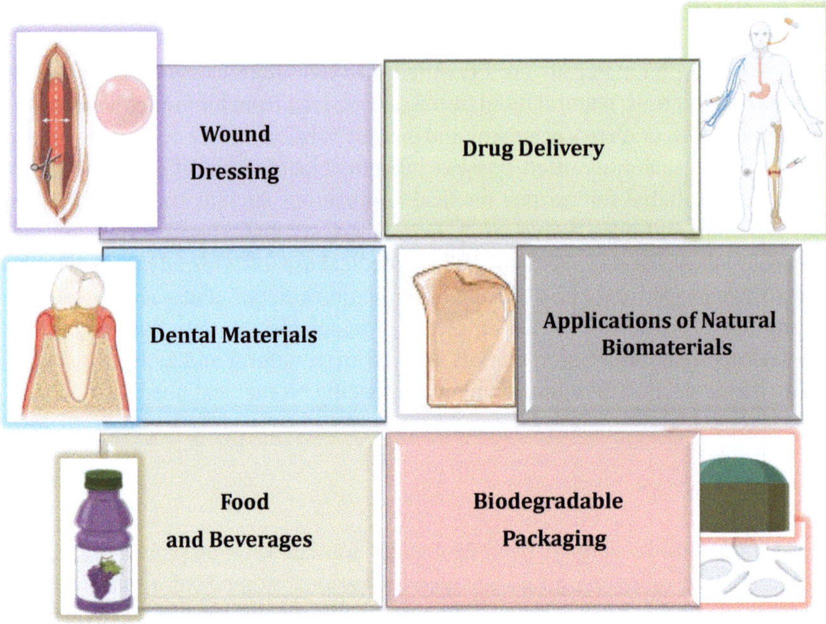


**Fig. 2.3** Various applications of natural biomaterials

innovation, offering sustainable and biocompatible solutions for addressing various healthcare challenges and improving patient outcomes.

Synthetic Biomaterials

Synthetic biomaterials, fabricated through chemical synthesis or manufacturing processes, are engineered materials designed specifically for biomedical applications. These materials offer several advantages, including precise control over composition, mechanical properties, degradation kinetics, and surface characteristics. Polymeric biomaterials, such as polyethylene glycol (PEG), poly (lactic acid) (PLA), and poly (lactic-co-glycolic acid) (PLGA), are among the most widely used synthetic biomaterials due to their versatility, biocompatibility, and tunable properties [12]. They are employed in various medical devices, drug delivery systems, and tissue engineering scaffolds, offering tailored solutions for specific clinical needs.

Moreover, synthetic biomaterials can be synthesized with controlled degradation rates, allowing for the design of bioresorbable implants and drug delivery systems that degrade over time within the body. This feature is particularly advantageous for applications requiring temporary support or localized drug release, such as sutures, stents, and implants. Synthetic biomaterials can be modified through surface functionalization or coating techniques to enhance their biocompatibility, promote cell adhesion, and modulate biological responses. Surface modifications enable the customization of synthetic biomaterials to interact favorably with biological tissues, reducing inflammation, improving integration, and enhancing therapeutic [12] efficacy. Synthetic biomaterials offer scalability, reproducibility, and consistency in manufacturing, ensuring uniformity and reliability in medical device production. Advanced manufacturing techniques, such as electrospinning, 3D printing, and micro molding, enable the fabrication of complex structures and patient-specific implants with precise control over geometry and architecture [13]. These capabilities facilitate the development of personalized medical devices and implants tailored to individual patient needs, enhancing treatment outcomes and patient satisfaction. Overall, synthetic biomaterials play a pivotal role in modern healthcare, driving innovation and advancements in medical device technology, drug delivery systems, and regenerative medicine.

#### 2.1.2.3 Based on Functionality

Based on functionality, biomaterials are categorized according to their intended purpose and role within biomedical applications. Structural biomaterials are designed to provide mechanical support and stability to tissues or organs, such as orthopedic implants for bone repair or dental materials for restoring tooth structure. These biomaterials must possess suitable mechanical properties to withstand physiological loads and ensure long-term durability within the body. In contrast, functional biomaterials are engineered to interact dynamically with biological systems, facilitating specific therapeutic or diagnostic functions [12, 13]. Examples include drug-eluting stents for controlled drug delivery, biosensors for real-time monitoring of physiological parameters, and tissue-engineered constructs for regenerative medicine applications. Understanding this classification based on functionality helps guide the selection and development of biomaterials tailored to meet specific clinical needs and applications.

Structural Biomaterials

Structural biomaterials are essential components in the field of biomedical engineering, providing mechanical support and stability to tissues and organs within the human body. These biomaterials are designed to mimic the properties of natural tissues while withstanding physiological loads and environmental stresses. Orthopedic implants, such as hip prostheses and bone plates, are common examples of structural biomaterials used to repair or replace damaged or diseased bone tissue. These implants must possess sufficient strength, durability, and biocompatibility to ensure successful integration with surrounding tissues and long-term functionality. One of the key requirements for structural biomaterials is mechanical properties that closely match those of the native tissue they are intended to support or replace. For instance, metallic implants like titanium alloys are favored for load-bearing applications in orthopedics due to their high strength-to-weight ratio and excellent biocompatibility. Similarly, ceramic materials such as hydroxyapatite-coated implants exhibit properties like natural bone, promoting osseointegration and minimizing the risk of implant failure [14]. Moreover, structural biomaterials play a critical role in various medical devices and surgical procedures beyond orthopedics. Dental biomaterials, including dental implants and prosthetic restorations, rely on structural biomaterials to restore lost or damaged tooth structure and support oral function. Similarly, cardiovascular biomaterials such as stents and heart valves are designed to provide mechanical support and functionality to the cardiovascular system, enhancing blood flow and cardiac function. The development of advanced structural biomaterials continues to drive innovations in medical technology, improving patient outcomes and quality of life. Structural biomaterials are indispensable components in biomedical engineering, providing essential mechanical support and stability to tissues and organs in the human body. Through careful design and engineering, these biomaterials offer solutions for repairing, replacing, or augmenting damaged or diseased tissues, improving patient mobility [15], comfort, and overall well-being. As research and technology continue to advance, the development of novel structural biomaterials holds promise for addressing evolving healthcare needs and challenges, furthering the progress of regenerative medicine, tissue engineering, and medical device innovation.

#### Functional Biomaterials

Functional biomaterials are a vital category in the field of biomedical engineering, designed to interact dynamically with biological systems to perform specific therapeutic or diagnostic functions. These biomaterials play a crucial role in advancing medical treatments and technologies by providing targeted solutions for a wide range of healthcare challenges. One example of functional biomaterials is drug-eluting stents, which are used to open blocked arteries and deliver medications directly to the site of injury [14, 15]. These stents release drugs over time, preventing restenosis and reducing the need for repeat interventions, thus improving patient outcomes. Another important application of functional biomaterials is in biosensors, which are devices capable of detecting and measuring biological signals or analytes within the body. Biosensors can be used for real-time monitoring of physiological parameters such as glucose levels in diabetic patients or the presence of specific biomarkers indicative of disease. By providing continuous, non-invasive monitoring, functional biomaterial-based biosensors offer valuable insights into patient health status, enabling timely intervention and personalized treatment strategies. Functional biomaterials also play a key role in tissue engineering, where they are used to create scaffolds and matrices that support cell growth and tissue regeneration. These biomaterials provide a temporary framework for cells to adhere, proliferate, and differentiate, ultimately leading to the formation of new functional tissues. For example, biodegradable polymers such as poly (lactic acid) (PLA) and poly (lactic-co-glycolic acid) (PLGA) are commonly used as scaffolds for tissue engineering applications due to their biocompatibility and ability to degrade over time, leaving behind regenerated tissue.

Furthermore, functional biomaterials are utilized in various drug delivery systems to control the release and distribution of therapeutic agents within the body [16]. These systems can be engineered to deliver drugs locally to specific target tissues or cells, minimizing systemic side effects and maximizing therapeutic efficacy. Functional biomaterial-based drug delivery systems offer precise control over drug release kinetics, dosage, and duration, allowing for customized treatment regimens tailored to individual patient needs. Overall, functional biomaterials represent a cornerstone of modern medical technology, enabling innovative approaches to diagnosis, treatment, and tissue regeneration, and ultimately improving patient care and quality of life.

#### 2.1.2.4 Based on Degradation Behavior

Based on degradation behavior, biomaterials are categorized into two main groups: bioresorbable and non-bioresorbable biomaterials. Bioresorbable biomaterials are designed to gradually degrade and be absorbed by the body over time, eliminating the need for surgical removal and minimizing long-term complications. These materials are often used in temporary implants, drug delivery systems, and tissue engineering scaffolds, where they support tissue regeneration and remodeling before gradually breaking down into non-toxic byproducts. In contrast, non-bioresorbable biomaterials remain stable and intact within the body without undergoing significant degradation [17]. These materials are commonly used in permanent implants and medical devices, such as dental implants, orthopedic screws, and cardiovascular stents, where long-term mechanical stability and durability are required. Understanding this classification based on degradation behavior is essential for selecting the most appropriate biomaterial for specific medical applications, ensuring optimal performance and patient outcomes.

##### Bioresorbable Biomaterials

Bioresorbable biomaterials represent a groundbreaking class of materials designed to degrade and be absorbed by the body over time, offering significant advantages in various medical applications. These biomaterials are engineered to provide temporary support or therapeutic functions while gradually breaking down into non-toxic byproducts that can be metabolized or excreted by the body. One of the primary benefits of bioresorbable biomaterials is their ability to eliminate the need for surgical removal, reducing the risk of complications and improving patient comfort and recovery. This feature makes them particularly well-suited for use in temporary implants, drug delivery systems, and tissue engineering scaffolds, where they support tissue regeneration and remodeling before being naturally replaced by newly formed tissue.

Polymeric biomaterials such as poly (lactic acid) (PLA), poly (glycolic acid) (PGA), and their copolymers (PLGA) are commonly used as bioresorbable materials due to their biocompatibility, tunable degradation kinetics, and ability to degrade into non-toxic metabolites. These polymers can be synthesized with precise control over molecular weight, degradation rate, and mechanical properties, allowing for customized solutions tailored to specific medical needs. Bioresorbable metals such as magnesium alloys have emerged as promising candidates for temporary implants, orthopedic screws, and cardiovascular stents [16, 17], offering mechanical strength and corrosion resistance while gradually degrading in the body. The use of bioresorbable biomaterials in medical devices and implants offers several advantages over permanent materials, including reduced risk of infection, inflammation, and long-term complications. By eliminating the need for implant removal surgeries, bioresorbable biomaterials minimize patient discomfort, hospitalization time, and healthcare costs. Moreover, bioresorbable materials support tissue regeneration and remodeling, promoting natural healing processes and improving long-term outcomes. However, challenges such as controlling degradation kinetics, ensuring mechanical integrity during the healing process, and optimizing tissue integration remain areas of active research and development. As advancements continue, bioresorbable biomaterials hold immense promise for revolutionizing healthcare by providing safe, effective, and patient-friendly solutions for a wide range of medical needs.

#### Non-bioresorbable Biomaterials

Non-bioresorbable biomaterials are a crucial category of materials used in various medical applications where long-term stability and durability are essential. Unlike bioresorbable biomaterials, which degrade and are absorbed by the body over time, non-bioresorbable biomaterials remain stable and intact within the body without undergoing significant degradation. These materials are commonly employed in permanent implants and medical devices, such as dental implants, orthopedic screws, and cardiovascular stents, where mechanical strength and longevity are paramount. Non-bioresorbable biomaterials offer the advantage of providing sustained structural support and functionality over extended periods, ensuring reliable performance, and reducing the need for frequent replacements or revisions. Metals such as titanium alloys, stainless steel, and cobalt-chromium alloys [18] are among the most used non-bioresorbable biomaterials due to their exceptional mechanical properties, corrosion resistance, and biocompatibility. These metals are favored for load-bearing applications in orthopedics, dentistry, and cardiovascular surgery, where high strength and long-term stability are critical requirements. Ceramics such as alumina and zirconia are utilized in dental and orthopedic implants for their biocompatibility, wear resistance, and inertness in biological environments. These non-bioresorbable materials provide durable solutions for restoring or augmenting damaged or diseased tissues, improving patient mobility, and enhancing quality of life. Non-bioresorbable biomaterials offer several advantages over bioresorbable materials, including superior mechanical properties, predictable degradation behavior, and long-term stability.

By maintaining their structural integrity over time, non-bioresorbable biomaterials provide sustained support and functionality, reducing the risk of implant failure and associated complications. Moreover, non-bioresorbable [19] materials offer versatility in design and fabrication, allowing for the development of customized implants tailored to individual patient needs and anatomical requirements. However, challenges such as potential inflammatory responses, stress shielding effects, and long-term biocompatibility issues remain areas of concern that require ongoing research and optimization. Overall, non-bioresorbable biomaterials play a crucial role in modern medicine, providing durable solutions for a wide range of medical devices and implants that improve patient outcomes and quality of life.

#### 2.1.2.5 Based on Applications

Based on application, biomaterials are categorized according to their intended use and functional role within specific medical contexts. This classification framework enables researchers and practitioners to identify biomaterials tailored to address specific clinical needs and challenges across various healthcare domains. For example, orthopedic biomaterials are designed to support and restore skeletal structures, including implants for joint replacement and fracture fixation. In contrast, cardiovascular biomaterials [19] are engineered to interact with the circulatory system, encompassing devices such as stents, heart valves, and vascular grafts. By organizing biomaterials based on their intended applications, this classification system facilitates targeted research, development, and implementation of biomaterial-based solutions to improve patient care and treatment outcomes.

##### Orthopedic Biomaterials

Orthopedic biomaterials play a crucial role in the treatment of musculoskeletal disorders and injuries, providing support, stability, and functionality to bones and joints. These biomaterials are extensively used in orthopedic surgeries for joint replacements, fracture fixation, and spinal fusion procedures. One of the most used biomaterials in orthopedics is titanium and its alloys, renowned for their high strength, corrosion resistance, and biocompatibility. Titanium implants, such as hip and knee prostheses, restore mobility and alleviate pain in patients suffering from degenerative joint diseases like osteoarthritis. Bioactive ceramics such as hydroxyapatite (HA) and calcium phosphate coatings promote bone integration and osseointegration, enhancing the stability and longevity of orthopedic implants. Moreover, advancements in biomaterials science have led to the development of innovative biomaterial-based solutions for orthopedic applications, such as biodegradable polymers and composite materials. These materials offer the advantage of gradually degrading over time, eliminating the need for implant removal surgeries, and reducing the risk of complications. Biodegradable implants, such as screws and plates made from poly (lactic acid) (PLA) or polyglycolic acid (PGA), provide temporary support during bone healing and eventually dissolve into non-toxic byproducts [20]. Composite materials combining polymers with ceramics or metals offer customizable solutions with enhanced mechanical properties, promoting faster healing and better patient outcomes in orthopedic surgeries. Overall, orthopedic biomaterials continue to drive innovation and advancements in musculoskeletal healthcare, improving mobility, function, and quality of life for millions of patients worldwide.

#### Cardiovascular Biomaterials

Cardiovascular biomaterials play a pivotal role in the management of cardiovascular diseases, which remain a leading cause of morbidity and mortality worldwide. These biomaterials are specifically designed to interact with the cardiovascular system, addressing conditions such as coronary artery disease, heart failure, and arrhythmias. One of the most significant applications of cardiovascular biomaterials is in the development of stents, which are small mesh-like tubes used to open narrowed or blocked arteries and restore blood flow to the heart. Drug-eluting stents, coated with medications that inhibit cell proliferation and prevent restenosis, have revolutionized the treatment of coronary artery disease, reducing the need for repeat interventions, and improving patient outcomes. Furthermore, cardiovascular biomaterials are instrumental in the development of cardiac implants such as heart valves and pacemakers, which restore or regulate cardiac function in patients with structural heart defects or arrhythmias. Bioprosthetic heart valves, made from biological tissues or biomaterials, offer durable and biocompatible solutions for replacing damaged or diseased heart valves, restoring normal blood flow, and reducing symptoms of heart valve disease. Similarly, pacemaker leads and electrodes made from biocompatible metals like titanium or platinum provide reliable electrical stimulation and monitoring of cardiac activity [21], ensuring proper heart function and rhythm control. Through ongoing research and innovation, cardiovascular biomaterials continue to advance the field of cardiology, offering safer, more effective, and patient-friendly solutions for managing cardiovascular diseases and improving patient quality of life.

#### Dental Biomaterials

Dental biomaterials play a crucial role in restorative dentistry, providing solutions for repairing, replacing, and enhancing the function and aesthetics of teeth and oral tissues. These biomaterials are designed to withstand the harsh oral environment, including mechanical forces, temperature variations, and chemical exposure, while maintaining biocompatibility and longevity. One of the most used dental biomaterials is dental amalgam, a metallic alloy composed of silver, mercury, tin, and copper. Dental amalgam offers high strength, durability, and affordability, making it suitable for restoring posterior teeth affected by cavities or dental caries. However, advancements in dental biomaterials have led to the development of tooth-colored alternatives such as composite resins and ceramic materials, which offer improved aesthetics and better biomechanical properties for anterior restorations. Moreover, dental biomaterials play a critical role in various dental procedures, including endodontic treatment, periodontal therapy, and dental implantology. Root canal sealers and obturation materials ensure effective disinfection and sealing of the root canal system, preventing bacterial reinfection, and preserving the natural tooth structure. Periodontal biomaterials, such as bone grafts and membranes, promote tissue regeneration and osseointegration, enhancing the success of periodontal surgeries [22] and dental implant placement. Dental implants made from biocompatible materials like titanium or zirconia offer a durable and aesthetic solution for replacing missing teeth, restoring oral function, and improving patient quality of life. Through ongoing research and innovation, dental biomaterials continue to advance the field of dentistry, enabling clinicians to provide safe, effective, and aesthetically pleasing treatments to patients.

#### Soft Tissue Biomaterials

Soft tissue biomaterials are critical components in the field of regenerative medicine, offering solutions for repairing, augmenting, and replacing soft tissues such as skin, cartilage, and ligaments. These biomaterials are engineered to mimic the properties of native soft tissues while providing structural support, promoting tissue regeneration, and facilitating wound healing. One example of soft tissue biomaterials is dermal substitutes, which are used in the treatment of severe burns, chronic wounds, and skin defects. These substitutes provide a scaffold for cell infiltration and tissue ingrowth, promoting re-epithelialization and wound closure while minimizing scarring and infection risk [23]. Soft tissue biomaterials play a crucial role in tissue engineering applications, where they are used to create scaffolds, matrices, and grafts for repairing or replacing damaged or diseased tissues. Biomaterial-based constructs, such as collagen scaffolds, hyaluronic acid hydrogels [19], and decellularized extracellular matrices, provide a supportive environment for cell adhesion, proliferation, and differentiation. These biomaterials facilitate the regeneration of functional soft tissues, such as cartilage in joints or ligaments in musculoskeletal injuries, offering potential solutions for improving mobility, function, and quality of life for patients. Through ongoing research and innovation, soft tissue biomaterials continue to advance the field of regenerative medicine, enabling clinicians to provide safe, effective, and long-lasting treatments for soft tissue injuries and defects.

**Exercise 2.**

**Fill Ups**

**(Degradation, Applications, Functionality, Composition, Origin)**

1. Biomaterials can be classified based on their ...................., such as metals, ceramics, polymers, and composites.

2. Another classification criterion for biomaterials is their ..............., which includes natural, synthetic, and hybrid materials.

3. Functional classification of biomaterials categorizes them based on their intended .................., such as structural support, tissue regeneration, or drug delivery.

4. Biomaterials can also be classified based on their .................. behavior, including bioresorbable, non-resorbable, and bioinert materials.

5. Classification of biomaterials based on .................. involves grouping them according to their specific applications, such as orthopedic implants, cardiovascular devices, or dental materials.

## *2.1.3 Biocompatibility and Bioresponse*

Biocompatibility and bioresponse are fundamental concepts in the field of biomaterials science, encompassing the interaction between biomaterials and living organisms. Biocompatibility refers to the ability of biomaterials to perform their intended function within the body without eliciting adverse reactions or responses from the host tissues. Achieving biocompatibility is essential for ensuring the safety and effectiveness of medical implants, devices, and drug delivery systems [7], as it minimizes the risk of inflammation, infection, and rejection. Bioresponse, on the other hand, involves the biological reactions and responses that occur when biomaterials come into contact with living tissues. Understanding and optimizing biocompatibility and bioresponse are critical for the development of biomaterials with tailored properties and functionalities that promote tissue integration, healing, and regeneration while minimizing adverse effects.

### 2.1.3.1 Host Response to Biomaterials

The host response to biomaterials is a complex and dynamic process that involves interactions between the implanted biomaterial and the body's immune system. Upon implantation, biomaterials elicit a series of cellular and molecular responses aimed at assessing and responding to the foreign material. Initially, the biomaterial surface is recognized by circulating immune cells such as macrophages and neutrophils, triggering an inflammatory response characterized by the release of cytokines, chemokines, and reactive oxygen species. This acute inflammatory phase serves to remove debris, recruit immune cells to the site of injury, and initiate tissue repair processes. Subsequently, the host response transitions into a chronic phase characterized by the formation of a fibrous capsule around the biomaterial [24], known as fibrous encapsulation. Fibrous encapsulation serves as a protective barrier isolating the biomaterial from surrounding tissues and preventing its migration or degradation. However, excessive fibrous encapsulation can impair the function of the implanted biomaterial, leading to complications such as implant failure or tissue necrosis. Understanding the host response to biomaterials is crucial for designing materials that elicit favorable immune reactions and promote tissue integration while minimizing adverse effects and enhancing the long-term performance of biomedical implants and devices.

#### 2.1.3.2 Immunological Reactions

Immunological reactions to biomaterials occur when the body's immune system identifies the implanted material as foreign and mounts an immune response. This response can vary depending on factors such as the composition, structure, and surface properties of the biomaterial, as well as individual patient factors. In some cases, biomaterials may trigger hypersensitivity reactions or allergic responses, characterized by symptoms such as inflammation, swelling, and itching at the implant site. These reactions are mediated by immune cells such as mast cells and eosinophils, which release inflammatory mediators such as histamine and cytokines in response to the presence of the biomaterial. Biomaterials may provoke immune responses that lead to the activation of specific immune cells, such as T lymphocytes, which can recognize and target the implanted material for destruction [24]. This process, known as cell-mediated immunity, can result in the formation of granulomas or foreign body giant cells around the biomaterial, contributing to fibrous encapsulation and potential implant rejection. Understanding immunological reactions to biomaterials is essential for predicting and mitigating adverse immune responses, informing the design of biocompatible materials, and improving the safety and efficacy of biomedical implants and devices.

#### 2.1.3.3 Inflammatory Responses

The inflammatory response to biomaterials is a crucial aspect of the body's reaction to foreign materials introduced into its tissues. Upon implantation, biomaterials can elicit an inflammatory reaction characterized by the recruitment of immune cells, release of inflammatory mediators, and activation of cellular pathways. Inflammatory cells such as neutrophils and macrophages are among the first responders to the presence of biomaterials, initiating the cascade of events that constitute the inflammatory response. These cells release cytokines, chemokines, and reactive oxygen species, which mediate inflammation and recruit additional immune cells to the site of implantation. The inflammatory response serves several purposes, including clearing debris, combating infection, and initiating tissue repair processes. However, excessive or prolonged inflammation can lead to adverse effects such as tissue damage, fibrous encapsulation, and implant rejection. Strategies to modulate the inflammatory response to biomaterials include surface modifications to enhance biocompatibility [25], incorporation of anti-inflammatory agents, and designing biomaterials with properties that mimic those of native tissues. By understanding and controlling the inflammatory response, researchers can develop biomaterials that promote favorable tissue interactions and improve the safety and efficacy of biomedical implants and devices.

### *2.1.4 Surface Modification Techniques*

Surface modification techniques are pivotal tools in the field of biomaterials engineering, offering methods to tailor the surface properties of materials to meet specific biological and clinical requirements. These techniques aim to enhance the biocompatibility, functionality, and performance of biomaterials by modifying surface chemistry, topography, and roughness. By altering the surface characteristics of biomaterials [26], researchers can improve cell adhesion, proliferation, and differentiation, as well as reduce inflammation, infection, and implant rejection. Surface modification techniques play a crucial role in various biomedical applications, including medical implants, drug delivery systems, and tissue engineering scaffolds, enabling the development of safer, more effective, and patient-friendly biomaterials for clinical use.

#### 2.1.4.1 Coating

Coating is a versatile surface modification technique widely used in biomaterials engineering to improve the biocompatibility, functionality, and performance of medical implants and devices. Coatings can be applied to biomaterial surfaces using various methods, including physical vapor deposition, chemical vapor deposition, and dip coating. These coatings serve multiple purposes, such as enhancing corrosion resistance, reducing friction and wear, promoting cell adhesion, and providing controlled release of therapeutic agents. For example, hydroxyapatite coatings are commonly applied to orthopedic implants to improve osseointegration and accelerate bone healing, while drug-eluting coatings on stents enable localized drug delivery to prevent restenosis and enhance vascular healing.

#### 2.1.4.2 Plasma Treatment

Plasma treatment is another effective surface modification technique utilized to modify the surface properties of biomaterials through exposure to plasma, a highly reactive gas containing ions, radicals, and electrons. Plasma treatment can alter surface chemistry, topography, and wettability, leading to improved cell adhesion, protein adsorption, and biocompatibility. Common plasma treatment methods include plasma etching, plasma deposition, and plasma activation, each offering unique advantages for specific biomaterial applications [26, 27]. For instance, plasma etching can create micro- or nanostructured surfaces on biomaterials, enhancing cell attachment and proliferation, while plasma deposition can coat biomaterial surfaces with thin films of functional molecules or polymers to improve bioactivity and surface properties.

#### 2.1.4.3 Functionalization

Functionalization is a surface modification strategy that involves attaching functional groups or molecules to biomaterial surfaces to impart specific properties or functionalities. Functionalization can be achieved through chemical reactions, physical adsorption, or biomolecule immobilization techniques. This technique enables the introduction of bioactive molecules, such as cell adhesion peptides, growth factors, or antimicrobial agents, onto biomaterial surfaces, enhancing cell-material interactions, tissue integration, and therapeutic efficacy. For example, covalent immobilization of RGD peptides on biomaterial surfaces can promote cell adhesion and spreading [26], facilitating tissue regeneration and wound healing. Moreover, functionalization can also be employed to introduce stimuli-responsive or smart properties to biomaterials, enabling the controlled release of drugs or bioactive molecules in response to specific environmental cues or physiological signals.

### *2.1.5 Biomaterials Characterization*

Biomaterials characterization is a crucial aspect of biomaterials science and engineering, aimed at understanding and evaluating the physical, chemical, and mechanical properties of biomaterials (Fig. 2.4). Microscopy techniques are commonly employed to visualize the surface morphology, structure, and interactions of biomaterials at the micro- and nanoscale. Scanning electron microscopy (SEM) and atomic force microscopy (AFM) are widely used microscopy techniques that provide high-resolution images of biomaterial surfaces, enabling researchers to observe features such as roughness, porosity, and topography. Transmission electron microscopy (TEM) allows for the visualization of biomaterials at even higher resolutions, providing insights into internal structures, particle sizes, and crystallinity. These

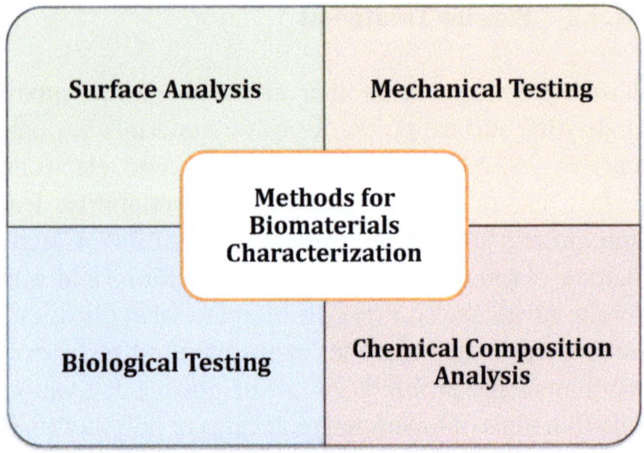


**Fig. 2.4** Various methods of biomaterial characterization

microscopy techniques are invaluable for characterizing biomaterials and understanding their structural properties, which influence cellular interactions and biological responses [28]. Spectroscopy techniques are utilized to analyze the chemical composition, molecular structure, and surface properties of biomaterials. Fourier transform infrared spectroscopy (FTIR) and Raman spectroscopy are commonly employed to identify functional groups, chemical bonds, and molecular vibrations present in biomaterials. These spectroscopic techniques provide valuable information about the chemical composition and molecular structure of biomaterials, aiding in the characterization of biomaterial surfaces, coatings, and interfaces. Additionally, X-ray photoelectron spectroscopy (XPS) and energy-dispersive X-ray spectroscopy (EDS) are used to analyze elemental compositions and quantify the presence of specific elements in biomaterials. Spectroscopy techniques are essential for elucidating the chemical properties and surface chemistry of biomaterials, which influence biocompatibility, bioactivity, and functionalization strategies.

Mechanical testing is critical for evaluating the mechanical properties and performance of biomaterials under physiological conditions. Tensile testing, compression testing, and flexural testing are commonly used mechanical testing methods to assess the tensile strength, compressive strength, and flexural modulus of biomaterials [1, 2, 4], respectively. These tests provide quantitative data on the mechanical behavior, stiffness, and elasticity of biomaterials, which are essential for predicting their performance in biomedical applications. Nanoindentation and microhardness testing techniques are employed to evaluate the hardness, indentation resistance, and surface mechanical properties of biomaterials at the micro- and nanoscale. Mechanical testing plays a crucial role in assessing the structural integrity, durability, and biomechanical compatibility of biomaterials, guiding the design and optimization of biomaterial-based implants, devices, and scaffolds.

Biomaterials characterization encompasses a range of techniques, including microscopy, spectroscopy, and mechanical testing, aimed at understanding and evaluating the properties of biomaterials. Microscopy techniques provide insights into the surface morphology and structure of biomaterials, while spectroscopy techniques analyze their chemical composition and surface chemistry. Mechanical testing evaluates the mechanical properties and performance of biomaterials under physiological conditions, guiding their design and optimization for biomedical applications [25]. Together, these characterization techniques play a crucial role in advancing biomaterials research and development, enabling the design of safer, more effective, and patient-friendly biomaterial-based therapies and devices.

### *2.1.6 Future Trends in Biomaterials*

Future trends in biomaterials are shaped by innovative approaches and emerging technologies that hold the potential to revolutionize the field of healthcare. Nanotechnology in biomaterials is one such trend, leveraging nanoscale materials and structures to engineer biomaterials with enhanced properties and functionalities. Nanomaterials [20] offer unique advantages such as high surface area-to-volume ratios, tunable mechanical properties, and versatile surface chemistry, making them ideal candidates for applications in drug delivery, tissue engineering, and medical diagnostics. By incorporating nanoparticles, nanofibers, and nanocomposites into biomaterials, researchers can create platforms for targeted drug delivery, controlled release of therapeutic agents, and enhanced tissue regeneration. Nanotechnology enables the development of diagnostic tools and imaging probes with unprecedented sensitivity and specificity, paving the way for early disease detection and personalized treatment strategies [28]. Bioinspired materials represent another promising trend in biomaterials research, drawing inspiration from nature's design principles and biological systems to develop biomimetic materials with advanced functionalities. By mimicking the hierarchical structures, properties, and functions found in natural tissues and organisms, bioinspired materials offer solutions for tissue engineering, regenerative medicine, and biomedical devices. For example, biomaterials designed to mimic the extracellular matrix can promote cell adhesion, migration, and differentiation, facilitating tissue regeneration and repair. Similarly, bioinspired materials inspired by the properties of spider silk, mussel adhesive proteins, and bone mineralization processes offer innovative approaches for designing biocompatible coatings, adhesives, and scaffolds. By harnessing nature's design strategies, bioinspired materials have the potential to address complex biomedical challenges and improve patient outcomes in diverse clinical settings. Personalized medicine approaches are reshaping the landscape of biomaterials research and development, aiming to tailor medical treatments and interventions to individual patient characteristics, preferences, and needs. Advances in genomics, proteomics, and biomarker technologies enable the identification of patient-specific biomarkers and molecular signatures, guiding the selection of biomaterials and treatment strategies. Personalized medicine approaches leverage this molecular information to design biomaterials that interact with the patient's biological systems in a targeted and customized manner. For example, biomaterial-based drug delivery [6, 7] systems can be engineered to deliver therapeutics at specific sites or in response to physiological cues, maximizing treatment efficacy while minimizing side effects. 3D printing technologies enable the fabrication of patient-specific implants and tissue scaffolds tailored to individual anatomical dimensions and pathology, offering personalized solutions for surgical interventions and tissue regeneration.

Future trends in biomaterials are driven by nanotechnology, bioinspired materials, and personalized medicine approaches, which hold immense promise for transforming healthcare and improving patient outcomes. Nanotechnology offers opportunities to engineer biomaterials with enhanced properties and functionalities for drug delivery [26], tissue engineering, and medical diagnostics. Bioinspired materials draw inspiration from nature's design principles to develop biomimetic materials with advanced functionalities for tissue regeneration, medical devices, and biomedical coatings. Personalized medicine approaches aim to tailor medical treatments and interventions to individual patient characteristics, leveraging molecular information to design biomaterials and treatment strategies customized to each patient's needs [29]. By embracing these innovative approaches, biomaterials researchers are poised to address complex biomedical challenges and pave the way for a new era of personalized and precision medicine.

**Exercise 3.**

**Choose the Correct Option**

1. Which of the following is NOT a category in the broad classification of biomaterials based on composition?
   a) Metals
   b) Ceramics
   c) Polysaccharides
   d) Polymers

2. Biomaterials can be classified based on origin into:
   a) Synthetic and natural
   b) Inorganic and organic
   c) Resorbable and non-resorbable
   d) Metals and ceramics

3. Functional biomaterials are designed to:
   a) Serve as structural components
   b) Mimic natural tissues and organs
   c) Prevent degradation in the body
   d) None of the above

4. Bioresorbable biomaterials:
   a) Remain unchanged in the body over time
   b) Are slowly broken down and absorbed by the body
   c) Are only used in dental applications
   d) Are completely inert in biological environments

5. Biomaterials used in orthopedic applications primarily focus on:
   a) Dental reconstruction
   b) Treatment of cardiovascular diseases
   c) Repair and replacement of bones and joints
   d) Brain stimulation therapy

6. Which of the following techniques is commonly used to analyze the surface morphology of biomaterials?
   a) Infrared spectroscopy
   b) X-ray diffraction
   c) Scanning electron microscopy (SEM)
   d) Nuclear magnetic resonance (NMR)

7. What does Fourier-transform infrared (FTIR) spectroscopy primarily measure in biomaterials characterization?
   a) Surface roughness
   b) Chemical composition
   c) Mechanical properties
   d) Electrical conductivity

8. Which of the following properties of biomaterials is crucial for their biocompatibility?
   a) Tensile strength
   b) Thermal conductivity
   c) Surface energy
   d) Electrical resistivity

9. The ability of a biomaterial to withstand mechanical stresses without breaking is known as:
   a) Biodegradability
   b) Flexibility
   c) Toughness
   d) Elasticity

10. Hydroxyapatite is commonly used in biomaterials due to its:
   a) Electrical conductivity
   b) Antibacterial properties
   c) Biodegradability
   d) Bioactivity

### 2.1.7 Summary

Biomaterials are specialized materials designed to interact with biological systems, ranging from tissues to cells, for medical purposes. They play a crucial role in various biomedical applications, including medical implants, tissue engineering, drug delivery systems, and diagnostic tools. The chapter begins by introducing the broad classification of biomaterials based on their composition and properties. These classifications typically include metals, polymers, ceramics, and composite materials. Each class offers unique advantages and disadvantages, making them suitable for specific applications. The selection of biomaterials [1, 2] is a critical aspect of biomaterials science and engineering. This chapter explores the criteria and considerations involved in selecting the most appropriate biomaterial for a particular application. Key factors include biocompatibility, mechanical properties, degradation behavior, surface characteristics, and manufacturing feasibility [4, 5]. Biocompatibility, in particular, refers to the ability of a material to perform its desired function within a specific biological environment without causing adverse reactions or harm to the host organism. Mechanical properties play a significant role in determining the suitability of a biomaterial for load-bearing applications, such as orthopedic implants or dental prostheses. Understanding the mechanical behavior of biomaterials under different physiological conditions is essential for ensuring their long-term performance and stability within the body. Moreover, degradation behavior is critical for biomaterials designed for temporary implants or drug delivery systems, where controlled degradation over time is desired to avoid long-term complications. Surface characteristics heavily influence the interaction between biomaterials and biological systems. Surface modifications, such as coatings or functionalization, can enhance biocompatibility, promote cell adhesion, and modulate the immune response to implanted devices. The chapter discusses various surface modification techniques and their impact on the performance and biocompatibility of biomaterials [15, 16]. Additionally, it highlights the importance of manufacturing processes in controlling the structure, properties, and performance of biomaterials, emphasizing the need for precise and reproducible fabrication methods.

Overall, this chapter serves as a comprehensive introduction to the fundamentals of biomaterials, providing readers with a solid understanding of the principles governing the design, selection, and application of biomaterials in medicine and biotechnology.

# Points to Remember

- **Nature has perfected the art of mineralization in organisms like sea sponges and diatoms, which produce intricate silica structures. Researchers draw inspiration from these natural processes to create biomaterials with unique properties, such as strength, porosity, and lightness.**

- **As humanity sets its sights on space colonization, biomaterials offer innovative solutions for sustaining life beyond Earth. Researchers are exploring bioregenerative life support systems that use biomaterials to recycle waste, produce food, and generate oxygen in closed-loop environments.**

- **Bacterial cellulose, produced by microorganisms like Acetobacter xylinum, is a versatile biomaterial with applications ranging from wound dressings to food packaging. Its unique properties, including high strength, biocompatibility, and transparency, make it an attractive alternative to traditional materials.**

- **Bioluminescent organisms, such as fireflies and certain marine species, produce light through biochemical reactions. Researchers are harnessing this natural phenomenon to develop bioluminescent biomaterials for applications in bioimaging, biosensing, and environmental monitoring.**

- **Biomimicry extends beyond biological structures to include natural processes like origami. Origami-inspired biomaterials leverage principles of folding and unfolding to create dynamic, shape-shifting materials with applications in drug delivery, soft robotics, and tissue engineering.**

- **Biomaterials can be designed to self-assemble at the molecular level, driven by specific interactions such as hydrogen bonding and hydrophobic interactions. This bottom-up approach allows for precise control over material properties and nanostructure formation, paving the way for advanced materials with tailored functionalities.**

## References

1. Baltacis K, Bystrov V, Bystrova A, Dekhtyar Y, Freivalds T, Raines J, Rozenberga K, Sorokins H, Zeidaks M (2020) Physical fundamentals of biomaterials surface electrical functionalization. Materials (Basel, Switzerland) 13(20):4575. https://doi.org/10.3390/ma13204575
2. Leali PT, Merolli A (2009) Fundamentals of biomaterials. In: Merolli A, Joyce TJ (eds) Biomaterials in hand surgery. Springer, Milano. https://doi.org/10.1007/978-88-470-1195-3_1
3. Rathore R, Suhag D, Wan F, Verma R, Thakur A, Thakur P (2023) Functionalization of biomaterials. In: Suhag D, Thakur A, Thakur P (eds) Integrated nanomaterials and their applications. Springer, Singapore. https://doi.org/10.1007/978-981-99-6105-4_11
4. Troy E, Tilbury MA, Power AM, Wall JG (2021) Nature-based biomaterials and their application in biomedicine. Polymers 13(19):3321. https://doi.org/10.3390/polym13193321
5. Lemons JE, Lucas LC (1986) Properties of biomaterials. J Arthroplasty 1(2):143–147. https://doi.org/10.1016/s0883-5403(86)80053-5
6. Rathore R, Suhag D, Wan F, Thakur A, Thakur P (2023) Toxicology and toxicity studies of nano-biomaterials. In: Suhag D, Thakur A, Thakur P (eds) Integrated nanomaterials and their applications. Springer, Singapore. https://doi.org/10.1007/978-981-99-6105-4_20
7. Suhag D, Bhatia R, Das S, Shakeel A, Ghosh AB, Singh A et al (2015) Physically cross-linked pH-responsive hydrogels with tunable formulations for controlled drug delivery. RSC Adv 5(66):53963–53972. https://doi.org/10.1039/c5ra07424j
8. Agrawal R, Kumar A, Mohammed MKA, Singh S (2023) Biomaterial types, properties, medical applications, and other factors: a recent review [最新综述:生物材料类型、性能、医疗应用及其他因素]. J Zhejiang Univ. Sci 1–16 (Advance online publication). https://doi.org/10.1631/jzus.A2200403
9. Wang L, Wang C, Wu S, Fan Y, Li X (2020) Influence of the mechanical properties of biomaterials on degradability, cell behaviors and signaling pathways: current progress and challenges. Biomater Sci 8(10):2714–2733. https://doi.org/10.1039/d0bm00269k
10. Nakayama K (1976) Iyo denshi to seitai kogaku. Jpn J Med Electron Biol Eng 14(1):58–63
11. Binyamin G, Shafi BM, Mery CM (2006) Biomaterials: a primer for surgeons. Semin Pediatr Surg 15(4):276–283. https://doi.org/10.1053/j.sempedsurg.2006.07.007
12. Williams DF (1976) Biomaterials and biocompatibility. Med Prog Technol 4(1–2):31–42
13. Bhardwaj A, Pandey, LM (2022) Biomaterials: types and applications. In: Nanoscale engineering of biomaterials: properties and applications (pp. 89–114). https://doi.org/10.1007/978-981-16-3667-7_4
14. Shirvan AR, Nouri A, Wen C (2021) Structural polymer biomaterials. In: Elsevier eBooks (pp. 395–439). https://doi.org/10.1016/b978-0-12-818831-6.00010-0
15. Ali MA, Bhuiyan MH (2021) Types of biomaterials useful in brain repair. Neurochem Int 146:105034. https://doi.org/10.1016/j.neuint.2021.105034
16. Hench LL (1980) Biomaterials. Science (New York, N.Y.) 208(4446):826–831. https://doi.org/10.1126/science.6246576
17. Verma D, Okhawilai M, Goh KL, Sharma M, Pasbakhsh P, Ramakrishna S (2024) Introduction to bioresorbable polymers: types and biomedical applications. In: Elsevier eBooks (pp. 3–22). https://doi.org/10.1016/b978-0-443-18915-9.00020-3
18. Springer IN, Fleiner B, Jepsen S, Acil Y (2001) Culture of cells gained from temporomandibular joint cartilage on non-absorbable scaffolds. Biomaterials 22(18):2569–2577
19. Todros S, Todesco M, Bagno A (2021) Biomaterials and their biomedical applications: from replacement to regeneration. Processes 9(11):1949. https://doi.org/10.3390/pr9111949
20. Shakeel A, Singh A, Das S et al (2017) Synthesis and morphological insight of new biocompatible smart hydrogels. J Polym Res 24:113. https://doi.org/10.1007/s10965-017-1267-7
21. Jaganathan SK, Supriyanto E, Murugesan S, Balaji A, Asokan MK (2014) Biomaterials in cardiovascular research: applications and clinical implications. BioMed Res Int 2014:1–11. https://doi.org/10.1155/2014/459465

22. Bayne SC (2005) Dental biomaterials: where are we and where are we going? J Dent Educ 69(5):571–585. https://doi.org/10.1002/j.0022-0337.2005.69.5.tb03943.x
23. Cornwell KG, Landsman A, James KS (2009) Extracellular matrix biomaterials for soft tissue repair. Clin Podiatr Med Surg 26(4):507–523. https://doi.org/10.1016/j.cpm.2009.08.001
24. Wei F, Liu S, Chen M, Tian G, Zha K, Yang Z, Jiang S, Li M, Sui X, Chen Z, Guo Q (2021) Host response to biomaterials for cartilage tissue engineering: key to remodeling. Front Bioeng Biotechnol 9:664592. https://doi.org/10.3389/fbioe.2021.664592
25. Tang L, Eaton JW (1995) Inflammatory responses to biomaterials. Am J Clini Pathol 103(4):466–471. https://doi.org/10.1093/ajcp/103.4.466
26. Raval N, Kalyane D, Maheshwari R, Tekade RK (2019) Surface modifications of biomaterials and their implication on biocompatibility. In: Elsevier eBooks (pp. 639–674). https://doi.org/10.1016/b978-0-12-814427-5.00017-2
27. Chu P (2002) Plasma-surface modification of biomaterials. Mater Sci Eng R Rep 36(5–6):143–206. https://doi.org/10.1016/s0927-796x(02)00004-9
28. Oliveira JPR, Melendez-Ortiz HI, Bucio E, Alves PT, Lima MIS, Goulart LR, Mathor MB, Varca GHC, Lugao AB (2018) Current methods applied to biomaterials—Characterization approaches, safety assessment and biological international standards. Curr Top Med Chem 18(4):256–274. https://doi.org/10.2174/1568026618666180410151518
29. Ratner BD (1983) Surface characterization of biomaterials by electron spectroscopy for chemical analysis. Ann Biomed Eng 11(3–4):313–336. https://doi.org/10.1007/BF02363290

# Chapter 3
# Biodegradable Biomaterials

## 3.1 Introduction

Biodegradable biomaterials have emerged as a vital area of research and development at the intersection of materials science, biology, and engineering. These materials possess the unique ability to degrade over time into harmless byproducts, making them environmentally friendly alternatives to conventional non-biodegradable materials. The significance of biodegradable biomaterials lies in their diverse applications across various fields, including biomedicine, environmental protection, and sustainable packaging. At the core of biodegradable biomaterials is their inherent ability to undergo degradation through natural processes [1]. Unlike traditional materials that persist in the environment for extended periods, biodegradable biomaterials break down into simpler compounds through enzymatic, hydrolytic, or microbial actions. This degradation process not only eliminates the accumulation of waste but also offers the potential for recycling nutrients back into the ecosystem. The key advantage of biodegradable biomaterials is their compatibility with biological systems, making them ideal candidates for biomedical applications [2]. These materials can be engineered to mimic the properties of native tissues, facilitating integration with the body, and minimizing adverse reactions. Biodegradable scaffolds for tissue engineering, drug delivery carriers, and implantable medical devices are just a few examples of their transformative impact on healthcare. Biodegradable biomaterials play a crucial role in addressing environmental challenges associated with plastic pollution and resource depletion. By offering sustainable alternatives to conventional plastics, these materials contribute to reducing the burden on landfills and mitigating the adverse effects of plastic waste on ecosystems and wildlife. From biodegradable packaging materials to compostable agricultural films, the versatility of biodegradable biomaterials offers promising solutions for a more sustainable future.

The development of biodegradable biomaterials is driven by ongoing advancements in materials science and engineering [3]. Researchers are continuously

 
D. Suhag et al., *Handbook of Biomaterials for Medical Applications, Volume 1*,
Biomedical Materials for Multi-functional Applications,
https://doi.org/10.1007/978-981-97-4818-1_3

exploring novel biomaterials and fabrication techniques to enhance their properties and expand their applications. From biopolymers derived from renewable resources to bioactive nanoparticles with tailored functionalities, the landscape of biodegradable biomaterials continues to evolve with innovative solutions.

## *3.1.1 Types of Biodegradable Biomaterials*

Biodegradable biomaterials encompass a diverse range of materials with the common characteristic of being capable of undergoing degradation into non-toxic byproducts in a biological environment. These materials can be broadly classified into natural polymers, synthetic polymers, and composite biomaterials, each offering unique advantages and applications. Natural polymers, such as collagen, chitosan, and alginate, are derived from renewable sources and exhibit excellent biocompatibility, making them ideal for biomedical applications. Synthetic polymers, including polylactic acid (PLA), polyglycolic acid (PGA), and polycaprolactone (PCL), offer tunable mechanical properties and degradation kinetics, allowing for precise control over material performance [1, 2]. Composite biomaterials combine the advantages of natural and synthetic components to achieve tailored properties and functionalities, paving the way for innovative solutions in tissue engineering, drug delivery, and regenerative medicine.

### 3.1.1.1 Natural Polymers

Natural polymers are biodegradable biomaterials derived from renewable sources such as plants, animals, and microorganisms, offering a sustainable alternative to synthetic materials. One of the key advantages of natural polymers lies in their inherent biocompatibility, which arises from their chemical similarity to biological macromolecules found in the human body. Examples of natural polymers include collagen, elastin, chitosan, alginate, and hyaluronic acid, each exhibiting unique properties and functionalities that make them suitable for a wide range of biomedical applications. Collagen, the most abundant protein in the human body, is widely used in tissue engineering and regenerative medicine due to its excellent biocompatibility and ability to promote cell adhesion and proliferation. Elastin, another essential structural protein, provides elasticity and resilience to tissues such as skin, blood vessels, and lungs, making it valuable for applications requiring flexibility and durability. Chitosan, derived from chitin found in the exoskeleton of crustaceans, possesses antimicrobial properties and is utilized in wound dressings, drug delivery systems, and tissue scaffolds. Alginate, extracted from brown seaweed, forms hydrogels capable of encapsulating cells and bioactive molecules, making it suitable for controlled drug delivery and tissue engineering applications. Hyaluronic acid, a major component of the extracellular matrix, exhibits viscoelastic properties and plays a crucial role in tissue hydration and lubrication [3, 4]. Its biocompatibility and biodegradability make it an attractive candidate for dermal fillers, ophthalmic solutions, and osteoarthritis treatments.

Natural polymers offer versatility in terms of processing techniques, allowing for the fabrication of scaffolds, films, fibers, and nanoparticles tailored to specific biomedical applications. Biodegradable and non-toxic, these materials minimize the risk of adverse reactions and foreign body responses, making them suitable for implantable devices and therapeutic formulations [5]. The abundance and diversity of natural polymers in nature provide ample opportunities for innovation and the development of sustainable biomaterials with tailored properties and functionalities.

#### 3.1.1.2 Synthetic Polymers

Synthetic polymers represent a diverse class of biodegradable biomaterials engineered through chemical synthesis, offering precise control over their composition, structure, and properties (Fig. 3.1). These materials play a crucial role in biomedical applications, providing tailored solutions for tissue engineering, drug delivery, and medical device fabrication. Key examples of synthetic polymers include polylactic acid (PLA), polyglycolic acid (PGA), polycaprolactone (PCL), and polyethylene glycol (PEG), each with unique characteristics that make them suitable for specific applications. Polylactic acid (PLA) is a biocompatible and biodegradable polymer derived from renewable resources such as corn starch or sugarcane [4, 5]. It exhibits good mechanical properties and degradation kinetics, making it an attractive choice for sutures, tissue scaffolds, and drug delivery systems. Polyglycolic acid (PGA), known for its high strength and rapid degradation, is commonly used in absorbable sutures and tissue engineering scaffolds due to its ability to provide temporary mechanical support before being metabolized by the body. Polycaprolactone (PCL) is a semi-crystalline polyester with a slow degradation rate, making it suitable for long-term implantable devices such as drug-eluting stents and bone fixation screws. Its flexibility and biocompatibility allow for the fabrication of porous scaffolds for tissue regeneration and controlled release formulations for drug delivery. Polyethylene glycol (PEG) is a hydrophilic polymer widely used in biomaterials due to its ability to improve biocompatibility, reduce protein adsorption, and enhance drug solubility [6]. It is utilized in formulations for drug delivery, tissue engineering, and surface modification of medical devices.

Synthetic polymers offer tunable properties and functionalities through modifications in molecular weight, chemical structure, and copolymerization with other monomers. This versatility enables the design of biomaterials tailored to specific applications and performance requirements. Synthetic polymers can be processed using a variety of techniques, including extrusion, injection molding, and electrospinning, to create complex architectures and controlled release systems. The biodegradability of synthetic polymers can be tailored to match the desired degradation profile for a given application, ranging from weeks to months or even years. By adjusting the polymer's molecular weight and crystallinity [6], researchers can control the rate of degradation and the release kinetics of encapsulated drugs or bioactive molecules.

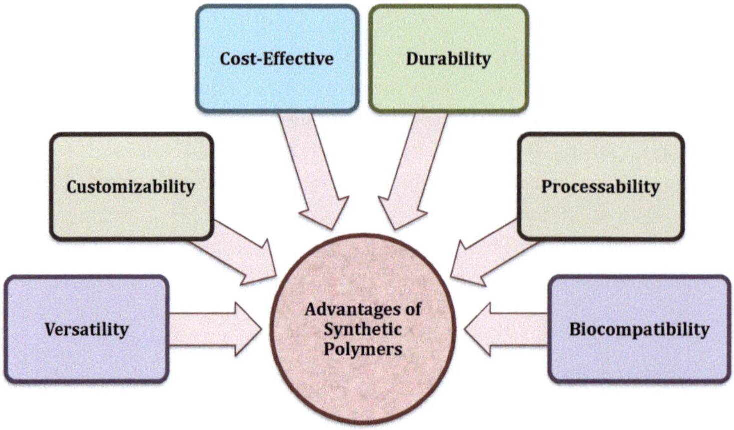


**Fig. 3.1** Advantages of synthetic polymers

This level of control is essential for optimizing the efficacy and safety of biomedical implants and drug delivery systems.

#### 3.1.1.3 Composite Biomaterials

Composite biomaterials represent a class of biodegradable materials composed of two or more distinct components, each contributing unique properties and functionalities to the overall material. By combining the advantages of different materials, composite biomaterials offer enhanced mechanical strength, biological performance, and versatility compared to individual components alone. These materials play a pivotal role in various biomedical applications, including tissue engineering, drug delivery, and regenerative medicine. One of the key advantages of composite biomaterials is their ability to mimic the complex structure and properties of native tissues, providing an ideal microenvironment for cell growth and tissue regeneration. By incorporating bioactive components such as growth factors, peptides, or nanoparticles into a biocompatible matrix, composite biomaterials can modulate cellular behavior and promote tissue regeneration [7]. This ability to tailor material properties at the molecular level enables precise control over tissue-specific functions and therapeutic outcomes. Examples of composite biomaterials include polymer-ceramic composites, polymer-hydrogel hybrids, and nanocomposites, each designed to meet specific requirements for tissue regeneration and drug delivery. Polymer-ceramic composites, for instance, combine the flexibility of polymers with the bioactivity of ceramics, making them suitable for bone tissue engineering applications. By incorporating ceramic nanoparticles or microparticles into a polymer matrix, these composites can enhance mechanical properties and stimulate bone formation. Polymer-hydrogel hybrids offer another versatile platform for composite biomaterials, combining the water-retaining properties of hydrogels with the mechanical strength of polymers. These hybrids can be tailored to mimic the extracellular matrix and provide a supportive scaffold for cell growth and tissue regeneration. Moreover, nanocomposites composed of polymer matrices reinforced with nanoparticles such as carbon nanotubes or graphene oxide exhibit superior mechanical properties and electrical conductivity [8], making them promising candidates for neural tissue engineering and biosensing applications. Composite biomaterials enable the incorporation of multiple therapeutic agents or bioactive molecules within a single platform, allowing for synergistic effects and controlled release kinetics. By encapsulating drugs, growth factors, or genes within a biodegradable matrix, composite biomaterials can deliver therapeutic payloads locally and sustainably, minimizing systemic side effects and improving therapeutic efficacy. This targeted delivery approach is particularly advantageous for treating localized diseases or promoting tissue regeneration in specific anatomical sites. Composite biomaterials offer tunable degradation kinetics and mechanical properties, allowing for customization according to specific application requirements. By adjusting the composition, ratio, and processing parameters, researchers can tailor the degradation rate, stiffness, and porosity of composite biomaterials to match the dynamic needs of the host tissue environment. This flexibility enables the design of biomimetic scaffolds and implants with optimal performance and compatibility for clinical translation.

### *3.1.2 Biodegradation Mechanisms*

Biodegradation mechanisms are fundamental processes through which biodegradable biomaterials undergo breakdown and transformation into simpler compounds in response to biological environments. Understanding these mechanisms is essential for designing biomaterials with tailored degradation profiles and predicting their behavior in vivo. Biodegradation can occur through various pathways, including enzymatic degradation, hydrolytic degradation, and microbial degradation, each influenced by factors such as material composition, structure, and environmental conditions [8]. By elucidating the mechanisms underlying biodegradation, researchers can develop biocompatible and environmentally sustainable biomaterials for diverse applications in healthcare, environmental protection, and beyond.

#### 3.1.2.1 Enzymatic Degradation

Enzymatic degradation is a prominent pathway through which biodegradable biomaterials undergo breakdown in biological environments, facilitated by the action of enzymes produced by living organisms. Enzymes are biological catalysts that accelerate chemical reactions, including the hydrolysis of polymer chains into smaller fragments. In the context of biodegradable biomaterials, enzymes such as proteases, lipases, and carbohydrases play key roles in breaking down protein, lipid, and carbohydrate-based polymers, respectively [9]. The specificity of enzymes enables the selective degradation of biomaterials based on their chemical composition, leading to a controlled and site-specific breakdown. Proteases are enzymes that catalyze the hydrolysis of peptide bonds in proteins, leading to the cleavage of protein chains into amino acid fragments. In the context of biodegradable biomaterials, proteases play a crucial role in degrading protein-based polymers such as collagen, gelatin, and silk fibroin. These materials are commonly used in tissue engineering scaffolds and drug delivery systems, where controlled degradation is essential for tissue regeneration and therapeutic release. Proteases produced by cells in the surrounding tissue or infiltrating immune cells can catalyze the breakdown of implanted biomaterials, allowing for gradual integration and remodeling of the tissue-engineered constructs [10]. Similarly, lipases are enzymes that catalyze the hydrolysis of ester bonds in lipids, leading to the degradation of lipid-based polymers such as polyesters and polylactones. These materials, including polylactic acid (PLA) and polycaprolactone (PCL), are widely used in biomedical applications due to their biocompatibility and biodegradability. Lipases secreted by cells in the local microenvironment or present in bodily fluids can initiate the degradation of implanted biomaterials, resulting in the release of degradation products that are metabolized or excreted by the body.

Carbohydrases, including amylases and cellulases, catalyze the hydrolysis of glycosidic bonds in carbohydrates, leading to the degradation of carbohydrate-based polymers such as cellulose, chitosan, and starch. These materials are utilized in various biomedical applications, including wound dressings, drug delivery systems, and tissue engineering scaffolds. Carbohydrases produced by microorganisms in the surrounding environment or by host cells can initiate the degradation of implanted biomaterials, enabling their clearance and remodeling over time. The rate and extent of enzymatic degradation can be modulated through various strategies, including the incorporation of enzyme-sensitive moieties into the polymer backbone or surface modifications to enhance enzyme accessibility [11]. By engineering biomaterials to respond to specific enzymes present in the target tissue or biological fluid, researchers can achieve precise control over degradation kinetics and degradation products. This tunability enables the design of biomaterials with tailored degradation profiles for applications requiring sustained release of therapeutic agents or controlled tissue regeneration.

Enzymatic degradation of biodegradable biomaterials is influenced by factors such as pH, temperature, and the presence of cofactors or inhibitors. Variations in these environmental conditions can affect enzyme activity and substrate specificity, thereby modulating the rate and mechanism of degradation. Understanding the interplay between enzymatic activity and environmental factors is essential for optimizing the performance and longevity of biomaterial implants in vivo.

#### 3.1.2.2 Hydrolytic Degradation

Hydrolytic degradation is a predominant pathway by which biodegradable biomaterials undergo breakdown in aqueous environments, involving the cleavage of chemical bonds through the addition of water molecules. This process is particularly relevant for hydrolytically sensitive polymers such as polyesters, polycarbonates, and polyamides, which contain ester, carbonate, or amide linkages susceptible to hydrolysis. Hydrolytic degradation occurs through nucleophilic attack on the polymer backbone, leading to chain scission and the formation of soluble degradation products. Polyesters, including polylactic acid (PLA), polyglycolic acid (PGA), and their copolymers, are widely utilized in biomedical applications due to their biocompatibility and biodegradability [9, 10]. Hydrolytic degradation of polyesters occurs through the cleavage of ester bonds, resulting in the formation of carboxylic acids and alcohols as degradation products. The rate of hydrolytic degradation is influenced by factors such as polymer crystallinity, molecular weight, and chemical structure, with amorphous regions and higher molecular weights generally exhibiting faster degradation kinetics. Polycarbonates, such as poly (trimethylene carbonate) (PTMC) and poly (ε-caprolactone-co-trimethylene carbonate) (P(TMC-CL)), are another class of biodegradable polymers susceptible to hydrolytic degradation. These materials contain carbonate linkages that undergo hydrolysis under physiological conditions, leading to the formation of hydroxy acids and carbon dioxide as degradation byproducts. The degradation rate of polycarbonates can be modulated by adjusting the ratio of carbonate to other monomers in copolymer formulations, enabling control over degradation kinetics and mechanical properties [11]. Hydrolytic degradation of biodegradable biomaterials is influenced by environmental factors such as pH, temperature, and humidity, which can affect the rate and mechanism of degradation. For example, acidic pH and elevated temperatures accelerate hydrolytic degradation by increasing the rate of hydrolysis reactions, whereas alkaline pH and low temperatures may slow down degradation kinetics. Understanding the interplay between environmental conditions and polymer chemistry is essential for predicting the behavior of biodegradable biomaterials in vivo and optimizing their performance for specific applications. Hydrolytic degradation can be exploited for the controlled release of encapsulated drugs or bioactive molecules from biodegradable matrices. By incorporating hydrolytically sensitive linkages into polymer chains or drug delivery systems, researchers can achieve programmable release profiles tailored to the therapeutic requirements of a given application [12]. This targeted release approach minimizes systemic side effects and maximizes the efficacy of therapeutic agents, offering advantages for treatments requiring sustained drug delivery or localized therapy. Hydrolytic degradation products generated during the breakdown of biodegradable biomaterials are generally metabolizable and non-toxic, minimizing the risk of adverse effects on surrounding tissues or systemic circulation. The gradual release of degradation products allows for their clearance by the body's natural detoxification pathways, ensuring biocompatibility and safety. This feature is particularly important for biomedical implants and drug delivery systems intended for long-term use in vivo.

#### 3.1.2.3 Microbial Degradation

Microbial degradation is a significant pathway through which biodegradable biomaterials undergo breakdown in natural environments, facilitated by the activity of microorganisms such as bacteria, fungi, and algae. These microorganisms produce enzymes capable of catalyzing the degradation of organic materials, including biodegradable polymers, into simpler compounds that can be assimilated as nutrients. Microbial degradation plays a crucial role in the biogeochemical cycling of carbon and other elements, contributing to the recycling of organic matter and the maintenance of ecosystem balance [13]. Biodegradable polymers such as polyesters, polycarbonates, and polysaccharides are susceptible to microbial degradation under appropriate environmental conditions. Microorganisms colonize the surface of biodegradable biomaterials and secrete extracellular enzymes that hydrolyze polymer chains into monomers or oligomers, which can be transported into microbial cells and metabolized as carbon and energy sources. This process leads to the gradual breakdown of biomaterials and the release of degradation products into the surrounding environment. Bacteria are among the most common microorganisms involved in microbial degradation of biodegradable biomaterials, possessing a wide range of metabolic capabilities and adaptability to diverse environmental conditions. Bacterial species such as Pseudomonas, Bacillus, and Clostridium have been identified as capable of degrading various biopolymers, including polyhydroxyalkanoates (PHA), polylactic acid (PLA), and cellulose [3]. These bacteria produce extracellular enzymes such as lipases, proteases, and cellulases that catalyze the breakdown of polymer chains into smaller fragments suitable for microbial assimilation. Fungi also play a significant role in microbial degradation, particularly in environments rich in organic matter such as soil, compost, and aquatic ecosystems. Fungal species such as Aspergillus, Penicillium, and Trichoderma are known for their ability to degrade lignocellulosic materials such as cellulose and lignin. Fungi secrete a diverse array of enzymes, including cellulases, hemicellulases, and ligninases, which act synergistically to break down complex plant polymers into simpler sugars and aromatic compounds [3, 13]. Microbial degradation of biodegradable biomaterials is influenced by environmental factors such as temperature, moisture, pH, and nutrient availability, which affect the composition and activity of microbial communities. Optimal conditions for microbial degradation vary depending on the specific polymers and microorganisms involved, with some polymers being more readily degraded under aerobic conditions, while others are degraded preferentially under anaerobic conditions. Microbial degradation of biodegradable biomaterials can be harnessed for waste management and environmental remediation purposes. Biodegradable polymers such as polyhydroxyalkanoates (PHA) and polylactic acid (PLA) have been explored as alternatives to conventional plastics for applications such as compostable packaging and biodegradable mulches [14]. By designing polymers that are specifically targeted by microbial communities present in composting facilities or soil environments, researchers can accelerate the degradation process and minimize the environmental impact of plastic waste. Microbial degradation products generated during the breakdown of biodegradable biomaterials can serve as valuable resources for microbial growth and metabolism. These degradation products, including sugars, organic acids, and amino acids, can support the proliferation of microbial communities involved in nutrient cycling and soil fertility [15]. By promoting microbial activity and diversity, biodegradable biomaterials contribute to the sustainability of agricultural and environmental systems.

**Exercise 1**

**Fill ups**

**(Rate, Degradation, Release, Biodegradation, Bulk)**

1. ............................... is the process by which materials are broken down into simpler compounds by natural biological processes.
2. Enzymes play a crucial role in catalyzing the .................. of biodegradable polymers.
3. Biodegradable materials can undergo ............... degradation, where they break down into smaller fragments.
4. Environmental factors such as temperature and pH can influence the ............ of biodegradable materials.
5. Biodegradable mechanisms are important in applications such as drug delivery, where controlled .................. of the carrier is desired.

### *3.1.3 Fabrication Techniques*

Fabrication techniques are essential processes in the production of biodegradable biomaterials, enabling the creation of complex structures and tailored properties for various applications in healthcare, environmental protection, and beyond (Fig. 3.2). These techniques encompass a diverse array of methodologies, including solution casting, electrospinning, and 3D printing, each offering unique advantages in terms of scalability, precision, and versatility. Fabrication techniques play a crucial role in translating the design concepts of biodegradable biomaterials into functional devices and scaffolds with controlled morphology, mechanical properties, and degradation kinetics [16]. Understanding the principles and capabilities of different fabrication techniques is essential for optimizing the performance and efficacy of biodegradable biomaterials in biomedical applications, environmental remediation, and sustainable manufacturing processes.

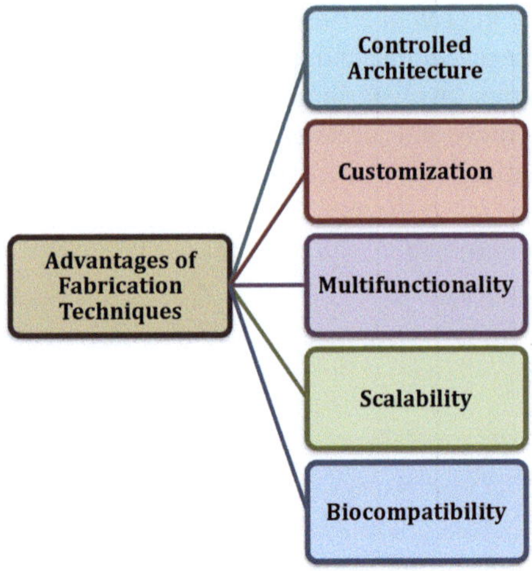


**Fig. 3.2** Advantages of fabrication techniques

#### 3.1.3.1 Solution Casting

Solution casting is a versatile fabrication technique widely used in the production of biodegradable biomaterials, offering precise control over material composition, morphology, and properties. In solution casting, a polymer is dissolved in a solvent to form a homogeneous solution, which is then cast into a mold or onto a substrate to create thin films, coatings, or 3D structures. This technique is particularly well-suited for biodegradable polymers such as polylactic acid (PLA), polyglycolic acid (PGA), and polycaprolactone (PCL), which can be dissolved in organic solvents such as chloroform, dichloromethane, or tetrahydrofuran. The versatility of solution casting lies in its ability to produce thin films with controlled thickness and uniformity, making it suitable for applications such as drug delivery systems, tissue engineering scaffolds, and barrier coatings. By adjusting parameters such as polymer concentration, solvent type, casting temperature, and drying conditions, researchers can tailor the properties of solution-cast films to meet specific requirements for mechanical strength [14], degradation rate, and biocompatibility. Additionally, solution casting allows for the incorporation of additives such as drugs, growth factors, or nanoparticles into the polymer matrix to impart desired functionalities to the final product. Solution casting offers scalability and cost-effectiveness compared to other fabrication techniques such as electrospinning or 3D printing, making it suitable for large-scale production of biodegradable biomaterials [16]. Batch-to-batch consistency and reproducibility are achievable through careful control of processing parameters and quality assurance measures, ensuring reliability and uniformity in the final product. Solution casting can be combined with other processing techniques such as solvent evaporation-induced phase separation or templating to create hierarchical structures or porous architectures with enhanced properties for specific applications.

Solution casting enables the fabrication of multicomponent systems and composite materials by blending different polymers or incorporating reinforcing agents such as fibers or nanoparticles into the solution. This approach allows for the creation of hybrid materials with synergistic properties, such as improved mechanical strength, bioactivity, or controlled release kinetics. By optimizing the formulation and processing conditions, researchers can engineer solution-cast biomaterials with tailored properties and functionalities for diverse biomedical and environmental applications [17]. Solution casting offers advantages in terms of environmental sustainability, as it enables the use of biodegradable solvents and reduces waste generation compared to solvent-intensive techniques. Solvent recycling and purification techniques can be employed to minimize environmental impact and enhance resource efficiency in the fabrication process. This eco-friendly aspect makes solution casting an attractive option for the production of biodegradable biomaterials aligned with principles of green chemistry and sustainable manufacturing [17]. Solution casting is a versatile and cost-effective fabrication technique for producing biodegradable biomaterials with tailored properties and functionalities. Its ability to create thin films, coatings, and 3D structures with precise control over material composition and morphology makes it suitable for a wide range of biomedical and environmental applications.

#### 3.1.3.2 Electrospinning

Electrospinning is a versatile and widely used fabrication technique for producing nanofibrous structures from biodegradable polymers, offering precise control over fiber morphology, diameter, and alignment. In electrospinning, a polymer solution or melt is extruded from a syringe needle or spinneret under the influence of an electric field, resulting in the formation of ultrafine fibers that are collected on a grounded substrate or rotating drum. This process leverages electrostatic forces to stretch and elongate the polymer solution into fibers with diameters ranging from nanometers to micrometers, making it suitable for applications such as tissue engineering scaffolds, drug delivery carriers, and filtration membranes. The unique architecture of electrospun nanofibers, characterized by a high surface area-to-volume ratio and interconnected porous structure, offers several advantages for biomedical and environmental applications. These include enhanced cell adhesion, proliferation, and differentiation in tissue engineering scaffolds, as well as high drug loading capacity and controlled release kinetics in drug delivery systems. Moreover, the tunable porosity and surface chemistry of electrospun nanofibers enable selective adsorption and separation of molecules or contaminants, making them promising candidates for water purification and environmental remediation.

Electrospinning offers unparalleled versatility in the fabrication of biodegradable biomaterials, as it allows for the incorporation of bioactive molecules, nanoparticles, or other additives into the polymer matrix to impart desired functionalities to the resulting nanofibers. By adjusting parameters such as polymer concentration, solution viscosity, electric field strength, and processing conditions, researchers can tailor the properties of electrospun nanofibers to meet specific requirements for mechanical strength, degradation rate, and bioactivity [18]. Additionally, the use of co-electrospinning techniques enables the production of composite nanofibers with synergistic properties derived from multiple components. Electrospinning offers scalability and adaptability to various substrate geometries, making it suitable for the fabrication of 3D scaffolds, aligned fiber arrays, and hierarchical structures. Advanced electrospinning setups, such as multi-jet electrospinning and coaxial electrospinning, enable precise control over fiber alignment, patterning, and encapsulation, allowing for the creation of biomimetic architectures that mimic the complexity and functionality of native tissues. This capability is particularly valuable for applications requiring spatially controlled release of therapeutic agents or directional guidance of cell growth and tissue regeneration. Electrospinning holds promise for the development of next-generation biomaterials with advanced functionalities, such as stimuli-responsive or conductive nanofibers for targeted drug delivery or electrical stimulation of cells [18]. By incorporating responsive polymers or conductive additives into electrospun nanofibers, researchers can engineer materials that respond to external stimuli such as pH, temperature, or electric fields, enabling on-demand release of drugs or modulation of cellular behavior. These smart biomaterials have potential applications in personalized medicine, regenerative therapies, and bioelectronics.

Electrospinning offers advantages in terms of material efficiency and waste reduction, as it enables precise control over fiber deposition and minimizes the use of solvents compared to other fabrication techniques.

#### 3.1.3.3 3D Printing

3D printing, also known as additive manufacturing, is a transformative fabrication technique that enables the layer-by-layer deposition of materials to create 3D objects with complex geometries. In the realm of biodegradable biomaterials [17], 3D printing offers unprecedented versatility and precision in the fabrication of scaffolds, implants, and drug delivery systems for biomedical applications. This technology allows for the customization of implants and devices according to patient-specific anatomical features, leading to improved treatment outcomes and reduced risk of complications. One of the key advantages of 3D printing is its ability to fabricate complex structures with high resolution and accuracy, making it ideal for applications requiring intricate architectures and precise control over material distribution. By precisely controlling the deposition of biodegradable polymers or composites, researchers can create scaffolds with tailored mechanical properties, degradation kinetics, and bioactive functionalities. This level of customization enables the design of scaffolds that mimic the hierarchical organization and mechanical properties of native tissues, promoting cell adhesion, proliferation, and tissue regeneration.

3D printing offers scalability and flexibility in the production of biodegradable biomaterials, allowing for rapid prototyping and iterative design optimization.

This capability accelerates the translation of research findings into clinical applications, facilitating the development of innovative therapies and medical devices [17]. Additionally, 3D printing can integrate multiple materials or components within a single scaffold, enabling the fabrication of complex structures with spatially controlled properties and functionalities. 3D printing enables the incorporation of bioactive molecules, cells, or growth factors into biodegradable scaffolds, facilitating controlled release and localized delivery for therapeutic applications. By encapsulating drugs or biological agents within the scaffold matrix, researchers can achieve sustained release profiles tailored to specific treatment regimens, minimizing systemic side effects, and maximizing therapeutic efficacy. This targeted delivery approach is particularly valuable for tissue engineering, regenerative medicine, and personalized medicine applications. 3D printing allows for the fabrication of patient-specific implants and medical devices, revolutionizing the field of personalized healthcare. By leveraging medical imaging data such as MRI or CT scans, clinicians can design implants that precisely match the anatomical geometry of individual patients, leading to improved fit, function, and comfort. This personalized approach reduces the risk of implant rejection and enhances patient satisfaction, highlighting the transformative potential of 3D printing in healthcare.

3D printing holds promise for advancing regenerative medicine and tissue engineering by enabling the fabrication of vascularized scaffolds and organoids. By incorporating microvascular networks or bioactive cues within 3D-printed scaffolds [19], researchers can promote angiogenesis and tissue innervation, enhancing the survival and functionality of engineered tissues and organs. This capability opens up new avenues for the development of implantable constructs for organ replacement and transplantation, addressing critical challenges in organ shortage and patient care.

3D printing offers environmental benefits by minimizing material waste and energy consumption compared to traditional subtractive manufacturing techniques. Additive manufacturing processes generate less waste by utilizing only the necessary materials for building the desired object, reducing material costs and environmental impact.

### *3.1.4 Properties of Biodegradable Biomaterials*

The properties of biodegradable biomaterials play a pivotal role in determining their performance and suitability for various biomedical applications. These materials exhibit a diverse range of characteristics, including mechanical strength, biocompatibility, degradation kinetics, and bioactivity, which influence their behavior in biological environments. Understanding and controlling these properties are essential for designing biomaterials that can safely and effectively interact with living tissues, promote tissue regeneration, and degrade at controlled rates [1, 4]. By tailoring the properties of biodegradable biomaterials to specific application requirements, researchers can develop innovative solutions for tissue engineering, drug delivery, medical devices, and environmental remediation.

#### 3.1.4.1 Mechanical Properties

The mechanical properties of biodegradable biomaterials are critical factors that influence their performance and functionality in biomedical applications. These properties include stiffness, strength, toughness, and elasticity, which collectively determine the material's ability to withstand mechanical forces and maintain structural integrity during use. In tissue engineering scaffolds, for example, mechanical properties play a crucial role in providing mechanical support to cells, guiding tissue formation, and facilitating integration with the surrounding tissue. Similarly, in drug delivery systems and medical devices, mechanical properties dictate the material's ability to withstand physiological stresses and maintain functionality over time. One of the key considerations in designing biodegradable biomaterials is achieving mechanical properties that match those of the native tissue or biological environment [20]. This ensures compatibility and minimizes the risk of mechanical mismatch, which can lead to inflammation, tissue damage, or implant failure. By selecting appropriate polymer compositions, processing techniques, and cross-linking methods, researchers can tailor the mechanical properties of biodegradable biomaterials to match the specific requirements of different tissues and applications. The mechanical properties of biodegradable biomaterials can be modulated over a wide range to accommodate diverse biomedical needs. For example, in bone tissue engineering, materials with high stiffness and strength are desirable to provide structural support and mimic the mechanical properties of bone tissue. In contrast, in soft tissue engineering or drug delivery applications, materials with lower stiffness and higher elasticity may be preferred to enhance biocompatibility and minimize mechanical trauma to surrounding tissues.

The degradation behavior of biodegradable biomaterials can significantly influence their mechanical properties over time [21]. As the material degrades, its structural integrity may weaken, leading to changes in stiffness, strength, and other mechanical properties. Understanding the degradation kinetics and degradation mechanisms of biodegradable biomaterials is essential for predicting their mechanical behavior and ensuring long-term performance in vivo. Strategies such as controlling polymer degradation rate, reinforcing the material with nanoparticles or fibers, or designing composite structures can help mitigate degradation-induced changes in mechanical properties [21]. The mechanical properties of biodegradable biomaterials can be tailored to accommodate dynamic physiological environments and biological processes. For example, in wound healing applications, materials with tunable mechanical properties can adapt to changes in tissue mechanics during the healing process, promoting optimal tissue regeneration and remodeling. Similarly, in drug delivery systems, materials with stimuli-responsive mechanical properties can release therapeutic agents in response to specific physiological cues, enhancing treatment efficacy and reducing side effects. Advancements in fabrication techniques such as 3D printing and electrospinning enable precise control over the mechanical properties of biodegradable biomaterials at the micro- and nanoscale levels. By designing complex architectures, hierarchical structures, and gradient compositions, researchers can create biomaterials with tailored mechanical properties optimized for specific applications. This level of control opens new possibilities for engineering biomimetic scaffolds, implants, and drug delivery systems that closely mimic the mechanical behavior of native tissues and organs.

The mechanical properties of biodegradable biomaterials are critical determinants of their performance and functionality in biomedical applications. By understanding the relationship between material composition, processing techniques, degradation behavior, and mechanical properties, researchers can design biomaterials that meet the mechanical requirements of diverse tissues and applications [20].

#### 3.1.4.2 Degradation Kinetics

Degradation kinetics is a fundamental aspect of biodegradable biomaterials that governs the rate and mechanism of material breakdown over time in biological environments. Understanding degradation kinetics is essential for predicting the behavior of biomaterials in vivo and optimizing their performance for specific biomedical applications. The degradation process of biodegradable biomaterials involves various chemical and biological mechanisms, including hydrolysis, enzymatic degradation, and microbial degradation, which collectively influence the overall degradation kinetics [13]. Hydrolysis is one of the primary degradation mechanisms for biodegradable polymers, involving the cleavage of chemical bonds by water molecules. The rate of hydrolysis depends on factors such as polymer chemistry, molecular weight, crystallinity, and environmental conditions such as pH and temperature. Hydrolytic degradation kinetics typically follow first-order or pseudo-first-order kinetics, characterized by an initial rapid degradation phase followed by a slower degradation rate as the polymer chains degrade into smaller fragments. Enzymatic degradation occurs through the action of enzymes produced by living organisms, such as proteases, lipases, and carbohydrases, which catalyze the breakdown of biodegradable polymers into soluble degradation products. The rate of enzymatic degradation is influenced by factors such as enzyme concentration, specificity, and substrate accessibility [13]. Enzymatic degradation kinetics can vary depending on the type of enzyme involved and the chemical structure of the polymer substrate, leading to differential degradation rates for different biomaterials. Microbial degradation involves the activity of microorganisms such as bacteria, fungi, and algae, which produce extracellular enzymes capable of degrading biodegradable polymers into simpler compounds. Microbial degradation kinetics depend on factors such as microbial diversity, activity, and environmental conditions such as temperature, moisture, and nutrient availability. Microbial degradation can accelerate the breakdown of biodegradable biomaterials in natural environments such as soil, water, or compost, leading to complete mineralization or assimilation of the material by microorganisms.

The degradation kinetics of biodegradable biomaterials can be modulated through various strategies, including material composition, structure, and processing techniques. For example, altering the chemical composition of polymers, incorporating hydrolytically or enzymatically sensitive linkages, or controlling the degree of cross-linking can influence degradation kinetics and degradation products. Additionally, adjusting processing parameters such as polymer concentration, solvent type, and curing conditions can affect the degradation behavior of biomaterials. Degradation kinetics play a crucial role in determining the lifespan and stability of biodegradable biomaterials in biomedical applications. For temporary implants or drug delivery systems, controlling degradation kinetics is essential to ensure that the biomaterial degrades at a rate compatible with tissue healing or therapeutic release requirements. Conversely, for long-term implants or tissue engineering scaffolds, modulating degradation kinetics is necessary to achieve gradual remodeling and integration with the surrounding tissue without compromising mechanical integrity [13, 21]. Degradation kinetics can be tailored to match the dynamic needs of specific biological environments or physiological processes.

#### 3.1.4.3 Biocompatibility

Biocompatibility is a fundamental characteristic of biomaterials that refers to their ability to perform their intended function without eliciting harmful effects on biological systems. In biomedical applications, biocompatibility is essential to ensure that biomaterials interact safely and effectively with living tissues, cells, and organs. Biocompatible biomaterials should not trigger adverse immune responses, inflammation, cytotoxicity, or other deleterious reactions when implanted or used in contact with biological fluids [22]. Achieving biocompatibility requires careful consideration of material properties, surface characteristics, and interactions with the host environment. One of the key aspects of biocompatibility is the surface properties of biomaterials, including surface chemistry, topography, and charge. Surface modification techniques such as plasma treatment, chemical functionalization, or biomimetic coatings can alter the surface properties of biomaterials to enhance biocompatibility and promote favorable interactions with biological entities. For example, the immobilization of bioactive molecules or cell-adhesive peptides onto biomaterial surfaces can facilitate cell attachment, proliferation, and tissue integration, leading to improved biocompatibility and tissue regeneration. The degradation of products of biodegradable biomaterials can influence their biocompatibility and biological responses. Biodegradable polymers such as polylactic acid (PLA), polyglycolic acid (PGA), and polycaprolactone (PCL) degrade into metabolizable byproducts that are cleared from the body through natural metabolic pathways [13, 22]. However, the degradation process can release acidic or toxic degradation products that may affect cell viability, inflammatory responses, or tissue remodeling. Understanding the degradation kinetics and biocompatibility of biodegradable biomaterials is essential for designing implants, scaffolds, and drug delivery systems with optimal performance and safety profiles.

The mechanical properties of biomaterials play a crucial role in their biocompatibility and suitability for biomedical applications. Biomaterials should possess mechanical properties that match those of the surrounding tissues to prevent mechanical mismatches and minimize stress concentrations that could lead to tissue damage or implant failure. Flexible and elastic biomaterials are preferred for applications such as soft tissue engineering and wound healing, where dynamic mechanical forces are present.

The immunogenicity of biomaterials is a critical factor in determining their biocompatibility and host response. Biomaterials that induce immune responses or foreign body reactions may trigger inflammation, fibrosis, or rejection by the host immune system. Strategies to minimize immunogenicity include using biocompatible materials, optimizing surface properties to reduce protein adsorption and immune cell activation, and incorporating immunomodulatory agents or anti-inflammatory drugs into biomaterial formulations [23]. Biocompatibility testing is an essential step in the development and evaluation of biomaterials for biomedical applications. In vitro assays, such as cell viability assays, cytotoxicity assays, and inflammation assays, are commonly used to assess the biocompatibility of biomaterials in controlled laboratory settings. In vivo studies, including animal models and preclinical trials, provide valuable insights into the biocompatibility, biodegradation, and tissue responses of biomaterials in more complex biological environments.

The biocompatibility of biomaterials can be influenced by factors such as sterilization methods, storage conditions, and manufacturing processes. Sterilization techniques such as gamma irradiation, ethylene oxide gas, or autoclaving should be carefully selected to ensure that biomaterials retain their structural integrity and biocompatibility while eliminating microbial contaminants. Additionally, proper storage conditions, such as temperature, humidity, and packaging, should be maintained to prevent degradation or contamination of biomaterials prior to use.

### *3.1.5 Applications in Biomedical Engineering*

In biomedical engineering, the applications of advanced materials play a pivotal role in revolutionizing healthcare and improving patient outcomes. From tissue engineering and regenerative medicine to drug delivery systems and medical devices, biomaterials contribute to innovative solutions for addressing complex medical challenges [24]. These materials are designed to interact seamlessly with biological systems, providing structural support, promoting tissue regeneration, and delivering therapeutic agents with precision and efficiency. By harnessing the unique properties of biomaterials, biomedical engineers can develop customized implants, scaffolds, and devices tailored to individual patient needs, advancing the frontiers of diagnosis, treatment, and patient care.

#### 3.1.5.1 Tissue Engineering Scaffolds

Tissue engineering scaffolds represent a cornerstone in regenerative medicine, offering a promising approach to repair or replace damaged or diseased tissues and organs (Fig. 3.3). These scaffolds serve as temporary 3D frameworks that support cell attachment, proliferation, and differentiation, facilitating the formation of new tissue

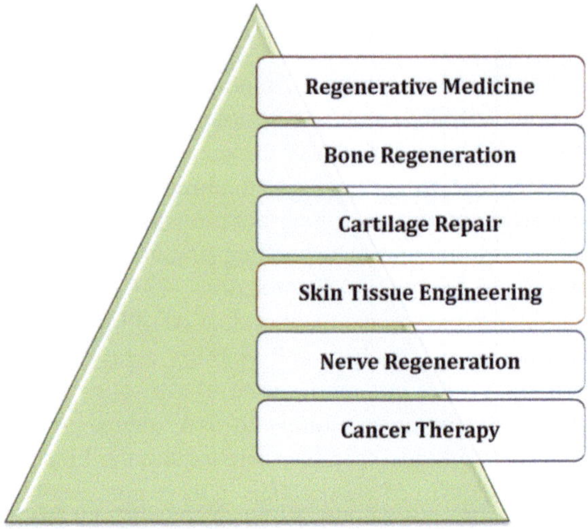


**Fig. 3.3** Applications of tissue engineering scaffolds using biodegradable materials

in vitro or in vivo. Tissue engineering scaffolds are designed to mimic the architecture, mechanical properties, and biological cues of native extracellular matrices, providing a conducive environment for tissue regeneration and integration. By leveraging the principles of biomaterials science, scaffold design, and cell biology, tissue engineering scaffolds hold immense potential for addressing critical healthcare needs and advancing personalized medicine. One of the key considerations in tissue engineering scaffold design is the selection of biomaterials with suitable properties for the target tissue or organ [24, 25]. Biodegradable polymers such as polylactic acid (PLA), polyglycolic acid (PGA), and polycaprolactone (PCL) are commonly used due to their biocompatibility, tunable mechanical properties, and ability to degrade into non-toxic byproducts. Natural polymers such as collagen, fibrin, and hyaluronic acid are also utilized for their inherent bioactivity and ability to mimic the composition and structure of native extracellular matrices. Tissue engineering scaffolds can be fabricated using various techniques, including electrospinning, 3D printing, and particulate leaching, to achieve desired architectural features and mechanical properties. Electrospinning allows for the production of nanofibrous scaffolds with high surface area and porosity, which promote cell adhesion and tissue integration. 3D printing enables the precise control of scaffold geometry and internal structure, facilitating the creation of complex tissue-like architectures with spatially controlled properties. Particulate leaching involves the incorporation of sacrificial particles into a polymer matrix, which are subsequently removed to create interconnected porous structures suitable for cell infiltration and nutrient exchange. Tissue engineering scaffolds can be functionalized with bioactive molecules, growth factors, or signaling peptides to enhance cell attachment, proliferation, and differentiation. Surface modification techniques such as covalent immobilization, physical adsorption, or biomimetic coating allow for the incorporation of bioactive cues into scaffold matrices, promoting specific cellular responses and guiding tissue regeneration.

Additionally, scaffolds can be engineered to release therapeutic agents in a controlled manner, providing localized delivery of drugs or growth factors to promote tissue healing and regeneration.

The mechanical properties of tissue engineering scaffolds play a crucial role in determining their performance and suitability for different applications. Scaffolds should possess mechanical properties that match those of the native tissue or organ to provide adequate support and mechanical cues for cell growth and tissue formation. Scaffolds should be biocompatible and degrade at a rate compatible with tissue regeneration, allowing for gradual remodeling and integration with the surrounding tissue without compromising structural integrity. Tissue engineering scaffolds can be tailored to specific tissue types and anatomical sites, allowing for personalized approaches to regenerative medicine. By adjusting scaffold composition, architecture, and biofunctionalization strategies, researchers can develop customized implants and devices optimized for individual patient needs [26]. This personalized approach holds promise for addressing challenges such as tissue defects, organ failure, and age-related degenerative diseases, offering new avenues for enhancing quality of life and extending lifespan.

#### 3.1.5.2 Drug Delivery Systems

Drug delivery systems play a pivotal role in the context of biodegradable biomaterials, offering innovative solutions for targeted and controlled drug release while minimizing adverse effects and optimizing therapeutic outcomes. Biodegradable biomaterials, such as polymers and hydrogels, serve as carriers or matrices for encapsulating drugs and delivering them to specific sites in the body. These biomaterials degrade over time into non-toxic byproducts, eliminating the need for surgical removal and reducing the risk of long-term complications associated with permanent implants. One of the key advantages of drug delivery systems based on biodegradable biomaterials is their ability to provide sustained and localized drug release, ensuring therapeutic concentrations of drugs at the target site while minimizing systemic exposure and side effects. Biodegradable polymers such as polylactic acid (PLA), polyglycolic acid (PGA), and their copolymers can be formulated into nanoparticles, microparticles [24, 25], or implants to encapsulate drugs and release them in a controlled manner over extended periods. This sustained release of kinetics improves patient compliance and reduces the frequency of drug administration, leading to better treatment outcomes.

Drug delivery systems based on biodegradable biomaterials offer versatility in terms of drug loading capacity, release kinetics, and route of administration, making them suitable for a wide range of therapeutic applications. These systems can be engineered to accommodate different types of drugs, including small molecules, proteins, peptides, and nucleic acids, allowing for personalized and tailored treatment regimens. Biodegradable biomaterials can be formulated into various dosage forms, such as injectable hydrogels, implantable scaffolds, or transdermal patches, to meet specific patient needs and clinical requirements. Drug delivery systems based on biodegradable biomaterials enable the co-delivery of multiple drugs or therapeutic agents, allowing for synergistic effects, combination therapies, or sequential release profiles. By encapsulating multiple drugs within the same carrier system, clinicians can achieve enhanced therapeutic efficacy, overcome drug resistance, and minimize adverse drug interactions. This approach is particularly beneficial for treating complex diseases such as cancer, infectious diseases, and chronic inflammatory conditions, where multiple therapeutic agents may be required for optimal treatment outcomes. Drug delivery systems based on biodegradable biomaterials can be tailored to target specific tissues, organs, or cells within the body, enhancing drug delivery efficiency and minimizing off-target effects [27]. Surface modification techniques, such as ligand conjugation, nanoparticle functionalization, or hydrogel patterning, allow for the precise targeting of drug-loaded carriers to diseased or inflamed tissues. This targeted delivery approach improves drug bioavailability, reduces systemic toxicity, and enhances therapeutic efficacy, making it ideal for precision medicine and personalized treatment strategies.

Drug delivery systems based on biodegradable biomaterials offer advantages in terms of biocompatibility, safety, and biodegradability compared to conventional drug delivery platforms. These biomaterials are inherently biocompatible and degrade into non-toxic byproducts that are metabolized and eliminated from the body through natural pathways [27]. As a result, they minimize the risk of immune responses, inflammation, or tissue damage associated with foreign body reactions, enhancing patient tolerance and acceptance of the treatment.

Drug delivery systems based on biodegradable biomaterials contribute to environmental sustainability by reducing the accumulation of non-biodegradable waste and minimizing the use of harmful chemicals or additives. Biodegradable polymers derived from renewable resources or waste materials further enhance the sustainability profile of these systems, supporting the transition toward greener and more environmentally friendly healthcare practices.

#### 3.1.5.3 Implants and Prosthetics

Implants and prosthetics represent transformative technologies in healthcare, restoring functionality and improving quality of life for individuals with physical impairments or injuries (Fig. 3.4). In the context of biodegradable biomaterials, these devices offer innovative solutions that address challenges associated with traditional permanent implants, such as long-term biocompatibility, risk of infection, and need for revision surgeries. Biodegradable biomaterials, including polymers and composites, serve as promising alternatives for implants and prosthetics, providing temporary support or replacement while facilitating tissue regeneration and eventual degradation. One of the key advantages of implants and prosthetics based on biodegradable biomaterials is their ability to degrade over time within the body [28], eliminating the need for surgical removal. Unlike permanent implants made from metals or ceramics, biodegradable implants gradually degrade into non-toxic byproducts that are metabolized and excreted by the body's natural processes. This eliminates the risk of long-term complications associated with permanent implants, such as stress shielding, corrosion, and foreign body reactions, leading to improved patient outcomes and satisfaction.

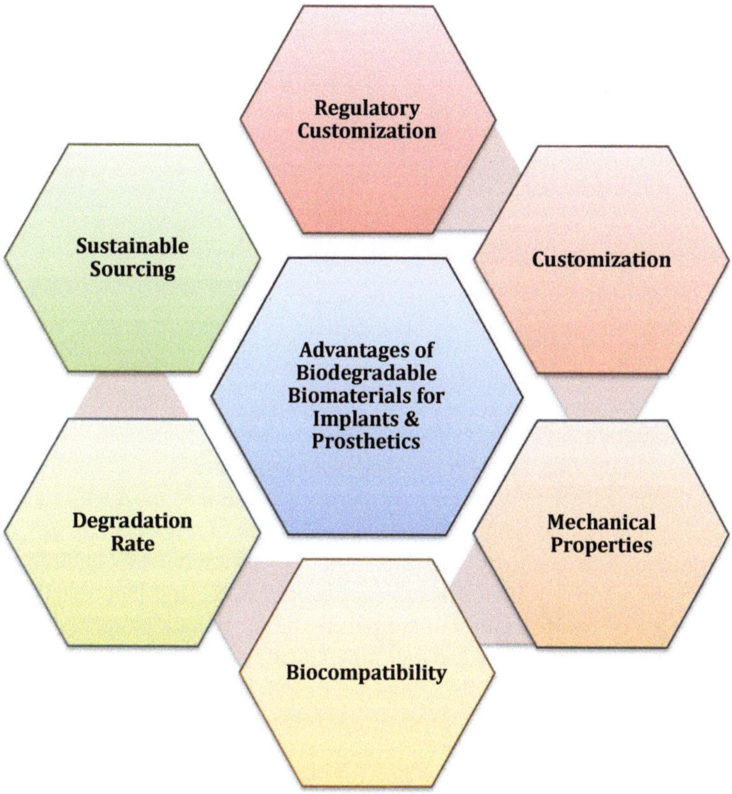


**Fig. 3.4** Advantages of biodegradable biomaterials for implants & prosthetics

**Exercise 2**

**Fill ups**

**(Cell growth, Biomedical implants, Tissue engineering, Efficacy, Restenosis)**

**1.** Biodegradable polymers are commonly used in ................................. applications for temporary support and eventual tissue regeneration.

**2.** Drug-eluting stents made from biodegradable materials can release medication to prevent ........................and promote vessel healing.

**3.** Biodegradable scaffolds provide a temporary framework for .................. and guide tissue regeneration in regenerative medicine.

**4.** Biodegradable nanoparticles can deliver drugs with controlled release kinetics, enhancing ..................and reducing side effects.

**5.** Biodegradable hydrogels are utilized in ........................ for controlled drug delivery and tissue engineering applications.

Implants and prosthetics based on biodegradable biomaterials offer tailored degradation profiles and mechanical properties that match those of native tissues, facilitating tissue integration and regeneration. Biodegradable polymers such as polylactic acid (PLA), polyglycolic acid (PGA), and their copolymers can be engineered to degrade at controlled rates, providing temporary structural support while promoting tissue healing and remodeling. This tailored degradation kinetics ensures that the implant maintains its integrity and functionality throughout the healing process before gradually resorbing and being replaced by new tissue. Implants and prosthetics based on biodegradable biomaterials support tissue regeneration and remodeling by providing a scaffold or matrix for cell attachment, proliferation, and differentiation. These biomaterials mimic the composition and structure of native extracellular matrices, creating a conducive environment for tissue ingrowth and vascularization.

Implants and prosthetics based on biodegradable biomaterials offer advantages in terms of surgical handling and customization, allowing for precise shaping, sizing, and placement to fit individual patient anatomy and clinical needs [28]. Biodegradable polymers can be fabricated using techniques such as injection molding, 3D printing, or electrospinning to create customized implants and prosthetics with complex geometries and patient-specific features. This customization improves implant-host integration, reduces surgical complications, and enhances patient comfort and satisfaction. Implants and prosthetics based on biodegradable biomaterials contribute to environmental sustainability by reducing the accumulation of non-biodegradable waste associated with traditional permanent implants. Biodegradable polymers derived from renewable resources or waste materials further enhance the sustainability profile of these devices, supporting the transition toward greener and more environmentally friendly healthcare practices. Implants and prosthetics based on biodegradable biomaterials offer opportunities for dynamic or temporary interventions in cases where permanent implants may not be suitable or necessary [26]. For example, biodegradable fixation devices for fracture repair or ligament reconstruction provide temporary stability and support during the healing process before gradually resorbing and being replaced by native tissue. Similarly, biodegradable scaffolds for tissue engineering and regenerative medicine applications enable the regeneration of damaged or diseased tissues without the need for permanent implants.

### *3.1.6 Environmental Impact and Sustainability*

Environmental impact and sustainability are pivotal considerations that shape the development, utilization, and disposal of these materials in various industries, including healthcare, agriculture, and packaging. Biodegradable biomaterials offer the potential to reduce the environmental footprint of products and processes by providing alternatives to conventional materials that persist in the environment for extended periods. These biomaterials degrade naturally into harmless byproducts through biological processes, minimizing pollution, resource depletion, and waste accumulation. Environmental impact assessments of biodegradable biomaterials consider factors such as raw material sourcing, manufacturing processes, product performance, end-of-life disposal, and ecological interactions to ensure that they contribute positively to sustainability goals.

#### **3.1.6.1 Life Cycle Assessment**

Life cycle assessment (LCA) serves as a valuable tool for evaluating the environmental impact of these materials throughout their entire life cycle, from raw material extraction to end-of-life disposal. LCA takes into account the various stages of a product's life cycle, including material production, manufacturing, use, and disposal, to assess resource consumption, energy use, emissions, and waste generation. For biodegradable biomaterials, LCA helps quantify the environmental benefits of using materials that degrade naturally, reducing the overall environmental burden compared to non-biodegradable alternatives. One of the key advantages of using LCA in the context of biodegradable biomaterials is its ability to provide a comprehensive understanding of the environmental implications of different material choices and processing methods [29]. By comparing the environmental impacts of biodegradable biomaterials to those of conventional materials, stakeholders can make informed decisions that prioritize sustainability and minimize ecological harm. LCA enables the identification of hotspots in the life cycle of biodegradable biomaterials, allowing for targeted interventions to improve environmental performance and resource efficiency.

LCA helps assess the potential benefits and trade-offs associated with biodegradable biomaterials in terms of energy use, greenhouse gas emissions, water consumption, and land use. By quantifying these environmental indicators, LCA provides insights into the overall sustainability of biodegradable biomaterials and their suitability for specific applications. For example, LCA can compare the environmental impact of using biodegradable polymers derived from renewable resources to that of using petroleum-based polymers, helping guide decision-making toward more sustainable material choices. LCA considers the fate of biodegradable biomaterials at the end of their life cycle, including their potential to degrade in various environmental conditions and the impacts of degradation byproducts on ecosystems. By assessing the environmental implications of biodegradation, LCA helps ensure that biodegradable biomaterials do not contribute to pollution or harm natural habitats. LCA can evaluate the potential for recycling or composting biodegradable biomaterials at the end of their useful life, further minimizing waste and resource consumption. LCA facilitates the development of eco-design strategies that optimize the environmental performance of biodegradable biomaterials throughout their life cycle. By integrating environmental considerations into the design process, manufacturers can identify opportunities to reduce material use, energy consumption, and emissions while maximizing resource efficiency and recyclability [11]. LCA-driven eco-design approaches help create more sustainable products and processes that minimize environmental impact and align with circular economy principles. LCA supports transparency and accountability in the use of biodegradable biomaterials by providing stakeholders with objective data and metrics to evaluate environmental performance. By conducting LCA studies and communicating results to consumers, policymakers, and industry partners, companies can demonstrate their commitment to sustainability and foster trust in their products and practices. Transparent LCA data also enables informed decision-making and facilitates collaboration toward shared sustainability goals.

LCA contributes to the development of regulations, standards, and certification schemes for biodegradable biomaterials, ensuring that they meet environmental criteria and perform as intended throughout their life cycle.

#### 3.1.6.2 Eco-Friendly Processing Methods

Eco-friendly processing methods are essential for minimizing environmental impact and promoting sustainable manufacturing practices. These methods prioritize resource efficiency, waste reduction, and pollution prevention throughout the production process, from raw material sourcing to end-of-life disposal. By adopting eco-friendly processing methods, manufacturers can minimize the ecological footprint of biodegradable biomaterials and contribute to a more sustainable and circular economy. One of the key eco-friendly processing methods for biodegradable biomaterials is green chemistry, which focuses on the design and synthesis of chemicals and materials that minimize or eliminate the use of hazardous substances and byproducts. Green chemistry principles emphasize the use of renewable feedstocks, non-toxic solvents, energy-efficient reactions, and waste prevention strategies to reduce environmental pollution and human health risks [10]. By incorporating green chemistry principles into the production of biodegradable biomaterials, manufacturers can minimize the use of harmful chemicals and reduce the generation of hazardous waste. Eco-friendly processing methods for biodegradable biomaterials include the use of renewable energy sources such as solar, wind, and hydroelectric power to reduce reliance on fossil fuels and minimize greenhouse gas emissions. By investing in renewable energy technologies and infrastructure, manufacturers can reduce their carbon footprint and mitigate the environmental impact of biodegradable biomaterial production. Energy-efficient processing technologies and practices, such as low-temperature synthesis, microwave heating, and process optimization, further reduce energy consumption and environmental pollution.

Eco-friendly processing methods prioritize water conservation and wastewater treatment to minimize water use and pollution in biodegradable biomaterial production. By implementing water-saving technologies, recycling and reusing process water, and treating wastewater to remove contaminants, manufacturers can reduce their water footprint and protect freshwater resources. Water-efficient processing methods not only reduce environmental impact but also lower operational costs and enhance regulatory compliance. Eco-friendly processing methods for biodegradable biomaterials focus on waste reduction, reuse, and recycling to minimize the generation of solid waste and promote a circular economy. By implementing waste minimization strategies, such as lean manufacturing [30], material substitution, and product redesign, manufacturers can reduce waste generation, conserve resources, and decrease disposal costs. Moreover, by recycling and reusing materials and byproducts, manufacturers can minimize the need for virgin raw materials, reduce energy consumption, and lower greenhouse gas emissions. Eco-friendly processing methods for biodegradable biomaterials prioritize sustainable sourcing and supply chain management practices to ensure the responsible procurement of raw materials and components. By sourcing materials from sustainable and ethical suppliers, manufacturers can minimize environmental and social risks throughout the supply chain and support sustainable resource management practices. Eco-friendly packaging and transportation practices, such as using recycled or biodegradable materials and optimizing logistics to minimize emissions, further contribute to overall sustainability efforts. Eco-friendly processing methods for biodegradable biomaterials encompass the use of environmentally friendly additives, coatings, and treatments to enhance product performance while minimizing environmental impact [17].

#### 3.1.6.3 Waste Management

Waste management of biodegradable biomaterials is crucial for minimizing environmental impact and promoting sustainability throughout the product life cycle. Biodegradable biomaterials offer the advantage of natural degradation, reducing the burden of long-term waste accumulation compared to non-biodegradable materials. However, effective waste management strategies are still essential to ensure that biodegradable biomaterials degrade properly and do not cause harm to the environment [10, 30]. One of the key aspects of waste management for biodegradable biomaterials is proper end-of-life disposal. Unlike traditional materials that persist in the environment for extended periods, biodegradable biomaterials naturally degrade into harmless byproducts through biological processes. However, it is essential to ensure that biodegradable waste is disposed of in environments conducive to degradation, such as composting facilities or anaerobic digesters, to facilitate efficient decomposition and minimize environmental impact. Waste management for biodegradable biomaterials involves educating consumers and stakeholders about proper disposal practices to prevent contamination and ensure effective degradation. Providing clear labeling and instructions on product packaging can help consumers understand how to dispose of biodegradable products responsibly. Additionally, public awareness campaigns and outreach efforts can raise awareness about the benefits of biodegradable materials and promote proper waste management practices. Waste management for biodegradable biomaterials includes monitoring and regulation to ensure compliance with environmental standards and regulations. Regulatory agencies and waste management authorities play a critical role in overseeing the collection, transportation, and disposal of biodegradable waste to prevent pollution and protect public health [31]. By enforcing regulations and promoting best practices, authorities can ensure that biodegradable biomaterials are managed in a manner that minimizes environmental impact and maximizes resource recovery.

Waste management for biodegradable biomaterials involves the development of infrastructure and facilities for composting, anaerobic digestion, and other forms of organic waste treatment. Composting facilities, for example, provide controlled environments for the decomposition of biodegradable waste into compost, a valuable soil amendment that enriches soil fertility and promotes plant growth. Anaerobic digestion facilities harness the natural decomposition process to generate biogas, a renewable energy source, from organic waste. Waste management for biodegradable biomaterials includes research and innovation to improve waste treatment technologies and optimize decomposition processes. Advances in composting techniques, microbial inoculants, and process monitoring systems can enhance the efficiency and effectiveness of biodegradable waste management, reducing processing times, increasing degradation rates [31], and improving resource recovery.

Waste management for biodegradable biomaterials encompasses the promotion of circular economy principles and sustainable resource management practices. By integrating biodegradable biomaterials into closed-loop systems that prioritize waste reduction, reuse, and recycling, stakeholders can minimize environmental impact and maximize resource efficiency. For example, organic waste streams from agricultural, food, and forestry industries can be diverted from landfills and converted into biodegradable biomaterials or bio-based products through bioconversion processes [31].

### 3.1.7 Emerging Technologies

Emerging technologies are playing a pivotal role in advancing the field of biodegradable biomaterials, offering innovative solutions to address pressing environmental and societal challenges. These technologies encompass a wide range of approaches, including materials science, manufacturing techniques, and biotechnological processes, aimed at developing biodegradable materials with enhanced properties, functionality, and performance. As the demand for sustainable alternatives to traditional materials grows, emerging technologies hold promise for revolutionizing industries such as healthcare, agriculture, packaging, and textiles while promoting environmental stewardship and resource conservation. One area of innovation in biodegradable biomaterials is the development of advanced polymers and composites with tailored properties and degradation profiles [27, 30]. Emerging technologies such as bio-based polymers, nanotechnology, and additive manufacturing enable the synthesis and fabrication of biodegradable materials with precise control over structure, composition, and functionality. These materials exhibit improved mechanical strength, biocompatibility, and degradation kinetics, making them suitable for a wide range of applications [32], including medical implants, drug delivery systems, and sustainable packaging.

Emerging technologies are driving advancements in manufacturing processes for biodegradable biomaterials, leading to increased efficiency, scalability, and cost-effectiveness. Techniques such as 3D printing, electrospinning, and microfluidics enable the rapid prototyping and mass production of complex biodegradable structures with customizable geometries and properties. By streamlining production processes and reducing waste, these technologies facilitate the widespread adoption of biodegradable biomaterials across industries, contributing to a more sustainable and circular economy. Biotechnological approaches are revolutionizing the production of biodegradable biomaterials through fermentation, enzyme engineering, and metabolic engineering. Microorganisms such as bacteria, fungi, and algae can be engineered to produce biodegradable polymers and specialty chemicals from renewable feedstocks [33], such as agricultural residues, industrial waste, and carbon dioxide. These bio-based production methods offer environmentally friendly alternatives to traditional petrochemical-based processes, reducing dependence on fossil resources and minimizing greenhouse gas emissions.

Nanotechnology is driving innovation in biodegradable biomaterials by enabling the design and fabrication of nanostructured materials with unique properties and functionalities. Nanomaterials such as nanoparticles, nanofibers, and nanostructured surfaces exhibit enhanced mechanical strength, surface area, and bioactivity compared to their bulk counterparts, making them promising candidates for biomedical and environmental applications [11].

#### 3.1.7.1 Overcoming Barriers to Adoption

Overcoming barriers to the adoption of biodegradable biomaterials requires a multifaceted approach that addresses technological, economic, regulatory, and societal challenges. As the demand for sustainable alternatives to traditional materials grows, it is essential to identify and address barriers that hinder the widespread adoption of biodegradable biomaterials in various industries, including healthcare, packaging, agriculture, and textiles. By implementing targeted strategies and fostering collaboration among stakeholders, barriers to adoption can be overcome, unlocking the full potential of biodegradable biomaterials and accelerating the transition toward a more sustainable and circular economy [33]. One of the key barriers to the adoption of biodegradable biomaterials is the limited availability of cost-effective and scalable manufacturing processes. Many biodegradable materials require specialized production techniques and equipment, leading to higher manufacturing costs and limited scalability compared to traditional materials. Overcoming this barrier requires investment in research and development to optimize production processes, improve efficiency, and reduce costs. Additionally, collaboration between academia, industry, and government agencies can facilitate technology transfer and commercialization, accelerating the adoption of biodegradable biomaterials. Supply chain constraints and infrastructure limitations can hinder the adoption of biodegradable biomaterials, particularly in regions with underdeveloped waste management systems or limited access to recycling facilities. Overcoming these barriers involves investing in infrastructure for waste collection, sorting, and processing, as well as promoting the development of composting and recycling facilities for biodegradable materials [31]. Collaboration between government, industry, and non-profit organizations can help address supply chain constraints and infrastructure limitations, creating an enabling environment for the adoption of biodegradable biomaterials. Technological barriers such as limited material performance, durability, and functionality pose challenges to the adoption of biodegradable biomaterials in certain applications, particularly those requiring high strength, temperature resistance, or barrier properties.

### *3.1.8 Case Studies and Success Stories*

Case studies and success stories in the realm of biodegradable biomaterials showcase the transformative potential of these innovative materials across various industries. These real-world examples highlight the practical applications and tangible benefits of biodegradable biomaterials, demonstrating their ability to address pressing environmental challenges while meeting performance requirements and regulatory standards. From healthcare to packaging to agriculture, case studies illuminate the diverse ways in which biodegradable biomaterials are being utilized to reduce waste, minimize environmental impact, and promote sustainability.

#### 3.1.8.1 Biodegradable Packaging Materials

Biodegradable packaging materials represent a promising solution to the environmental challenges posed by conventional packaging materials, such as plastic. These materials are designed to degrade naturally into harmless byproducts through biological processes, reducing the accumulation of waste in landfills and ecosystems. Biodegradable packaging materials encompass a wide range of options, including biopolymers derived from renewable resources, such as corn starch, cellulose, and PLA (polylactic acid), as well as compostable materials that break down into organic matter under controlled conditions. By replacing non-biodegradable packaging materials with biodegradable alternatives, companies can minimize their environmental footprint and contribute to a more sustainable and circular economy. One of the key advantages of biodegradable packaging materials is their ability to reduce pollution and environmental degradation by breaking down into natural compounds without leaving behind harmful residues [30, 31]. Unlike traditional plastics, which can persist in the environment for hundreds of years, biodegradable packaging materials degrade relatively quickly, minimizing their impact on ecosystems and wildlife. Biodegradable packaging materials offer opportunities for composting and organic waste management, providing a closed-loop solution that returns organic matter to the soil and reduces the need for landfill disposal. Biodegradable packaging materials can be tailored to meet specific performance requirements, such as barrier properties, mechanical strength, and shelf life, making them suitable for a wide range of packaging applications. Advances in materials science and manufacturing techniques have led to the development of biodegradable packaging materials with properties comparable to traditional plastics, enabling their use in food packaging, cosmetics, pharmaceuticals, and other industries. By providing effective protection and preservation of products, biodegradable packaging materials help maintain product quality and safety while minimizing environmental impact. Biodegradable packaging materials offer opportunities for innovation and differentiation in the marketplace, as consumers increasingly demand sustainable and eco-friendly packaging solutions. Companies that adopt biodegradable packaging materials can enhance their brand reputation, attract environmentally conscious consumers, and differentiate themselves from competitors.

#### 3.1.8.2 Biopolymer-Based Medical Devices

Biopolymer-based medical devices are revolutionizing the healthcare industry by offering innovative solutions that are both biocompatible and environmentally sustainable. These devices are made from biodegradable polymers derived from renewable resources, such as plant starches, cellulose, and proteins, making them safe for use in medical applications while reducing dependence on fossil fuels. Biopolymer-based medical devices encompass a wide range of products, including sutures, implants, drug delivery systems, wound dressings, and tissue scaffolds, which provide critical functions in diagnosis, treatment, and patient care [28]. One of the key advantages of biopolymer-based medical devices is their biocompatibility, which minimizes the risk of adverse reactions and complications when implanted or used in contact with biological tissues and fluids. Unlike synthetic materials, biopolymers are inherently compatible with the body's natural processes, reducing the likelihood of inflammation, rejection, or immune response. This biocompatibility is particularly beneficial for applications such as implantable medical devices, where long-term compatibility and integration with host tissues are essential for therapeutic efficacy and patient safety. Biopolymer-based medical devices offer advantages in terms of controlled drug release and tissue regeneration, thanks to their tunable properties and degradation kinetics. Biopolymers can be engineered to degrade at a controlled rate, releasing drugs or therapeutic agents in a sustained manner to achieve targeted therapeutic outcomes [28]. Biopolymer scaffolds provide a 3D structure that supports cell adhesion, proliferation, and differentiation, facilitating tissue regeneration and repair. This ability to tailor biopolymer properties and functionality to specific medical needs enables the development of personalized and precision medicine approaches in healthcare.

Biopolymer-based medical devices contribute to environmental sustainability by reducing reliance on non-renewable resources and minimizing waste generation. Biopolymers are derived from natural sources and can be produced using environmentally friendly processes, such as fermentation or enzymatic conversion, which consume less energy and produce fewer greenhouse gas emissions compared to petrochemical-based polymers. The biodegradability of biopolymers ensures that medical devices degrade naturally over time, reducing the burden on landfills and ecosystems and promoting a circular economy model in healthcare.

Biopolymer-based medical devices offer versatility in design and fabrication, allowing for the development of complex and customized products that meet diverse medical needs.

**Exercise 3**

**Multiple Choice Questions**

1. Which of the following is a key advantage of biodegradable biomaterials?
   a) Permanent implantation
   b) Long-term stability
   c) Minimal tissue response
   d) Degradation into non-toxic byproducts

2. What is the primary mechanism by which biodegradable biomaterials degrade?
   a) Mechanical breakdown
   b) Chemical oxidation
   c) Enzymatic hydrolysis
   d) Thermal decomposition

3. Which polymer is commonly used in biodegradable sutures and drug delivery systems?
   a) Polyethylene
   b) Polycarbonate
   c) Polyglycolic acid
   d) Polystyrene

4. Biodegradable biomaterials are often used in which medical application?
   a) Orthopedic implants
   b) Dental fillings
   c) Cardiovascular stents
   d) All of the above

5. What role do degradation products of biodegradable biomaterials play in the body?
   a) They trigger inflammation
   b) They promote tissue regeneration
   c) They inhibit cell proliferation
   d) They have no biological effect

6. Which factor does NOT influence the degradation rate of biodegradable biomaterials?
   a) Polymer composition
   b) Implant location
   c) pH of the surrounding environment
   d) Temperature

7. Biodegradable biomaterials are often used to avoid the need for:
   a) Surgical removal
   b) Immunosuppression
   c) Sterilization
   d) Allergy testing

8. Which of the following is a challenge associated with biodegradable biomaterials?
   a) Inflammation
   b) Risk of infection
   c) Limited mechanical strength
   d) All of the above

9. Which type of degradation mechanism occurs in biodegradable metallic implants?
   a) Hydrolysis
   b) Oxidation
   c) Corrosion
   d) Enzymatic cleavage

10. Biodegradable biomaterials play a crucial role in:
   a) Long-term tissue replacement
   b) Temporary medical interventions
   c) Permanent prosthetic devices
   d) Genetic engineering applications

## 3.2 Summary

The book chapter on biodegradable biomaterials provides a comprehensive overview of the latest advancements, applications, and challenges in the field. It covers a wide range of topics, including the types of biodegradable biomaterials, their fabrication techniques, degradation mechanisms, properties, and various applications in biomedical engineering. Through detailed discussions and case studies, the chapter highlights the significant role that biodegradable biomaterials play in addressing pressing environmental concerns, promoting sustainability, and advancing healthcare technologies. The chapter begins by introducing the concept of biodegradable biomaterials and their importance in reducing environmental pollution and resource depletion compared to traditional non-biodegradable materials. It discusses the different types of biodegradable biomaterials, including natural polymers, synthetic polymers, and composite materials, along with their properties, advantages, and limitations. Furthermore, the chapter explores various fabrication techniques for biodegradable biomaterials, such as solution casting, electrospinning, and 3D printing, and their implications for product design, customization, and scalability. The chapter delves into the mechanisms of biodegradation, including enzymatic degradation, hydrolytic degradation, and microbial degradation, elucidating the factors that influence degradation rates, products, and environmental interactions. It also discusses the properties of biodegradable biomaterials, such as mechanical properties, degradation kinetics, and biocompatibility, and their significance for specific applications in tissue engineering, drug delivery, implants, and prosthetics. Finally, the chapter concludes by outlining future perspectives, emerging technologies, regulatory considerations, and challenges in the field, emphasizing the need for continued research, collaboration, and innovation to realize the full potential of biodegradable biomaterials in promoting human health and environmental sustainability.

## Points to Remember

- **Biodegradable biomaterials can be sourced from natural materials like cellulose, starch, chitosan (from crustacean shells), or proteins like collagen.**
- **They can also be synthesized from synthetic polymers like polylactic acid (PLA), polyglycolic acid (PGA), or polyhydroxyalkanoates (PHA).**
- **Biodegradable biomaterials offer environmental benefits as they can decompose naturally, reducing pollution and waste accumulation.**
- **They are widely used in medical applications such as sutures, implants, drug delivery systems, and tissue engineering scaffolds due to their biocompatibility and ability to degrade in the body without causing harm.**
- **Biodegradable biomaterials are gaining popularity in the packaging industry as an eco-friendly alternative to traditional plastics, reducing the burden of non-biodegradable waste.**
- **Some biodegradable biomaterials are used in agricultural applications such as mulches and soil erosion control mats, providing benefits for soil health and sustainability.**
- **The time taken for biodegradable biomaterials to degrade varies depending on factors such as composition, environmental conditions, and microbial activity.**

## References

1. Geevarghese R, Sajjadi SS, Hudecki A, Sajjadi S, Jalal NR, Madrakian T, Ahmadi M, Włodarczyk-Biegun MK, Ghavami S, Likus W, Siemianowicz K, Łos MJ (2022) Biodegradable and non-biodegradable biomaterials and their effect on cell differentiation. Int J Mol Sci 23(24):16185. https://doi.org/10.3390/ijms232416185
2. Pişkin E (1995) Biodegradable polymers as biomaterials. J Biomater Sci Polym Ed 6(9):775–795. https://doi.org/10.1163/156856295x00175
3. Suhag D, Bhatia R, Das S, Shakeel A, Ghosh AB, Singh A et al (2015) Physically cross-linked pH-responsive hydrogels with tunable formulations for controlled drug delivery. RSC Adv 5(66):53963–53972. https://doi.org/10.1039/c5ra07424j
4. Vroman I, Tighzert L (2009) Biodegradable polymers. Materials 2(2):307–344. https://doi.org/10.3390/ma2020307

5. Oleksy M, Dynarowicz K, Aebisher D (2023) Advances in biodegradable polymers and biomaterials for medical applications-a review. Molecules (Basel, Switzerland) 28(17):6213. https://doi.org/10.3390/molecules28176213
6. Modrák M, Trebuňová M, Balogová AF, Hudák R, Živčák J (2023) Biodegradable materials for tissue engineering: development, classification and current applications. J Funct Biomater 3(3):159. https://doi.org/10.3390/jfb3030159
7. Aslam Khan MU, Abd Razak SI, Al Arjan WS, Nazir S, Sahaya Anand TJ, Mehboob H, Amin R (2021) Recent advances in biopolymeric composite materials for tissue engineering and regenerative medicines: a review. Molecules (Basel, Switzerland) 26(3):619. https://doi.org/10.3390/molecules26030619
8. Kaur N, Sharma AK, Shakeel A, Kumar V, Singh A, Gupta A et al (2017) Therapeutic implications of superoxide dismutase and its importance in kinase drug discovery. Curr Top Med Chem 17(22). https://doi.org/10.2174/1568026617666170307112837
9. Kikkawa Y, Tanaka S, Norikane Y (2017) Photo-triggered enzymatic degradation of biodegradable polymers. RSC Adv 7(88):55720–55724. https://doi.org/10.1039/c7ra10598c. PMID: 35287884
10. Elsawy MA, Kim K, Park J, Deep A (2017) Hydrolytic degradation of polylactic acid (PLA) and its composites. Renew Sustain Energ Rev 79:1346–1352. https://doi.org/10.1016/j.rser.2017.05.143
11. Cakmak OK (2023) Biodegradable polymers—a review on properties, processing, and degradation mechanism. Circ Econ Sustain/Circ Econ Sustain 4(1):339–362. https://doi.org/10.1007/s43615-023-00277-y
12. Hofmann D, Entrialgo-Castaño M, Kratz K, Lendlein A (2009) Knowledge-based approach towards hydrolytic degradation of polymer-based biomaterials. Adv Mater 21(32–33):3237–3245. https://doi.org/10.1002/adma.200802213
13. Pathak VM, Navneet N (2017) Review on the current status of polymer degradation: a microbial approach. Bioresour Bioprocess 4(1). https://doi.org/10.1186/s40643-017-0145-9
14. Suhag D, Thakur P, Thakur A (2023) Introduction to nanotechnology. In: Suhag D, Thakur A, Thakur P (eds) Integrated nanomaterials and their applications. Springer, Singapore. https://doi.org/10.1007/978-981-99-6105-4_1
15. Arutchelvi J, Sudhakar M, Arkatkar A, Doble M, Bhaduri S, Uppara PV (2008) Biodegradation of polyethylene and polypropylene
16. Kirillova A, Yeazel TR, Asheghali D, Petersen SR, Dort S, Gall K, Becker ML (2021) Fabrication of biomedical scaffolds using biodegradable polymers. Chem Rev 121(18):11238–11304. https://doi.org/10.1021/acs.chemrev.0c01200
17. Tolbert JW, Hammerstone DE, Yuchimiuk N, Seppala JE, Chow LW (2021) Solvent-cast 3D printing of biodegradable polymer scaffolds. Macromol Mater Eng 306(12). https://doi.org/10.1002/mame.202100442
18. Kai D, Liow SS, Loh XJ (2014) Biodegradable polymers for electrospinning: towards biomedical applications. Mater Sci Eng Biomimetic Mater Sens Syst 45:659–670. https://doi.org/10.1016/j.msec.2014.04.051
19. Begum NSA, Krishnan PSG, Kanny K (2023) Bio-based polymers: a review on processing and 3D printing. Polym Sci. https://doi.org/10.1134/s0965545x2360045x
20. Singh A, Suhag D (2023) Polymers and polymeric composites in nano/bio-medicine. In: Suhag D, Thakur A, Thakur P (eds) Integrated nanomaterials and their applications. Springer, Singapore. https://doi.org/10.1007/978-981-99-6105-4_8
21. Dintcheva NT (2024). Overview of polymers and biopolymers degradation and stabilization towards sustainability and materials circularity. Polymer 127136. https://doi.org/10.1016/j.polymer.2024.127136
22. Wang HMD, Thakur P, Thakur A (2023) Biocompatibility, bio-clearance, and toxicology. In: Suhag D, Thakur A, Thakur P (eds) Integrated nanomaterials and their applications. Springer, Singapore. https://doi.org/10.1007/978-981-99-6105-4_10
23. Rathore R, Suhag D, Wan F, Thakur A, Thakur P (2023) Toxicology and toxicity studies of nano-biomaterials. In: Suhag D, Thakur A, Thakur P (eds) Integrated nanomaterials and their applications. Springer, Singapore. https://doi.org/10.1007/978-981-99-6105-4_20

24. Jadoun S, Riaz U, Budhiraja V (2020) Biodegradable conducting polymeric materials for biomedical applications: a review. Med Devices Sensors 4(1). https://doi.org/10.1002/mds3.10141
25. Stevens B, Yang Y, Mohandas A, Stucker B, Nguyen KT (2008) A review of materials, fabrication methods, and strategies used to enhance bone regeneration in engineered bone tissues. J Biomed Mater Res Part B Appl Biomater 85(2):573–582. https://doi.org/10.1002/jbm.b.30962
26. Bhushan S, Singh S, Maiti TK, Sharma C, Dutt D, Sharma S, Li C, Tag Eldin EM (2022) Scaffold fabrication techniques of biomaterials for bone tissue engineering: a critical review. Bioengineering (Basel, Switzerland) 9(12):728. https://doi.org/10.3390/bioengineering9120728
27. Prajapati SK, Jain A, Jain A, Jain S (2019) Biodegradable polymers and constructs: a novel approach in drug delivery. Euro Polym J/Eur Polym J 120:109191. https://doi.org/10.1016/j.eurpolymj.2019.08.018
28. Li C, Guo C, Fitzpatrick V, Ibrahim A, Zwierstra MJ, Hanna P, Lechtig A, Nazarian A, Lin SJ, Kaplan DL (2019) Design of biodegradable, implantable devices towards clinical translation. Nat Rev Mater 5(1):61–81. https://doi.org/10.1038/s41578-019-0150-z
29. Yates MR, Barlow CY (2013) Life cycle assessments of biodegradable, commercial biopolymers—a critical review. Resour Conserv Recycl 78:54–66. https://doi.org/10.1016/j.resconrec.2013.06.010
30. Kuang T, Esmaeili A, Ehsani M (2022) Eco-friendly biodegradable polymers: sustainable future. Polym Renew Resour 13(1–2):71–79. https://doi.org/10.1177/20412479221109875
31. Ali GaM, Thalji MR, Makhlouf ASH (2022) Biodegradable materials: fundamentals, importance, and impacts. In: Springer eBooks (pp. 1–16). https://doi.org/10.1007/978-3-030-83783-9_74-1
32. Kumar A, Wang X, Nune KC, Misra R (2017) Biodegradable hydrogel-based biomaterials with high absorbent properties for non-adherent wound dressing. Int Wound J 3(6):1076–1087. https://doi.org/10.1111/iwj.12762
33. Kurowiak J, Klekiel T, Będziński R (2023) Biodegradable polymers in biomedical applications: a review—Developments, perspectives and future challenges. Int J Mol Sci 24(23):16952. https://doi.org/10.3390/ijms242316952

# Chapter 4
# Nanomaterials in Biomedicine

## 4.1 Introduction

Nano biomaterials, at the intersection of nanotechnology and biomedicine, represent a burgeoning field with promising implications for advancing healthcare. These materials, engineered at the nanoscale, offer unique properties that can revolutionize various aspects of medicine, from drug delivery and imaging to tissue engineering and cancer therapy. By harnessing the power of nanotechnology [1], scientists and engineers aim to develop innovative solutions to address pressing medical challenges, such as improving treatment efficacy, enhancing diagnostic accuracy, and promoting tissue regeneration. This chapter provides an accessible introduction to the exciting world of nano biomaterials, exploring their definition, historical evolution, and the significant role they play in shaping the future of healthcare.

### *4.1.1 Historical Development and Importance in Biomedicine*

The historical evolution of nano biomaterials traces back to the pioneering work of scientists and researchers who recognized the potential of nanotechnology in revolutionizing biomedical applications (Fig. 4.1). In the 1950s, Richard Feynman's visionary lecture "There's Plenty of Room at the Bottom" laid the groundwork for exploring manipulation and engineering at the atomic and molecular levels, planting the seeds for nanotechnology's emergence. Subsequent decades witnessed remarkable advancements in materials science, leading to the development of nanomaterials with tailored properties suitable for biomedical use. The importance of nano biomaterials in biomedicine cannot be overstated, as they offer unprecedented opportunities to address longstanding challenges in healthcare [1]. One of their key advantages lies in their ability to interact with biological systems at the molecular level, enabling precise targeting and manipulation for therapeutic and diagnostic purposes.

 
D. Suhag et al., *Handbook of Biomaterials for Medical Applications, Volume 1*,
Biomedical Materials for Multi-functional Applications,
https://doi.org/10.1007/978-981-97-4818-1_4

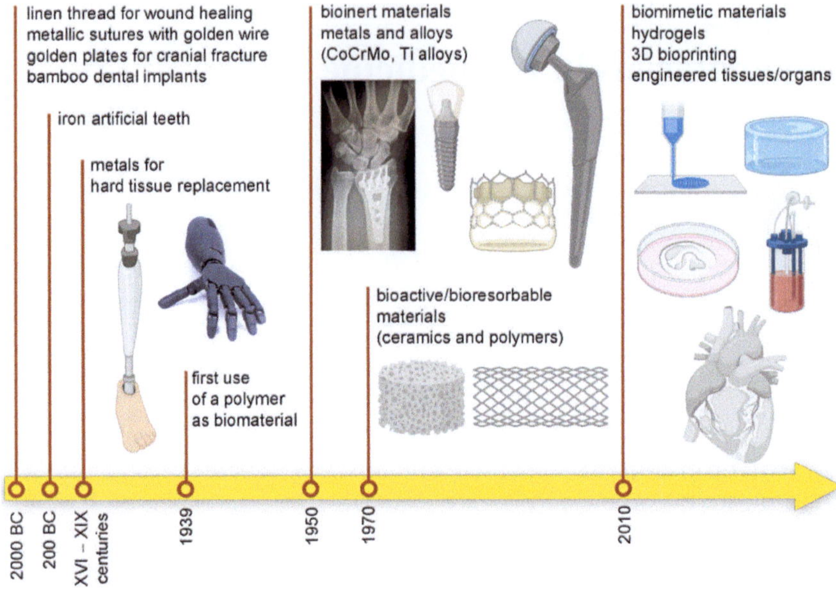


**Fig. 4.1** The history of biomaterials. https://doi.org/10.4490/pr9111949 [5]

For instance, nanoparticles engineered with specific surface properties can deliver therapeutic agents directly to diseased tissues while minimizing off-target effects, thus enhancing treatment efficacy, and reducing side effects.

Nano biomaterials play a pivotal role in advancing medical imaging technologies, offering enhanced contrast, sensitivity, and resolution for visualizing anatomical structures and physiological processes. Nanoparticle-based contrast agents, such as quantum dots and magnetic nanoparticles, enable non-invasive imaging modalities like magnetic resonance imaging (MRI) and fluorescence imaging, facilitating early disease detection and monitoring of treatment responses [2]. These imaging advancements hold tremendous promise for improving patient outcomes and refining personalized medicine approaches. In tissue engineering, nano biomaterials serve as building blocks for constructing biomimetic scaffolds that mimic the extracellular matrix's architecture and biochemical cues. By precisely controlling the nanoscale features of these scaffolds, researchers can regulate cell behavior and guide tissue regeneration, offering potential solutions for repairing damaged organs and facilitating regenerative medicine therapies. Moreover, nanotechnology-enabled biosensors and diagnostic devices empower clinicians with rapid and sensitive tools for detecting biomarkers of disease, enabling early intervention and personalized treatment strategies. The interdisciplinary nature of nano biomaterials research fosters collaboration between scientists, engineers, clinicians, and regulatory experts, driving innovation and translation from bench to bedside [2]. However, alongside their transformative potential, nano biomaterials also raise ethical and safety concerns regarding their long-term effects on human health and the environment. Thus, responsible development and thorough evaluation of these materials are imperative to ensure their safe and effective integration into clinical practice. In conclusion, nano biomaterials represent a paradigm shift in biomedicine, offering multifaceted solutions to improve healthcare outcomes and pave the way for a healthier future.

### *4.1.2 Nanoparticles for Drug Delivery*

Nanoparticles have emerged as promising vehicles for drug delivery, offering precise control over therapeutic payloads and enhancing their efficacy while minimizing systemic side effects. Engineered at the nanoscale, these particles exhibit unique properties that enable targeted delivery to specific tissues or cells, overcoming biological barriers that hinder conventional drug administration routes. By encapsulating drugs within nanoparticles, researchers can prolong circulation time, improve bioavailability, and facilitate sustained release kinetics, thus optimizing treatment outcomes [3]. This chapter explores the diverse range of nanoparticle platforms used in drug delivery, including liposomes, polymeric nanoparticles, and inorganic nanoparticles, highlighting their potential to revolutionize the field of medicine and improve patient care.

#### **4.1.2.1 Liposomes**

Liposomes, spherical vesicles composed of lipid bilayers, have garnered significant attention in drug delivery due to their biocompatibility and versatility. These microscopic structures offer an ideal platform for encapsulating hydrophobic and hydrophilic drugs within their aqueous core or lipid bilayers, allowing for tailored delivery to target tissues or cells. Liposomes can be engineered with precise control over size, surface charge, and membrane composition, enabling customization to suit specific therapeutic needs. Moreover, their ability to protect encapsulated drugs from degradation and clearance by the immune system enhances drug stability and bioavailability, prolonging their circulation time in the body. One of the key advantages of liposomal drug delivery is its potential to minimize systemic toxicity while maximizing therapeutic efficacy. By encapsulating drugs within liposomes, researchers can achieve controlled release profiles, ensuring sustained drug concentrations at the target site while reducing exposure to healthy tissues [4]. This targeted delivery approach is particularly advantageous in cancer therapy, where conventional chemotherapeutic agents often exhibit dose-limiting toxicity. Liposomal formulations allow for higher drug concentrations to be delivered to tumor tissues while sparing surrounding healthy cells, thereby enhancing treatment outcomes, and minimizing adverse effects. Liposomes offer a versatile platform for multifunctional drug delivery strategies, allowing for the co-delivery of multiple therapeutic agents or the conjugation of targeting ligands for site-specific delivery. Through surface modification with ligands such as antibodies, peptides, and aptamers, liposomes can selectively bind to receptors overexpressed on diseased cells, facilitating precise targeting and internalization of therapeutic cargo. This targeted approach enhances drug accumulation at the desired site, maximizing therapeutic efficacy while minimizing off-target effects, thus improving patient compliance and quality of life [5]. Despite their considerable promise, liposomal drug delivery systems also face challenges such as instability during storage, premature drug leakage, and batch-to-batch variability. Addressing these limitations requires ongoing research efforts to optimize liposomal formulations, improve manufacturing processes, and enhance scalability for clinical translation.

#### 4.1.2.2 Polymeric Nanoparticles

Polymeric nanoparticles have emerged as versatile carriers for drug delivery, offering numerous advantages over traditional delivery systems. These nanoparticles are typically composed of biocompatible and biodegradable polymers such as poly (lactic-co-glycolic acid) (PLGA), polyethylene glycol (PEG), and chitosan, among others [3]. Their tunable properties, including size, surface charge, and drug release kinetics, make them suitable for encapsulating a wide range of therapeutic agents and tailoring drug delivery to specific biological targets. Polymeric nanoparticles can be synthesized using various techniques, including emulsion/solvent evaporation, nanoprecipitation, and electrospraying, allowing for precise control over particle characteristics and drug loading efficiency. One of the key advantages of polymeric nanoparticles is their ability to protect encapsulated drugs from enzymatic degradation and immune clearance, thereby enhancing drug stability and bioavailability. Their small size and surface modifications enable prolonged circulation in the bloodstream, facilitating passive accumulation at disease sites through the enhanced permeability and retention (EPR) effect. This passive targeting mechanism is particularly advantageous in cancer therapy, where leaky vasculature and impaired lymphatic drainage in tumors result in preferential accumulation of nanoparticles within the tumor microenvironment, maximizing therapeutic efficacy while minimizing systemic toxicity.

Polymeric nanoparticles also offer opportunities for controlled and sustained drug release, allowing for the modulation of pharmacokinetics and minimizing the need for frequent dosing [6, 7]. By engineering nanoparticles with stimuli-responsive or biodegradable polymers, researchers can design systems that release drugs in response to specific triggers such as pH, temperature, enzymes, and light, enabling spatiotemporal control over drug delivery [6]. Polymeric nanoparticles can be surface-functionalized with targeting ligands or therapeutic moieties to enhance specificity and efficacy [8]. These multifunctional nanoparticles can selectively bind to receptors or antigens expressed on target cells, facilitating active targeting and internalization of therapeutic cargo while minimizing off-target effects, thus improving therapeutic outcomes. Despite their considerable promise, polymeric nanoparticles face challenges related to manufacturing scalability, reproducibility, and regulatory approval. Addressing these challenges requires interdisciplinary collaborations between scientists, engineers, and clinicians to optimize nanoparticle formulations, streamline production processes, and conduct rigorous preclinical and clinical evaluations.

#### 4.1.2.3 Inorganic Nanoparticles

Inorganic nanoparticles, including gold, silver, iron oxide, and others, have gained prominence in drug delivery and biomedical applications due to their unique physico-chemical properties and versatile functionalities. These nanoparticles exhibit distinct optical, magnetic, and catalytic properties, making them suitable candidates for a wide range of therapeutic and diagnostic applications. Gold nanoparticles, for example, are renowned for their excellent biocompatibility, optical properties, and ease of surface functionalization [8], allowing for precise control over drug loading and release. Similarly, silver nanoparticles possess antimicrobial properties, making them valuable for wound healing and infection control. Iron oxide nanoparticles, particularly magnetic nanoparticles such as magnetite ($Fe_4O_4$) and maghemite ($\gamma$-$Fe_2O_4$), offer opportunities for magnetic targeting and hyperthermia-based therapies. By applying an external magnetic field, these nanoparticles can be directed to specific disease sites, enabling localized drug delivery, or inducing hyperthermia to selectively destroy cancer cells. Additionally, iron oxide nanoparticles serve as contrast agents for magnetic resonance imaging (MRI), providing high-resolution anatomical and functional imaging with minimal toxicity. This dual functionality of iron oxide nanoparticles highlights their potential in theranostic applications, combining therapy and diagnostics within a single platform.

Inorganic nanoparticles also hold promise for enhancing the efficacy of conventional chemotherapy and radiotherapy through synergistic combination therapies [9]. For instance, gold nanoparticles can sensitize cancer cells to radiation therapy by enhancing radiation absorption and generating reactive oxygen species, thereby increasing treatment efficacy while minimizing damage to surrounding healthy tissues. Similarly, silver nanoparticles can potentiate the cytotoxic effects of chemotherapeutic drugs by promoting drug uptake and intracellular accumulation in cancer cells, offering a complementary approach to overcome drug resistance and improve treatment outcomes. Despite their immense potential, the clinical translation of inorganic nanoparticles faces challenges related to biocompatibility, toxicity [10], and regulatory approval. Rigorous preclinical evaluations are essential to assess the safety and efficacy of these nanoparticles in relevant animal models before advancing to human trials. Efforts to optimize nanoparticle design, surface modification, and targeting strategies are crucial to enhance their biocompatibility, minimize off-target effects, and maximize therapeutic benefits. Overall, inorganic nanoparticles represent a promising class of nanomaterials with multifaceted applications in biomedicine, holding the potential to revolutionize disease diagnosis, treatment, and management.

#### 4.1.2.4 Quantum Dots

Quantum dots, with their unique optical and physicochemical properties, are not only valuable in imaging but also hold promise as carriers in drug delivery systems. These semiconductor nanocrystals offer several advantages for drug delivery applications, including their tunable size, surface chemistry, and high payload capacity. Quantum dots can be functionalized by targeting ligands or therapeutic agents, allowing for precise delivery to specific tissues or cells. Their small size facilitates cellular uptake and intracellular trafficking, enhancing the efficiency of drug delivery to the desired site of action [11]. One of the key advantages of quantum dots in drug delivery is their ability to overcome biological barriers and improve drug stability and pharmacokinetics. Encapsulating drugs within quantum dots protects them from enzymatic degradation and clearance by the immune system, thereby increasing their circulation time in the bloodstream and enhancing their bioavailability. Moreover, quantum dots can be engineered to release drugs in a controlled manner, allowing for sustained drug release kinetics and prolonged therapeutic effects. This controlled release profile minimizes systemic toxicity and improves therapeutic efficacy by maintaining therapeutic drug concentrations within the therapeutic window. Quantum dots offer opportunities for multifunctional drug delivery strategies, enabling combination therapies or simultaneous imaging and therapy within a single platform. By conjugating imaging agents or diagnostic probes to quantum dots, researchers can monitor drug distribution and pharmacokinetics in real time, providing valuable insights into treatment efficacy and patient response. Additionally, quantum dots can be engineered to respond to external stimuli, such as light or magnetic fields, enabling triggered drug release at the target site, further enhancing therapeutic precision and efficacy.

Despite their considerable promise, the clinical translation of quantum dots in drug delivery faces challenges related to biocompatibility, toxicity, and regulatory approval. Addressing these challenges requires rigorous preclinical evaluation of quantum dot formulations to assess their safety, pharmacokinetics, and efficacy in relevant animal models. Efforts to optimize quantum dot design, surface modification, and targeting strategies are essential to enhance their biocompatibility and minimize off-target effects.

**Exercise 1.**

**Fill Ups**

**(Potential, Nanoparticles, Small size, Biological barriers, Drugs)**

**1.** Nanoparticles enhance drug delivery by encapsulating ............and targeting specific cells or tissues.

**2.** The ....................enables efficient penetration of biological barriers.

**3.** Surface modifications allow ..................to recognize and bind to specific markers on target cells.

**4.** Overcoming......................, nanoparticles can deliver drugs to previously inaccessible sites, such as the brain.

**5.** Nanoparticles offer ................for personalized medicine by tailoring treatments to individual patient characteristics.

### *4.1.3 Nanomaterials for Imaging Applications*

Nanomaterials have emerged as revolutionary tools in biomedical imaging, offering unprecedented capabilities for visualizing biological structures and processes with high precision and sensitivity. These materials, engineered at the nanoscale, exhibit unique optical, magnetic, and acoustic properties that enable enhanced contrast, resolution, and multiplexing in various imaging modalities. From quantum dots and gold nanoparticles to superparamagnetic iron oxide nanoparticles, nanomaterials offer diverse functionalities that cater to different imaging needs, ranging from fluorescence imaging and magnetic resonance imaging (MRI) to photoacoustic imaging and computed tomography (CT). This chapter explores the exciting potential of nanomaterials in imaging applications, highlighting their versatility and transformative impact on biomedical research and clinical diagnostics.

#### 4.1.3.1 Contrast Agents

Contrast agents play a pivotal role in biomedical imaging by enhancing the visibility and distinction of anatomical structures or pathological tissues in imaging modalities such as MRI, CT, ultrasound, and molecular imaging. These agents, typically composed of nanomaterials or small molecules, exploit different physical properties to generate contrast, including differences in magnetic susceptibility, X-ray attenuation, acoustic impedance, and optical properties [12]. In MRI, contrast agents containing paramagnetic or superparamagnetic ions, such as gadolinium or iron oxide nanoparticles, alter the relaxation times of surrounding water molecules, resulting in enhanced signal intensity and contrast between tissues. Similarly, in CT imaging, contrast agents containing high atomic number elements, such as iodine or gold nanoparticles, attenuate X-rays to a greater extent than surrounding tissues [9], leading to increased image contrast and improved visualization of blood vessels, organs, or tumors.

In ultrasound imaging, contrast agents composed of microbubbles or nanoparticles enhance the reflection of ultrasound waves, allowing for improved detection of blood flow and tissue perfusion. These agents provide valuable information about tissue morphology, vascularity, and function, aiding in the diagnosis and monitoring of cardiovascular diseases, cancer, and inflammatory conditions. Contrast agents used in molecular imaging leverage specific molecular interactions to target and visualize disease biomarkers or cellular processes with high sensitivity and specificity. For example, targeted contrast agents functionalized with ligands or antibodies can selectively bind to receptors or antigens overexpressed on cancer cells, enabling non-invasive detection and characterization of tumors at the molecular level. The development of novel contrast agents with improved biocompatibility, stability, and targeting specificity is a rapidly evolving field in biomedical imaging research. Researchers are exploring innovative nanomaterial-based formulations, including liposomes, polymer nanoparticles, and hybrid nanoparticles, to overcome existing limitations and enhance imaging performance. Moreover, efforts to optimize contrast agent design, surface functionalization, and pharmacokinetics are aimed at improving imaging sensitivity, reducing background noise, and enabling real-time monitoring of disease progression and treatment response.

#### 4.1.3.2 Fluorescent Probes

Fluorescent probes are indispensable tools in biomedical imaging, offering researchers and clinicians the ability to visualize and study biological processes with exceptional sensitivity and specificity. These probes, typically small molecules, or nanomaterials, emit fluorescent light upon excitation by a specific wavelength of light, allowing for real-time imaging of cellular structures, molecular interactions, and physiological events. Fluorescent probes cover a wide spectrum of applications, from basic research in cell biology and neuroscience to clinical diagnostics and therapeutics. One of the key advantages of fluorescent probes is their versatility, as they can be tailored to target specific molecules or organelles within cells, providing insights into cellular dynamics and function [13]. In cellular imaging, fluorescent probes enable researchers to label and track biomolecules such as proteins, nucleic acids, and lipids in living cells or tissues, facilitating the study of cellular processes such as protein trafficking, signal transduction, and gene expression. By selecting probes with appropriate spectral properties, researchers can perform multiplexed imaging experiments, allowing for simultaneous visualization of multiple targets within the same sample. Moreover, advancements in super-resolution microscopy techniques, such as stimulated emission depletion (STED) microscopy and stochastic optical reconstruction microscopy (STORM), have further expanded the capabilities of fluorescent probes, enabling visualization of cellular structures at nanometer-scale resolution.

Fluorescent probes also play a crucial role in molecular imaging, where they are used to detect and monitor disease biomarkers or molecular targets in vivo. For example, targeted fluorescent probes functionalized with ligands or antibodies can selectively bind to tumor-specific antigens or receptors, enabling non-invasive imaging and characterization of tumors in preclinical models and clinical patients. Fluorescent probes can be employed in fluorescence-guided surgery, where they help surgeons identify tumor margins and metastatic lesions in real-time during surgical procedures, improving surgical outcomes and patient prognosis. The development of novel fluorescent probes with improved brightness, photostability, and targeting specificity is a rapidly advancing field in biomedical research. Researchers are exploring innovative strategies to design and synthesize fluorescent probes with enhanced optical properties, including quantum dots [11], organic dyes, and genetically encoded fluorescent proteins. Efforts to develop activatable probes that respond to specific biological stimuli or environmental changes are aimed at improving imaging sensitivity and reducing background signals, enabling more accurate detection of disease biomarkers and cellular processes.

#### 4.1.3.3 Magnetic Resonance Imaging (MRI) Agents

Magnetic Resonance Imaging (MRI) agents play a crucial role in enhancing the quality and specificity of MRI scans, enabling detailed visualization of anatomical structures and physiological processes within the body [14]. These agents, typically composed of contrast-enhancing materials, interact with the magnetic fields generated during MRI scans to produce clearer and more informative images. One common type of MRI agent is superparamagnetic iron oxide nanoparticles (SPIONs), which are highly effective contrast agents due to their ability to generate strong magnetic signals. When administered into the body, SPIONs accumulate in tissues of interest, such as tumors or areas of inflammation, leading to enhanced contrast and improved delineation of pathological regions in MRI images. Another type of MRI agent is gadolinium-based nanoparticles, which contain gadolinium ions chelated within a nanoparticle core. These nanoparticles exhibit high relaxivity, allowing for increased sensitivity and contrast enhancement in MRI scans. Gadolinium-based nanoparticles offer advantages such as tunable size and surface functionalization, enabling customization for specific imaging applications and targeted delivery to diseased tissues. Their controlled release of gadolinium ions prolongs the enhancement of MRI contrast, providing valuable information about tissue morphology, vascularity, and function. The use of MRI agents extends beyond anatomical imaging to functional and molecular imaging applications, where they provide insights into disease pathology and treatment response. By targeting specific molecular markers or biological processes, MRI agents can help clinicians diagnose diseases such as cancer, cardiovascular disease, and neurological disorders with greater accuracy and specificity. Moreover, MRI agents can be engineered to respond to external stimuli or molecular interactions, enabling dynamic imaging of physiological processes such as blood flow, tissue perfusion, and cellular metabolism [14]. Overall, MRI agents represent indispensable tools in modern medical imaging, offering valuable diagnostic information and contributing to advancements in personalized medicine and precision healthcare.

#### 4.1.3.4 Photoacoustic Imaging Nanoprobes

Photoacoustic imaging nanoprobes represent an innovative approach to biomedical imaging, offering high resolution and deep tissue penetration capabilities for visualizing biological structures and functions. This emerging imaging modality combines the advantages of optical and ultrasound imaging, utilizing laser-induced acoustic waves to detect and map the distribution of light-absorbing molecules within tissues. Photoacoustic nanoprobes, typically composed of nanomaterials such as gold nanoparticles, carbon nanotubes, and organic dyes, absorb laser energy and convert it into acoustic signals, which are then detected by ultrasound transducers to generate detailed images [12]. One of the key advantages of photoacoustic imaging nanoprobes is their ability to provide functional and molecular information about biological tissues with high sensitivity and specificity. These nanoprobes can be functionalized by targeting ligands or contrast agents, enabling specific detection of disease biomarkers or molecular targets within tissues. Moreover, their small size and surface modifications facilitate efficient delivery and accumulation at the target site, enhancing imaging contrast and diagnostic accuracy. Photoacoustic nanoprobes offer potential applications in cancer imaging, neuroimaging, cardiovascular imaging, and molecular imaging, among others, providing valuable insights into disease pathology and treatment response.

Gold nanoparticles are among the most widely studied nanomaterials for photoacoustic imaging applications due to their strong optical absorption properties in the near-infrared region, where biological tissues exhibit minimal light attenuation [15]. These nanoparticles can be engineered with precise control over size, shape, and surface chemistry, enabling optimization for specific imaging applications and biological targets. Gold nanoparticles offer excellent biocompatibility, stability, and ease of surface functionalization, making them attractive candidates for targeted molecular imaging and image-guided therapy.

Carbon-based nanomaterials, such as carbon nanotubes and graphene oxide, have also shown promise as photoacoustic imaging nanoprobes due to their unique optical and acoustic properties. These nanomaterials exhibit strong light absorption and efficient conversion of optical energy into acoustic signals, enabling sensitive detection of molecular targets and cellular processes within tissues. Carbon-based nanomaterials offer advantages such as biocompatibility, low toxicity, and versatility in surface functionalization, making them suitable for a wide range of biomedical imaging applications.

### 4.1.4 *Nanotechnology in Tissue Engineering*

Nanotechnology has revolutionized tissue engineering by providing innovative solutions for the design, fabrication, and manipulation of biomaterials at the nanoscale to mimic the complex structures and functions of native tissues (Fig. 4.2). This interdisciplinary field merges principles from materials science, engineering, biology, and medicine to create advanced scaffolds, growth factors, and drug delivery systems tailored to specific tissue regeneration applications [16]. Nano biomaterials offer unique properties such as high surface area, tunable mechanical properties, and precise control over cellular interactions, enabling enhanced biocompatibility, bioactivity, and tissue integration. By harnessing the power of nanotechnology, tissue engineers can overcome key challenges in tissue regeneration, including vascularization, innervation, and immune response modulation, leading to the development of next-generation therapies for tissue repair and regeneration.

#### 4.1.4.1 Scaffold Design and Fabrication

Scaffold design and fabrication are critical aspects of tissue engineering, providing the architectural framework necessary for guiding cell growth, differentiation, and tissue formation. The design of scaffolds involves careful consideration of various factors, including material properties, structure, porosity, and mechanical strength, to mimic the native extracellular matrix (ECM) and create an optimal microenvironment for cell adhesion, proliferation, and differentiation. Biomaterials commonly used for scaffold fabrication include natural polymers such as collagen, fibrin, and hyaluronic acid, as well as synthetic polymers like poly (lactic-co-glycolic acid)

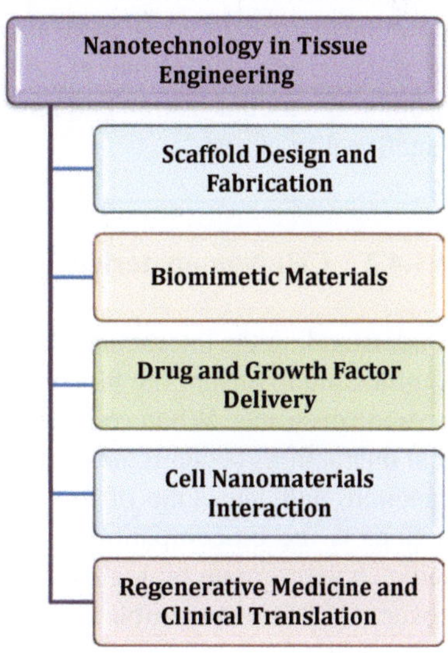


**Fig. 4.2** Nanotechnology in tissue engineering

(PLGA), polyethylene glycol (PEG), and polycaprolactone (PCL) [16]. These materials offer tunable properties that can be tailored to specific tissue types and applications, allowing for the development of customized scaffolds with desired characteristics. Fabrication techniques for scaffolds range from traditional methods such as solvent casting, particulate leaching, and electrospinning to advanced technologies like 4D bioprinting and microfluidics. Solvent casting involves dissolving the polymer in a solvent, casting it into a mold, and evaporating the solvent to create a solid scaffold with interconnected pores. Particulate leaching utilizes sacrificial particles that are mixed with the polymer solution, allowing for the creation of porous scaffolds with controlled pore sizes and interconnectivity. Electrospinning involves the extrusion of polymer fibers from a charged spinneret onto a collector, resulting in fibrous scaffolds with high surface area and porosity. These techniques offer precise control over scaffold morphology, porosity, and mechanical properties, enabling customization for specific tissue engineering applications.

In recent years, 4D bioprinting has emerged as a powerful tool for scaffold fabrication, allowing for the precise deposition of biomaterials, cells, and bioactive molecules layer-by-layer to create complex 3D structures [17]. This technology enables the incorporation of multiple materials and cell types within scaffolds, mimicking the hierarchical organization and heterogeneity of native tissues. Moreover, 4D bioprinting offers spatial control over cell distribution and scaffold architecture, facilitating the development of tissue constructs with precise anatomical features and functionality. Microfluidic-based approaches also hold promise for scaffold fabrication, leveraging microscale channels and fluid flow to control the spatial organization of cells and biomaterials within scaffolds, enabling the creation of biomimetic tissue models with high throughput and reproducibility. Advancements in scaffold design and fabrication have led to the development of tissue-engineered constructs for a wide range of applications, including bone, cartilage, skin, cardiac, and neural tissue regeneration. These engineered tissues hold potential for clinical translation, offering alternative therapies for patients with tissue damage or organ failure [16]. However, challenges remain in optimizing scaffold properties, promoting vascularization and innervation, and achieving long-term integration and functionality in vivo.

#### 4.1.4.2 Cell-Nanomaterials Interactions

Understanding the interactions between cells and nanomaterials is essential in tissue engineering to ensure the biocompatibility, functionality, and effectiveness of engineered constructs. When cells encounter nanomaterials, various physical and chemical interactions occur at the cellular and molecular levels, influencing cell behavior, function, and fate. One of the primary considerations in cell-nanomaterial interactions is the biocompatibility of the nanomaterial, which refers to its ability to elicit an appropriate biological response without causing adverse effects on cells or tissues [18]. Biocompatible nanomaterials promote cell adhesion, proliferation, and differentiation, facilitating tissue regeneration and integration within the host environment. Surface properties of nanomaterials, such as topography, roughness, and chemistry, play a crucial role in mediating cell-nanomaterial interactions. Nanomaterials with appropriate surface modifications, such as functional groups or biomolecular coatings, can enhance cell adhesion and spread by promoting cell-surface interactions and signaling pathways. Nanomaterials can influence cell behavior through mechano transduction mechanisms, where mechanical cues from the nanomaterial substrate are converted into biochemical signals that regulate cell function and fate. These mechano transduction processes are critical for modulating cell morphology, cytoskeletal organization, and gene expression in response to nanomaterial cues.

In addition to surface properties, the size, shape, and stiffness of nanomaterials also impact cell-nanomaterial interactions. Nanomaterials with nanoscale dimensions and high aspect ratios can mimic the structural features of the native extracellular matrix (ECM), facilitating cell attachment, migration, and differentiation. Furthermore, the mechanical properties of nanomaterials, such as stiffness or elasticity, can influence cellular behavior and lineage commitment by modulating cellular tension and cytoskeletal dynamics. Understanding how these physical cues affect cell behavior is essential for designing biomimetic scaffolds and engineering functional tissues with desired mechanical properties and functionality. Nanomaterials can serve as carriers for bioactive molecules, growth factors, and genetic materials, enabling spatiotemporal control over cell signaling and tissue regeneration. Functionalized nanomaterials can deliver therapeutic agents to target cells or tissues, modulate cellular responses, and promote tissue repair and regeneration [16]. By harnessing the synergistic interactions between cells and nanomaterials, researchers can develop advanced tissue engineering strategies for regenerating complex tissues and organs, with applications in regenerative medicine, drug delivery, and disease modeling. However, further research is needed to elucidate the mechanisms underlying cell-nanomaterial interactions and optimize their design and implementation in tissue engineering applications.

#### 4.1.4.3 Biomimetic Materials

Biomimetic materials represent a revolutionary approach in tissue engineering, aiming to replicate the structural and functional properties of native tissues and organs. Inspired by biological systems, biomimetic materials are designed to mimic the complex hierarchical organization, mechanical properties, and biochemical cues of the extracellular matrix (ECM) [19]. These materials offer numerous advantages, including enhanced biocompatibility, cell adhesion, and tissue integration, making them ideal candidates for regenerative medicine applications. By closely resembling the native tissue microenvironment, biomimetic materials promote cell attachment, proliferation, and differentiation, facilitating tissue regeneration and functional restoration. One of the key strategies in biomimetic materials design is the incorporation of bioactive molecules, such as growth factors, peptides, and extracellular matrix proteins, to provide spatiotemporal cues for cell behavior and tissue development. These bioactive molecules play critical roles in cell adhesion, migration, and differentiation, guiding tissue morphogenesis and remodeling. By immobilizing these molecules onto biomimetic scaffolds, researchers can create environments that promote specific cellular responses, such as stem cell differentiation toward desired lineages or enhanced tissue regeneration in damaged or diseased tissues. Another important aspect of biomimetic materials is their ability to recapitulate the mechanical properties of native tissues, including elasticity, stiffness, and viscoelasticity. These mechanical cues influence cell behavior and fate, regulating processes such as cell spreading, cytoskeletal organization, and gene expression. Biomimetic materials with tunable mechanical properties enable researchers to create scaffolds that match the mechanical characteristics of target tissues, providing optimal support and guidance for cell growth and tissue development.

Biomimetic materials offer versatility in fabrication techniques, allowing for the creation of complex 3D structures with precise control over composition, architecture, and functionality. Advanced manufacturing methods, such as 4D bioprinting [17], electrospinning, and microfluidics, enable the fabrication of biomimetic scaffolds with biomimetic features and spatial organization, mimicking the native tissue microenvironment [19]. These biomimetic materials hold great promise for tissue engineering applications, offering innovative solutions for regenerating damaged or diseased tissues and organs, with potential applications in regenerative medicine, organ transplantation, and disease modeling.

### *4.1.5 Nano Biomaterials for Cancer Therapy*

Nano biomaterials have emerged as promising tools in cancer therapy, offering innovative approaches to improve treatment efficacy while minimizing adverse effects on healthy tissues (Fig. 4.3). By combining nanotechnology with biomaterials, researchers can design and develop targeted drug delivery systems, photothermal therapies, and diagnostic agents tailored to the unique characteristics of cancer cells and tumors. These nano biomaterials offer several advantages, including enhanced tumor targeting, prolonged circulation time, and controlled drug release, leading to improved therapeutic outcomes and reduced systemic toxicity [20]. With their ability to overcome biological barriers, such as the blood–brain barrier, and selectively deliver therapeutic payloads to tumor sites, nano biomaterials hold great promise for revolutionizing cancer treatment and personalized medicine approaches in oncology.

#### **4.1.5.1 Targeted Drug Delivery Systems**

Targeted drug delivery systems represent a groundbreaking approach in cancer therapy, aiming to improve the efficacy and reduce the toxicity of anticancer drugs by selectively delivering them to tumor sites while sparing healthy tissues. These systems utilize various nano biomaterials, such as liposomes, polymer nanoparticles, and dendrimers, as carriers for encapsulating and delivering therapeutic agents to cancer cells. One of the key advantages of targeted drug delivery systems is

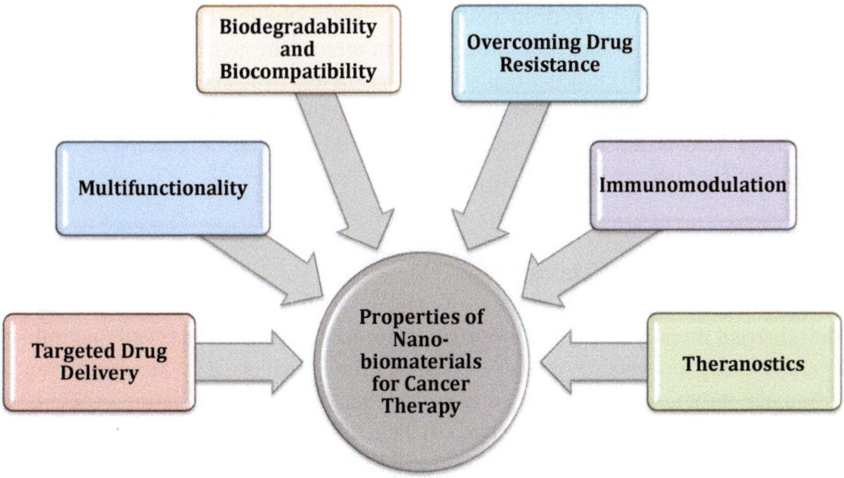


**Fig. 4.3** Different properties of nano biomaterials for cancer therapy

their ability to exploit unique features of tumors, such as overexpressed receptors or leaky vasculature, to achieve selective accumulation and retention of drugs within the tumor microenvironment, enhancing therapeutic efficacy. Liposomes are among the most widely studied nanocarriers for targeted drug delivery, offering advantages such as biocompatibility, versatility, and tunable properties [21]. These lipid-based vesicles can encapsulate hydrophilic and hydrophobic drugs within their aqueous core or lipid bilayer, respectively, allowing for efficient drug loading and controlled release kinetics. Moreover, liposomes can be surface-functionalized with targeting ligands, such as antibodies or peptides, to enhance tumor specificity and facilitate receptor-mediated endocytosis in cancer cells. By encapsulating cytotoxic drugs within targeted liposomes, researchers can achieve higher drug concentrations at the tumor site, leading to improved therapeutic outcomes and reduced off-target effects. Polymer nanoparticles are another class of nanocarriers widely explored for targeted drug delivery in cancer therapy. These nanoparticles offer advantages such as biodegradability, stability, and tunable drug release profiles, making them suitable for delivering a wide range of anticancer agents, including chemotherapeutic drugs, nucleic acids, and small molecules. Polymer nanoparticles can be engineered with surface modifications, such as polyethylene glycol (PEG) or ligand conjugation, to improve their circulation time, evade immune detection, and enhance tumor targeting efficiency [8]. Stimuli-responsive polymer nanoparticles can be designed to release drugs in response to specific triggers, such as pH, temperature, and enzymatic activity, further enhancing their therapeutic efficacy and selectivity.

Dendrimers, highly branched synthetic polymers with precise molecular structures, offer unique advantages for targeted drug delivery due to their well-defined size, shape, and surface functionality. These nanocarriers can be engineered with multiple functional groups for drug conjugation and targeting ligand attachment, enabling precise control over drug loading and release kinetics. Dendrimers can penetrate biological barriers and accumulate within tumors through an enhanced permeability and retention (EPR) effect, allowing for efficient drug delivery to cancer cells. With their multifunctional capabilities and ability to overcome biological barriers, dendrimers hold promise for developing advanced targeted drug delivery systems for personalized cancer therapy.

#### 4.1.5.2 Photothermal Therapy

Photothermal therapy (PTT) is a promising approach in cancer treatment that utilizes light-absorbing materials, typically nanomaterials such as gold nanoparticles or carbon nanotubes, to convert light energy into heat, selectively ablating cancer cells while sparing surrounding healthy tissues. This minimally invasive therapy relies on the principle of photothermal conversion, where the absorption of light by nanomaterials generates localized heating, leading to thermal damage and cell death within the tumor. One of the key advantages of PTT is its specificity, as the therapeutic effect is confined to regions where light is delivered, allowing for precise targeting of tumors with high spatial resolution and minimal collateral damage to adjacent tissues. Gold nanoparticles (GNPs) are widely used as photothermal agents in PTT due to their strong light absorption in the near-infrared (NIR) region, where biological tissues exhibit minimal light absorption and scattering [22]. GNPs can be functionalized and targeted to specific cancer cells or tissues using surface modifications, such as antibodies, peptides, or aptamers, enhancing tumor specificity and uptake. Upon irradiation with NIR light, GNPs efficiently convert photon energy into heat, leading to localized hyperthermia within the tumor microenvironment, which induces cancer cell death through protein denaturation, membrane disruption, and apoptosis. GNPs offer excellent biocompatibility, stability, and surface tunability, making them ideal candidates for PTT applications in cancer therapy.

Carbon-based nanomaterials, such as carbon nanotubes (CNTs) and graphene oxide (GO), have also shown promise as photothermal agents for cancer therapy. These nanomaterials exhibit strong light absorption in the NIR region, enabling efficient conversion of light energy into heat for selective tumor ablation. CNTs and GO can be functionalized by targeting ligands or therapeutic agents to enhance tumor specificity and therapeutic efficacy. Moreover, their unique physicochemical properties, including high surface area, mechanical strength, and optical transparency [1], make them attractive candidates for multifunctional applications in PTT, including imaging, drug delivery, and biosensing.

PTT offers several advantages over traditional cancer therapies, including non-invasiveness, minimal side effects, and rapid treatment response. PTT can be combined with other therapeutic modalities, such as chemotherapy, immunotherapy, and radiotherapy, to achieve synergistic effects and enhance therapeutic outcomes [23]. However, challenges remain in optimizing PTT parameters, such as light dosage, wavelength, and penetration depth, to maximize treatment efficacy while minimizing off-target effects and tissue damage. Ongoing research efforts aimed at developing novel photothermal agents, optimizing treatment protocols, and improving imaging and monitoring techniques are crucial for advancing PTT as a safe and effective modality for cancer therapy.

#### 4.1.5.3 Photodynamic Therapy

Nanomaterials have emerged as promising candidates in photodynamic therapy (PDT), offering innovative approaches to enhance the efficacy and specificity of this cancer treatment modality [24]. PDT relies on the administration of photosensitizing agents, which can be activated by light of a specific wavelength to generate reactive oxygen species (ROS) that induce tumor cell death. Nanomaterials, such as metal nanoparticles, carbon-based nanomaterials, and polymeric nanoparticles, can serve as carriers for photosensitizers, improving their solubility, stability, and tumor-targeting capabilities. By encapsulating photosensitizers within nanocarriers, researchers can enhance their accumulation within tumor tissues and achieve controlled release kinetics, leading to improved therapeutic outcomes. One of the key advantages of nanomaterial-based PDT is its ability to overcome biological barriers and selectively target tumor cells while sparing healthy tissues. Nanocarriers can be engineered with surface modifications, such as targeting ligands or stimuli-responsive moieties, to enhance their specificity for cancer cells and minimize off-target effects. Nanomaterials can facilitate the delivery of photosensitizers [24] to intracellular compartments or subcellular organelles, such as mitochondria or lysosomes, where ROS generation can induce selective cytotoxicity and apoptosis in cancer cells. Furthermore, nanomaterials offer opportunities for multimodal imaging and therapy by combining PDT with other treatment modalities, such as chemotherapy or photothermal therapy, within a single nanoplatform.

Nanomaterials can overcome limitations associated with conventional photosensitizers, such as poor water solubility, low photostability, and limited tissue penetration depth. Nanoparticles can be designed to enhance the photophysical properties of photosensitizers, such as their absorption and emission spectra, quantum yield, and singlet oxygen generation efficiency, leading to improved photodynamic efficacy [1, 24]. The unique physicochemical properties of nanomaterials, such as their high surface area, optical transparency, and biocompatibility, make them suitable candidates for theranostic applications, enabling real-time imaging and monitoring of PDT response in vivo.

#### 4.1.5.4 Radiotherapy Enhancement

Nanomaterials have garnered significant interest in the field of radiotherapy enhancement, offering novel strategies to improve the efficacy of radiation therapy in treating cancer [25]. One approach involves the use of high-Z (atomic number) nanoparticles, such as gold nanoparticles (GNPs), to enhance the radiation dose delivered to tumor cells while minimizing damage to surrounding healthy tissues. GNPs have a strong interaction with ionizing radiation, leading to increased energy deposition and enhanced production of secondary electrons, which can induce DNA damage and cell death. By selectively accumulating within tumor tissues via passive or active targeting mechanisms, GNPs can augment the therapeutic effect of radiation therapy, leading to improved tumor control and survival outcomes. Nanomaterials can be engineered to enhance the radiosensitivity of cancer cells by modulating cellular responses to ionizing radiation. For instance, nanoparticle-based radiosensitizers, such as cerium oxide nanoparticles and manganese oxide nanoparticles, can act as free radical scavengers or catalytic antioxidants, mitigating radiation-induced oxidative stress and DNA damage [25]. Nanomaterials can facilitate the delivery of radiosensitizing agents, such as chemotherapeutic drugs or nucleic acids, to tumor cells, enhancing their response to radiation therapy. By exploiting the unique physicochemical properties of nanomaterials, researchers can develop multifunctional platforms for synergistic combination therapies that target multiple pathways involved in cancer progression and treatment resistance. Moreover, nanomaterials offer opportunities to overcome biological barriers and improve the delivery and distribution of radiation therapy. For instance, nanocarriers, such as liposomes, polymeric nanoparticles, and dendrimers, can encapsulate and protect radiosensitizing agents from degradation, prolonging their circulation time and enhancing tumor accumulation. Surface modifications of nanomaterials with targeting ligands or stimuli-responsive moieties enable site-specific delivery of radiation enhancers to tumor tissues, minimizing off-target effects and systemic toxicity. Nanomaterial-based imaging agents, such as gold nanoprobes and iron oxide nanoparticles, can provide real-time visualization of tumor response to radiation therapy, enabling personalized treatment planning and monitoring of treatment efficacy.

Nanomaterials hold great promise for enhancing the therapeutic efficacy of radiation therapy in cancer treatment through various mechanisms, including radiation dose enhancement, radiosensitization, and targeted delivery of radiosensitizing agents [25]. With continued advancements in nanotechnology and interdisciplinary collaborations, nanomaterial-based radiotherapy enhancement strategies have the potential to revolutionize cancer treatment by improving tumor control, minimizing side effects, and overcoming treatment resistance. However, further research is needed to optimize the design, synthesis, and clinical translation of nanomaterial-based radiotherapy enhancement therapies for widespread clinical use.

**Exercise 2.**

**Fill Ups**

**(Cancer cells, Cancer treatments, Therapeutic, Drug penetration, Nano biomaterials)**

**1.** .................... hold promise in cancer therapy by delivering drugs directly to tumors while sparing healthy tissues.

**2.** The unique properties of biomaterials enable precise targeting and controlled release of .................agents.

**3.** Surface functionalization allows nano biomaterials to selectively bind to .................and evade immune detection.

**4.** Overcoming tumor microenvironment challenges, nano biomaterials can enhance ......................and efficacy.

**5.** Nano biomaterials offer potential for personalized and targeted ........................tailored to individual patient needs.

### *4.1.6 Nanotechnology in Diagnosis and Biosensing*

Nanotechnology has revolutionized the field of diagnosis and biosensing by providing innovative tools and platforms for the detection and monitoring of various diseases with high sensitivity, specificity, and efficiency (Fig. 4.4). Nanomaterials, such as nanoparticles, nanowires, and quantum dots, offer unique properties, including large surface area-to-volume ratio, tunable optical and electronic properties, and biocompatibility, making them ideal candidates for biosensing applications. By harnessing the capabilities of nanotechnology, researchers can develop ultrasensitive diagnostic assays, point-of-care devices, and imaging contrast agents for early detection, prognosis, and personalized treatment of diseases such as cancer, infectious diseases, and neurological disorders. Nanotechnology-based biosensors enable rapid and multiplexed detection of biomarkers, DNA, proteins, and other analytes in complex biological samples, providing valuable insights into disease pathology and facilitating timely interventions for improved patient outcomes [26].

#### 4.1.6.1 Biosensors for Disease Detection

Biosensors play a crucial role in disease detection, offering rapid, sensitive, and cost-effective diagnostic tools for the early identification of various health conditions. These devices, often utilizing nanotechnology, detect specific biomolecules, such as proteins, nucleic acids, and metabolites, that serve as indicators of disease presence or progression. Nanomaterials, such as nanoparticles, nanowires, and carbon

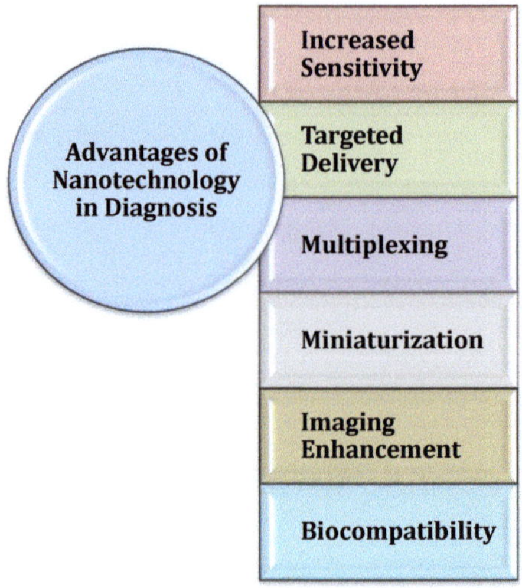


**Fig. 4.4** Advantages of nanotechnology in diagnosis and biosensing

nanotubes, are commonly incorporated into biosensors due to their unique physical, chemical, and biological properties, which enhance sensitivity and specificity [26]. Biosensors can be designed to detect biomarkers associated with a wide range of diseases, including cancer, infectious diseases, cardiovascular disorders, and neurological conditions, providing valuable information for disease diagnosis, prognosis, and treatment monitoring. One of the key advantages of biosensors is their ability to provide rapid and real-time detection of biomarkers with high sensitivity and specificity. By integrating nanomaterials with biomolecular recognition elements, such as antibodies, aptamers, and enzymes, biosensors can selectively capture and quantify target analytes in complex biological samples, such as blood, urine, and saliva. Advancements in microfluidics, miniaturization, and wireless communication enable the development of portable and point-of-care biosensing devices, facilitating decentralized testing and remote monitoring of disease biomarkers in resource-limited settings or home environments.

Nanotechnology-based biosensors offer multiplexing capabilities, allowing simultaneous detection of multiple biomarkers within a single assay. This capability is particularly valuable for diseases with complex pathophysiology or heterogeneous biomarker profiles [26], such as cancer, where the combination of multiple biomarkers can improve diagnostic accuracy and prognostic assessment. Nanomaterials can enhance signal amplification and signal-to-noise ratio in biosensing platforms, enabling the detection of low-abundance biomarkers and early-stage disease markers that may be challenging to detect using conventional methods [27]. Biosensors represent a powerful tool for disease detection and monitoring, offering advantages such as rapidity, sensitivity, specificity, and portability. With ongoing advancements in nanotechnology, biosensor development, and biomarker discovery, these devices hold promise for revolutionizing healthcare by enabling early and accurate diagnosis of diseases, facilitating timely interventions, and improving patient outcomes.

#### 4.1.6.2 Point-of-Care Devices

Point-of-care (POC) devices have emerged as valuable tools in healthcare, providing rapid and convenient diagnostic testing at or near the patient's location. These portable and easy-to-use devices enable healthcare providers to obtain immediate test results, facilitating timely decision-making and patient management. POC devices offer several advantages, including reduced turnaround time, improved patient outcomes, and increased accessibility to diagnostic testing, particularly in remote or underserved areas where traditional laboratory facilities may be limited. Nanotechnology plays a significant role in the development of POC devices, as it enables the miniaturization, integration, and enhancement of device performance, making them suitable for a wide range of diagnostic applications [28]. One of the key features of POC devices is their simplicity and user-friendliness, allowing non-specialized personnel, such as nurses or community health workers, to perform tests with minimal training. These devices typically utilize lateral flow assays, microfluidic cartridges, or paper-based sensors to detect target analytes, such as biomarkers, pathogens, and genetic material, in biological samples, such as blood, urine, and saliva. Nanomaterials, such as gold nanoparticles, quantum dots, and magnetic nanoparticles, are often incorporated into POC devices to enhance sensitivity, specificity, and signal amplification, enabling accurate and reliable detection of disease markers within minutes.

POC devices offer versatility in terms of diagnostic capabilities, allowing for the detection of a wide range of diseases, including infectious diseases, chronic conditions, and metabolic disorders [28]. For example, POC devices can be used for rapid screening and diagnosis of infectious pathogens, such as influenza viruses, HIV, and malaria parasites, enabling prompt initiation of appropriate treatment and infection control measures. POC devices can monitor disease progression and treatment response over time, providing valuable insights into disease management and personalized medicine approaches. POC devices have the potential to revolutionize healthcare delivery by shifting diagnostic testing from centralized laboratory facilities to the point of care, reducing healthcare costs, and improving patient outcomes. With ongoing advancements in nanotechnology, microfluidics, and biosensing technologies, POC devices continue to evolve, offering enhanced sensitivity, multiplexing capabilities, and connectivity for data sharing and remote monitoring. However, challenges such as regulatory approval, standardization, and integration into existing healthcare infrastructure need to be addressed to realize the full potential of POC devices in improving global health outcomes.

#### 4.1.6.3 Lab-on-a-Chip Technologies

Lab-on-a-chip (LOC) technologies represent a transformative approach to diagnostic testing, offering miniaturized and integrated platforms for performing multiple laboratory functions on a single device [29]. These microfluidic devices enable the manipulation, analysis, and detection of biological samples with high precision, throughput, and automation, making them ideal for a wide range of diagnostic applications. LOC devices typically consist of microchannels, chambers, valves, and sensors fabricated on a small chip, allowing for the integration of sample preparation, reaction, and detection steps within a compact and portable format. Nanotechnology plays a crucial role in LOC technologies, as it enables the fabrication of nanoscale features and the incorporation of nanomaterials for enhancing device performance and functionality.

One of the key advantages of LOC technologies is their ability to perform complex analytical tasks with minimal sample volume and reagent consumption, reducing costs and waste while increasing efficiency and throughput. These devices enable rapid and multiplexed analysis of biological samples, such as blood, saliva, and urine, for various diagnostic purposes, including disease diagnosis, monitoring, and personalized medicine. Furthermore, LOC devices offer high sensitivity and specificity, enabling the detection of low-abundance biomarkers or pathogens that may be challenging to detect using traditional methods. By integrating sample processing, amplification, and detection steps within a single device [29], LOC technologies streamline diagnostic workflows and minimize human error, leading to improved accuracy and reproducibility of test results. LOC technologies offer versatility in terms of applications, allowing for the detection of a wide range of analytes, including nucleic acids, proteins, cells, and small molecules. These devices can be customized and adapted for specific diagnostic needs, such as infectious disease screening, cancer detection, genetic testing, and environmental monitoring. LOC devices can be interfaced with external equipment, such as smartphones or computers, for data analysis, storage, and communication, enabling remote monitoring and telemedicine applications. With ongoing advancements in nanotechnology, microfluidics, and biosensing techniques, LOC technologies continue to evolve, offering new opportunities for improving healthcare delivery, disease management, and public health surveillance. LOC technologies hold promise for point-of-care testing, enabling decentralized diagnostic testing in resource-limited settings or remote areas where access to traditional laboratory facilities may be limited. These devices can be deployed in various settings, such as clinics, hospitals, field hospitals, and community health centers, providing rapid and reliable diagnostic results within minutes. By bringing diagnostic testing closer to the patient, LOC technologies empower healthcare providers to make timely clinical decisions, initiate appropriate treatment interventions, and improve patient outcomes.

### 4.1.7 Regulatory and Ethical Considerations

Regulatory and safety considerations are paramount in the development and translation of nano biomaterials from the laboratory to clinical applications. Safety assessment is a crucial step in evaluating the potential risks associated with nano biomaterials, including their biocompatibility, toxicity, and immunogenicity. Various in vitro and in vivo assays are employed to assess the cytotoxicity, genotoxicity, and immunotoxicity of nano biomaterials, providing valuable insights into their safety profile and potential adverse effects on biological systems. Moreover, nano biomaterials may undergo physicochemical characterization [30], stability testing, and biodegradation studies to assess their behavior in physiological environments and predict their long-term safety and efficacy in vivo. Regulatory approval processes play a critical role in ensuring the safety, efficacy, and quality of nano biomaterials for clinical use. Depending on the intended application and regulatory jurisdiction, nano biomaterials may be subject to different regulatory pathways, including premarket approval, 510(k) clearance, or investigational new drug (IND) application. Regulatory agencies, such as the Food and Drug Administration (FDA) in the United States or the European Medicines Agency (EMA) in Europe, review preclinical and clinical data to evaluate the safety and effectiveness of nano biomaterials, assess their manufacturing processes, and determine their labeling, indications, and contraindications for use.

Ethical considerations are also important in the development and use of nano biomaterials, particularly regarding their potential impact on human health, privacy, and societal values. Ethical implications may arise from issues such as informed consent, privacy protection, equitable access to healthcare, and the responsible conduct of research [30, 31]. Researchers and clinicians must adhere to ethical guidelines and principles, such as autonomy, beneficence, nonmaleficence, and justice, to ensure the ethical conduct of research involving nano biomaterials. Moreover, stakeholders, including patients, healthcare providers, policymakers, and the public, should be engaged in discussions about the ethical, social, and legal implications of nano biomaterials, fostering transparency, trust, and accountability in their development and use.

Considerations regarding environmental impact and sustainability are gaining increasing attention in the field of nano biomaterials. Nanoparticles and other nanomaterials may pose risks to the environment during their manufacturing, use, and disposal, potentially affecting ecosystems and human health. Therefore, it is essential to evaluate the environmental impacts of nano biomaterials throughout their lifecycle and implement measures to mitigate potential risks, such as green synthesis methods, recycling, and waste management strategies. Furthermore, efforts to promote sustainable practices, such as green chemistry principles, life cycle assessment, and eco-design strategies, can help minimize the environmental footprint of nano biomaterials and contribute to the development of environmentally friendly healthcare technologies.

Regulatory and safety considerations, along with ethical and environmental implications, are integral aspects of the development and translation of nano biomaterials for clinical applications. Rigorous safety assessment, compliance with regulatory requirements, and adherence to ethical guidelines are essential to ensure the safety, efficacy, and responsible use of nano biomaterials in healthcare. By addressing these considerations comprehensively and collaboratively, researchers, clinicians, policymakers, and stakeholders can advance the field of nano biomaterials while safeguarding human health, protecting the environment, and upholding ethical principles.

### *4.1.8 Emerging Trends in Nanomedicine*

Emerging trends in nanomedicine hold great promise for revolutionizing healthcare by offering innovative solutions for disease diagnosis, treatment, and personalized medicine. One notable trend is the development of multifunctional nanoplatforms capable of integrating diagnostic and therapeutic functionalities within a single nanocarrier. These multifunctional nanoparticles can simultaneously deliver therapeutic agents, such as drugs or nucleic acids, and imaging contrast agents to targeted sites, enabling real-time monitoring of treatment response and disease progression [32, 33]. Moreover, advances in nanotechnology have facilitated the engineering of smart nanomaterials with stimuli-responsive properties, allowing for triggered drug release in response to specific physiological cues, such as pH, temperature, and enzyme activity, further enhancing treatment efficacy and minimizing off-target effects. Another emerging trend in nanomedicine is the development of theranostic nanomedicines, which combine therapeutic and diagnostic capabilities for personalized medicine applications. Theranostic nanoparticles enable non-invasive imaging of disease biomarkers, such as molecular imaging or magnetic resonance imaging (MRI), to guide targeted drug delivery and monitor treatment response in real time. By tailoring treatment strategies based on individual patient characteristics and disease profiles, theranostic nanomedicines offer the potential for personalized and precision medicine approaches, optimizing therapeutic outcomes and minimizing adverse effects.

Nanotechnology-based approaches hold promise for overcoming biological barriers and improving drug delivery to target tissues and cells. Functionalized nanoparticles can bypass biological barriers, such as the blood-brain barrier or tumor microenvironment, and selectively accumulate within diseased tissues, enhancing drug bioavailability and therapeutic efficacy. Nanocarriers can be engineered to enhance intracellular uptake and trafficking, overcoming multidrug resistance mechanisms, and enabling targeted delivery of therapeutic agents to intracellular compartments, such as the nucleus or mitochondria, where they exert their pharmacological effects. Despite the significant potential of nanomedicine, several challenges must be addressed to facilitate its clinical translation and widespread adoption in healthcare. One major challenge is ensuring the safety and biocompatibility of nanomaterials, as well as understanding their long-term effects on biological systems and the environment. Rigorous safety assessment [1], standardized testing protocols, and regulatory guidelines are essential to evaluate the potential risks associated with nanomedicines and ensure their safe use in clinical settings. Scalability, reproducibility, and cost-effectiveness are important considerations in the development and manufacturing of nanomedicines, requiring innovative fabrication techniques and optimization of production processes to meet regulatory requirements and market demands.

Interdisciplinary collaborations and partnerships between academia, industry, and regulatory agencies are crucial for accelerating the clinical translation of nanomedicine and overcoming barriers to commercialization [34, 35]. Collaborative efforts can facilitate knowledge exchange, technology transfer, and regulatory approval processes, enabling a seamless transition from bench to bedside. Moreover, education and training programs are needed to increase awareness and understanding of nanomedicine among healthcare professionals [48], researchers, and the public, fostering a supportive ecosystem for the development and adoption of nanotechnology-based solutions in healthcare.

**Exercise 3.**

**Choose The Correct Option**

1. What is a common method used to fabricate nanoparticles for drug delivery?
   a) Injection molding
   b) 3D printing
   c) Solvent evaporation
   d) Laser cutting

2. Which of the following properties of nanoparticles makes them suitable for drug delivery?
   a) Large size
   b) Fast degradation
   c) High surface area to volume ratio
   d) Low stability

3. Which route of administration is often utilized for delivering drug-loaded nanoparticles?
   a) Topical
   b) Oral
   c) Intramuscular
   d) Intravenous

4. Which nanomaterial is commonly used for contrast enhancement in medical imaging?
   a) Carbon nanotubes
   b) Gold nanoparticles
   c) Silver nanoparticles
   d) Quantum dots

5. What imaging modality is often utilized in conjunction with nanomaterials for deep tissue imaging?
   a) X-ray imaging
   b) Magnetic resonance imaging (MRI)
   c) Positron emission tomography (PET)
   d) Near-infrared fluorescence imaging

6. Which of the following is a common strategy employed by nano biomaterials for targeted cancer therapy?
   a) Increasing tumor size
   b) Enhancing metastasis
   c) Passive targeting via enhanced permeability and retention (EPR) effect
   d) Inducing inflammation

7. Which type of nano biomaterial is often used to encapsulate chemotherapeutic drugs for targeted delivery to cancer cells?
   a) Liposomes
   b) Carbon nanotubes
   c) Silicon nanoparticles
   d) Polymer microspheres

8. What type of nanoparticles are commonly utilized in biosensing applications for detecting biomolecules?
   a) Iron nanoparticles
   b) Titanium nanoparticles
   c) Gold nanoparticles
   d) Silver nanoparticles

9. Which of the following techniques exploits the unique optical properties of nanoparticles for biosensing?
   a) Polymerase chain reaction (PCR)
   b) Surface plasmon resonance (SPR)
   c) Western blotting
   d) Gel electrophoresis

10. How do nanomaterial-based biosensors typically enhance detection sensitivity?
   a) By decreasing specificity
   b) By reducing detection speed
   c) By increasing surface area for biomolecule binding
   d) By increasing sample volume required

## 4.2 Summary

In this comprehensive chapter on "Nanomaterials in Biomedicine," various aspects of nanotechnology's applications in healthcare were explored, beginning with an introduction to nano biomaterials, highlighting their historical development and significance in biomedicine. The chapter delved into the diverse role nanomaterials play, particularly focusing on drug delivery, imaging, tissue engineering, cancer therapy, diagnosis, and biosensing. Nanoparticles for drug delivery were discussed extensively, including liposomes, polymeric nanoparticles, and inorganic nanoparticles like gold, silver, and iron oxide. These carriers offer improved drug delivery efficiency, specificity, and reduced side effects, revolutionizing therapeutic interventions [4, 5, 8]. The role of nanomaterials in imaging, such as contrast agents and fluorescent probes, was explored for their utility in visualizing and monitoring diseases with high precision. The chapter then delved into tissue engineering, illustrating how nanomaterials contribute to scaffold design, cell-material interactions, and the development of biomimetic materials. These advancements facilitate tissue regeneration and organ repair, paving the way for regenerative medicine. Nanotechnology's impact on cancer therapy was highlighted, particularly in targeted drug delivery, photothermal therapy, photodynamic therapy, and radiotherapy enhancement, offering novel strategies for precise and effective cancer treatment.

Nanotechnology's role in diagnosis and biosensing was discussed, emphasizing the development of biosensors and lab-on-a-chip technologies for rapid, sensitive, and portable disease detection. The chapter also addressed regulatory and ethical considerations surrounding nano biomaterials, emphasizing safety assessment, regulatory approval processes, and ethical implications for responsible research and clinical translation. Lastly, emerging trends in nanomedicine and strategies for overcoming barriers to clinical translation were outlined, underscoring the potential of nano biomaterials to revolutionize healthcare by offering personalized, precise, and minimally invasive diagnostic and therapeutic interventions.

## Points to Remember

- **Nanoparticles can be designed to specifically target tumors, delivering drugs directly to cancer cells while minimizing side effects on healthy tissues.**
- **Nanoparticles like quantum dots emit light when excited, making them valuable tools for sensitive and high-resolution medical imaging.**
- **Theranostic nanoparticles combine therapy and diagnostics in one package, allowing doctors to monitor treatment responses in real-time while delivering therapies directly to diseased cells.**
- **Nanostructured surfaces on implants promote better integration with surrounding tissues, improving the success rate of medical implants like joint replacements and dental implants.**
- **Nanomaterial-based biosensors offer rapid and sensitive detection of biomolecules, aiding in the diagnosis of diseases and monitoring of environmental pollutants.**
- **Nanoparticles serve as carriers for therapeutic genes, enabling gene therapy to treat genetic disorders and cancers by delivering genetic material directly into cells.**
- **Nanomaterials with antimicrobial properties, such as silver nanoparticles, are effective against drug-resistant pathogens, offering new solutions for combating infections in healthcare settings.**
- **Nanotechnology is revolutionizing organ-on-chip technology, allowing researchers to create miniature organs with lifelike functions for drug testing and disease modeling.**

## References

1. Patil-Sen Y (2021) Advances in nano-biomaterials and their applications in biomedicine. Emerg Top Life Sci 5(1):169–176. https://doi.org/10.1042/ETLS20200444
2. Jangid AK, Jain P, Pooja D, Kulhari H (2021) Introduction to nanomedicines: basic concept and applications. In: Singh S (eds) Emerging trends in nanomedicine. Springer, Singapore. https://doi.org/10.1007/978-981-15-9920-0_1
3. Afzal O, Altamimi ASA, Nadeem MS, Alzarea SI, Almalki WH, Tariq A, Mubeen B, Murtaza BN, Iftikhar S, Riaz N, Kazmi I (2022) Nanoparticles in drug delivery: from history to therapeutic applications. Nanomaterials (Basel, Switzerland) 12(24):4494. https://doi.org/10.3390/nano12244494

4. Guimarães D, Cavaco-Paulo A, Nogueira E (2021) Design of liposomes as drug delivery system for therapeutic applications. Int J Pharm 601:120571. https://doi.org/10.1016/j.ijpharm.2021.120571
5. Todros S, Todesco M, Bagno A. (2021) Biomaterials and their biomedical applications: from replacement to regeneration. Processes 9(11):1949. https://doi.org/10.3390/pr9111949
6. Shakeel A, Singh A, Das S et al (2017) Synthesis and morphological insight of new biocompatible smart hydrogels. J Polym Res 24:114. https://doi.org/10.1007/s10965-017-1267-7
7. Begines B, Ortiz T, Pérez-Aranda M, Martínez G, Merinero M, Argüelles-Arias F, Alcudia A (2020) Polymeric nanoparticles for drug delivery: recent developments and future prospects. Nanomaterials (Basel, Switzerland) 10(7):1403. https://doi.org/10.3390/nano10071403
8. Rathore R, Suhag D, Wan F, Verma R, Thakur A, Thakur P (2023) Functionalization of biomaterials. In: Suhag D, Thakur A, Thakur P (eds) Integrated nanomaterials and their applications. Springer, Singapore. https://doi.org/10.1007/978-981-99-6105-4_11
9. Paul W, Sharma CP (2020) Inorganic nanoparticles for targeted drug delivery. In: Elsevier eBooks (pp. 333–373). https://doi.org/10.1016/b978-0-08-102680-9.00013-5
10. Rathore R, Suhag D, Wan F, Thakur A, Thakur P (2023) Toxicology and toxicity studies of nano-biomaterials. In: Suhag D, Thakur A, Thakur P (eds) Integrated nanomaterials and their applications. Springer, Singapore. https://doi.org/10.1007/978-981-99-6105-4_20
11. Badıllı U, Mollarasouli F, Bakirhan NK, Ozkan Y, Ozkan SA (2020) Role of quantum dots in pharmaceutical and biomedical analysis, and its application in drug delivery. TrAC. Trends Anal Chem 131:116013. https://doi.org/10.1016/j.trac.2020.116013
12. Fu Q, Zhu R, Song J, Yang H, Chen X (2019) Photoacoustic imaging: contrast agents and their biomedical applications. Adv Mater (Deerfield Beach, Fla.) 31(6):e1805875. https://doi.org/10.1002/adma.201805875
13. Ueno T, Nagano T (2011b) Fluorescent probes for sensing and imaging. Nat Methods 8(8):642–645. https://doi.org/10.1038/nmeth.1663
14. Merbach AE, Toth E (2013) The chemistry of contrast agents in medical magnetic resonance imaging. Wiley eBooks. https://doi.org/10.1002/9781118503652
15. Li W, Chen X (2015) Gold nanoparticles for photoacoustic imaging. Nanomedicine 10(2):299–320. https://doi.org/10.2217/nnm.14.169
16. Kingsley JD, Ranjan S, Dasgupta N, Saha P (2013) Nanotechnology for tissue engineering: need, techniques and applications. J Pharm Res 7(2):200–204. https://doi.org/10.1016/j.jopr.2013.02.021
17. Naniz MA, Askari M, Zolfagharian A, Bodaghi M (2022) 4D bioprinting: fabrication approaches and biomedical applications. In: Elsevier eBooks (pp. 193–229). https://doi.org/10.1016/b978-0-12-824082-3.00023-4
18. Veranič P, Križaj I (2023): Interaction of Nanomaterials with cells and tissues. Int J Mol Sci 24(17):13667. https://doi.org/10.3390/ijms241713667
19. Shin H, Jo S, Mikos AG (2003) Biomimetic materials for tissue engineering. Biomaterials 24(24):4353–4364. https://doi.org/10.1016/s0142-9612(03)00339-9
20. Rathore R, Suhag D, Wan F, Thakur A, Thakur P (2024) Toxicology and toxicity studies of nano-biomaterials. In: Suhag D, Thakur A, Thakur P (eds) Integrated nanomaterials and their applications. Springer, Singapore. https://doi.org/10.1007/978-981-99-6105-4_20
21. Mansouri S, Alharbi Y, Alqahtani A (2024) Current status and prospects for improved targeted delivery approaches for cancer. Pathol Res Pract 253:154993. https://doi.org/10.1016/j.prp.2023.154993
22. Han HS, Choi KY (2021) Advances in nanomaterial-mediated photothermal cancer therapies: toward clinical applications. Biomedicines 9(3):305. https://doi.org/10.3390/biomedicines9030305
23. Kumari S, Sharma N, Sahi SV (2021) Advances in cancer therapeutics: conventional thermal therapy to nanotechnology-based photothermal therapy. Pharmaceutics 13(8):1174. https://doi.org/10.3390/pharmaceutics13081174

24. Jiang W, Liang M, Lei Q, Li G, Wu S (2023) The current status of photodynamic therapy in cancer treatment. Cancers 15(3):585. https://doi.org/10.3390/cancers15030585
25. Gerken LRH, Gerdes ME, Pruschy M, Herrmann IK (2023) Prospects of nanoparticle-based radioenhancement for radiotherapy. Mater Horiz 10(10):4059–4082. https://doi.org/10.1039/d3mh00265a
26. Malik S, Singh J, Goyat R, Saharan Y, Chaudhry V, Umar A, Ibrahim AA, Akbar S, Ameen S, Baskoutas S (2023b). Nanomaterials-based biosensor and their applications: A review. Heliyon 9(9):e19929. https://doi.org/10.1016/j.heliyon.2023.e19929
27. Haleem A, Javaid M, Singh RP, Suman R, Rab S (2021) Biosensors applications in medical field: a brief review. Sens Int 2:100100. https://doi.org/10.1016/j.sintl.2021.100100
28. Zarnegar A (2023) Point-of-care devices in healthcare: a public health perspective. In: Daimi K, Alsadoon A, Seabra Dos Reis S (eds) Current and future trends in health and medical informatics. Studies in Computational Intelligence, vol 1112. Springer, Cham. https://doi.org/10.1007/978-3-031-42112-9_4
29. Arshavsky-Graham S, Segal E (2020) Lab-on-a-chip devices for point-of-care medical diagnostics. In: Bahnemann J, Grünberger A (eds) Microfluidics in biotechnology. Advances in Biochemical Engineering/Biotechnology, vol 179. Springer, Cham. https://doi.org/10.1007/10_2020_127
30. Soares S, Sousa J, Pais A, Vitorino C (2018) Nanomedicine: principles, properties, and regulatory issues. Front Chem 6:360. https://doi.org/10.3389/fchem.2018.00360
31. Ali F, Neha K, Parveen S (2023) Current regulatory landscape of nanomaterials and nanomedicines: a global perspective. J Drug Deliv Sci Technol 80:104118. https://doi.org/10.1016/j.jddst.2022.104118
32. Parvanian S, Mostafavi SM, Aghashiri M (2017) Multifunctional nanoparticle developments in cancer diagnosis and treatment. Sens Bio-sens Res 13:81–87. https://doi.org/10.1016/j.sbsr.2016.08.002
33. Souto EB, Blanco-Llamero C, Krambeck K, Kiran NS, Yashaswini C, Postwala H, Severino P, Priefer R, Prajapati BG, Maheshwari R (2024) Regulatory insights into nanomedicine and gene vaccine innovation: safety assessment, challenges, and regulatory perspectives. Acta Biomater 180:1–17. https://doi.org/10.1016/j.actbio.2024.04.010
34. Zhang C, Yan L, Wang X, Zhu S, Chen C, Gu Z, Zhao Y (2020) Progress, challenges, and future of nanomedicine. Nano Today 35:101008. https://doi.org/10.1016/j.nantod.2020.101008
35. Csóka I, Ismail R, Jójárt-Laczkovich O, Pallagi E (2021) Regulatory considerations, challenges and risk-based approach in nanomedicine development. Curr Med Chem 28(36): 7461–7476. https://doi.org/10.2174/0929867328666210406115529

# Chapter 5
# Bioactive Polymers

## 5.1 Introduction

Bioactive polymers, at the nexus of polymer science and biomedical engineering, have garnered considerable attention for their unique ability to interact with biological systems. These versatile materials possess inherent or engineered properties that stimulate specific responses when in contact with living organisms, making them invaluable in various biomedical applications. From drug delivery systems to tissue engineering scaffolds [1], bioactive polymers play a pivotal role in advancing healthcare by providing tailored solutions for targeted delivery, controlled release, and tissue regeneration. Understanding the fundamentals of bioactive polymers is essential to harnessing their full potential in biomedical applications. This entails exploring their chemical structure, synthesis methods, and mechanisms of biological interaction. Additionally, considering factors such as biocompatibility, degradation kinetics, and mechanical properties is crucial for designing bioactive polymers that meet the specific requirements of diverse biomedical applications [1, 2]. Thus, this introductory section sets the stage for a comprehensive exploration of bioactive polymers, laying the foundation for subsequent discussions on their design, synthesis, characterization, and applications in healthcare.

### *5.1.1 Historical Development and Importance of Bioactive Polymers*

The historical development of bioactive polymers traces back to the mid-twentieth century when researchers began exploring the potential of synthetic materials in biomedical applications. Early efforts focused on developing biocompatible materials for implantable medical devices, such as pacemakers and artificial joints. However, it wasn't until the late twentieth century that the concept of bioactive



D. Suhag et al., *Handbook of Biomaterials for Medical Applications, Volume 1*,
Biomedical Materials for Multi-functional Applications,
https://doi.org/10.1007/978-981-97-4818-1_5

polymers truly took off with groundbreaking discoveries in tissue engineering and drug delivery [2, 3]. One significant milestone was the development of biodegradable polymers capable of delivering therapeutic agents to specific sites in the body, revolutionizing the field of controlled drug release. Concurrently, advancements in polymer chemistry and materials science led to the design of polymers with tailored bioactive properties, enabling precise interactions with biological systems. Today, bioactive polymers continue to evolve, with ongoing research focusing on enhancing their biocompatibility, functionality, and applicability in regenerative medicine and personalized healthcare. The importance of bioactive polymers in biomedical research and clinical practice cannot be overstated. These materials offer a myriad of advantages over traditional biomaterials, including enhanced biocompatibility [5], tunable mechanical properties, and the ability to promote tissue regeneration. In drug delivery applications, bioactive polymers enable targeted delivery of therapeutics, reducing systemic side effects and improving treatment efficacy. Moreover, bioactive polymers play a crucial role in tissue engineering, providing scaffolds that mimic the native extracellular matrix and support cell growth and differentiation [3, 5]. This capability holds immense promise for regenerating damaged tissues and organs, offering hope for patients suffering from a wide range of medical conditions. As research in bioactive polymers continues to advance, their importance in modern medicine is expected to grow, paving the way for innovative solutions to pressing healthcare challenges and improving the quality of life for countless individuals worldwide.

### *5.1.2 Fundamentals of Polymer Chemistry*

The fundamentals of polymer chemistry serve as the cornerstone for understanding the structure, properties, and behavior of polymers, which are ubiquitous in our everyday lives and play pivotal roles in numerous industries, including healthcare, electronics, and materials science. At its core, polymer chemistry delves into the molecular structure of polymers, elucidating how monomers, the building blocks of polymers, polymerize to form long chains or networks. Understanding polymerization mechanisms, such as addition and condensation polymerization, provides insight into the synthesis processes that yield polymers with tailored properties [5]. Moreover, polymer chemistry explores the relationships between polymer structure and properties, encompassing factors like chain length, branching, and chemical composition, which influence characteristics such as mechanical strength, flexibility, and thermal stability. By unraveling these fundamental principles, students can grasp the underlying mechanisms driving the behavior of polymers, laying the groundwork for their application in diverse fields and fostering innovations that address complex societal challenges.

#### 5.1.2.1 Polymer Structure and Composition

Polymer structure and composition are foundational concepts in understanding the behavior and properties of polymers, which are large molecules composed of repeating units called monomers. These monomers link together through covalent bonds to form long chains or networks, giving polymers their characteristic properties. The structure of a polymer can vary widely depending on the type of monomers used and the arrangement of these monomers within the polymer chain. For instance, polymers may have linear, branched, or cross-linked structures, each imparting distinct mechanical, thermal, and chemical properties to the material. Understanding polymer structure is essential for tailoring the properties of polymers to specific applications, whether it be designing flexible packaging materials or engineering durable automotive components. The composition of a polymer refers to the types of monomers present in its structure and their arrangement along the polymer chain [3, 5]. Homopolymers consist of repeating units derived from a single type of monomer, while copolymers contain two or more different types of monomers. Copolymers can be further classified based on the arrangement of monomer units, including alternating, random, block, and graft copolymers. The composition of a polymer profoundly influences its properties and performance characteristics. For example, copolymers with alternating monomer sequences may exhibit unique chemical reactivity or solubility properties, while block copolymers with distinct segments may self-assemble into ordered nanostructures with applications in drug delivery or membrane separation technologies.

Polymer composition also plays a crucial role in determining the material's compatibility with various environments and processing techniques. For instance, the presence of polar or nonpolar functional groups in the polymer backbone can influence its interaction with solvents or other materials, affecting factors such as swelling behavior or adhesion properties. The composition of a polymer can impact its processability, including melt viscosity, melt temperature, and crystallization behavior, which are critical considerations in manufacturing processes such as injection molding, extrusion, or blow molding [6]. Thus, a thorough understanding of polymer composition is essential for optimizing material properties and processing conditions to meet the requirements of specific applications while ensuring cost-effectiveness and sustainability.

Polymer structure and composition are fundamental aspects that govern the properties, performance, and processing of polymers in various industrial and technological applications.

#### 5.1.2.2 Polymerization Techniques

Polymerization techniques are fundamental processes in the synthesis of bioactive polymers, enabling the creation of materials with tailored properties and functionalities for biomedical applications. One commonly employed polymerization technique is solution polymerization, where monomers are dissolved in a solvent and initiated to form polymer chains. This method allows for the synthesis of bioactive polymers with precise control over molecular weight, composition, and architecture [5], making it suitable for applications such as drug delivery, tissue engineering, and medical device fabrication. Solution polymerization enables the incorporation of bioactive molecules or functional groups into polymer chains, enhancing the bioactivity and specificity of the resulting materials. Another widely used polymerization technique is emulsion polymerization, which involves dispersing monomer droplets in an aqueous phase containing surfactants and initiators. Emulsion polymerization enables the synthesis of bioactive polymers in colloidal form, with nanoparticles or microspheres being commonly used in drug delivery systems [6, 7] and biomedical coatings. This technique offers advantages such as high monomer conversion rates, narrow particle size distributions, and ease of scale-up, making it suitable for large-scale production of bioactive polymer materials for clinical applications. Moreover, emulsion polymerization allows for the encapsulation of hydrophobic drugs or bioactive molecules within polymer nanoparticles, providing controlled release and enhanced therapeutic efficacy.

Polymerization techniques such as suspension polymerization and bulk polymerization are utilized in the synthesis of bioactive polymers for specific applications. Suspension polymerization involves dispersing monomer droplets in a non-solvent medium and initiating polymerization to form polymer beads or microspheres, which find applications in drug delivery, tissue engineering, and medical device coatings. Bulk polymerization, on the other hand, involves polymerizing monomers in the absence of a solvent, resulting in the formation of dense polymer networks that are used in applications requiring high mechanical strength and stability, such as orthopedic implants and dental materials [8]. Overall, polymerization techniques play a crucial role in the design and synthesis of bioactive polymers with tailored properties and functionalities, enabling advancements in biomedicine and healthcare.

#### 5.1.2.3 Types of Polymerizations

Types of polymerizations encompass a diverse array of processes used to synthesize polymers with varying structures, properties, and applications. Addition polymerization is a prominent method wherein monomers undergo successive addition reactions to form polymer chains without the generation of any byproducts. This mechanism typically involves the initiation of polymerization by a catalyst or initiator, followed by the propagation of polymer chains through the addition of monomer units. Addition polymerization is widely utilized in the production of polyethylene, polypropylene, and polystyrene, among other polymers, and is favored for its simplicity, high conversion rates, and the ability to produce polymers with controlled molecular weights.

Condensation polymerization, in contrast, involves the sequential reaction of monomers with complementary functional groups, leading to the formation of covalent bonds and the release of small molecules such as water or alcohol. This process requires two or more different types of monomers and proceeds through a step-growth mechanism, resulting in the formation of linear or branched polymer chains. Condensation polymerization is commonly employed in the synthesis of polyesters, polyamides, and polyurethanes, offering versatility in producing polymers with tailored properties and functionalities. Living polymerization techniques represent a specialized class of polymerization methods that enable precise control over polymer chain growth and architecture. Unlike conventional polymerization processes, living polymerization proceeds in a controlled manner, allowing for the addition of monomers in a sequential fashion without termination [6, 7]. Examples of living polymerization techniques include controlled radical polymerization, ring-opening polymerization, and anionic polymerization, each offering unique advantages such as narrow molecular weight distributions and the ability to synthesize polymers with well-defined structures. These techniques find applications in areas such as polymer synthesis, macromolecular engineering, and the development of advanced materials with enhanced properties. Specialized polymerization methods such as emulsion polymerization and suspension polymerization are employed to produce polymer particles with controlled size, morphology, and dispersity. Emulsion polymerization involves the dispersion of monomers in an aqueous phase with the aid of surfactants, leading to the formation of polymer particles suspended in the reaction medium. Suspension polymerization, on the other hand, utilizes inert particles as a support matrix for polymerization, resulting in the formation of polymer beads or microspheres. These techniques are utilized in various industries for the production of latex paints, polymer composites, and microencapsulated materials [8], highlighting the importance of diverse polymerization methods in meeting the evolving needs of modern technology and industry.

#### 5.1.2.4 Polymer Properties

Polymer properties take on added significance in the realm of bioactive polymers, where the interaction between the polymer material and biological systems plays a pivotal role in determining its effectiveness in biomedical applications (Fig. 5.1). Mechanical properties, such as flexibility, elasticity, and strength, are critical considerations in bioactive polymers used for tissue engineering scaffolds or implantable medical devices. For instance, polymers with mechanical properties similar to natural tissues are desirable for scaffolds as they provide mechanical support while promoting cell attachment, proliferation, and tissue regeneration. Similarly, implantable medical devices require polymers with adequate strength and durability to withstand physiological stresses and ensure long-term biocompatibility. Thermal properties also play a crucial role in bioactive polymers, particularly in applications where sterilization or processing at elevated temperatures is necessary [9]. Polymers used in medical implants or drug delivery systems must maintain their structural integrity and bioactivity under sterilization conditions, such as autoclaving or gamma irradiation. Moreover, thermal stability is essential for ensuring the safe and effective performance of bioactive polymers during manufacturing processes and in vivo applications, where exposure to elevated temperatures may occur. Electrical

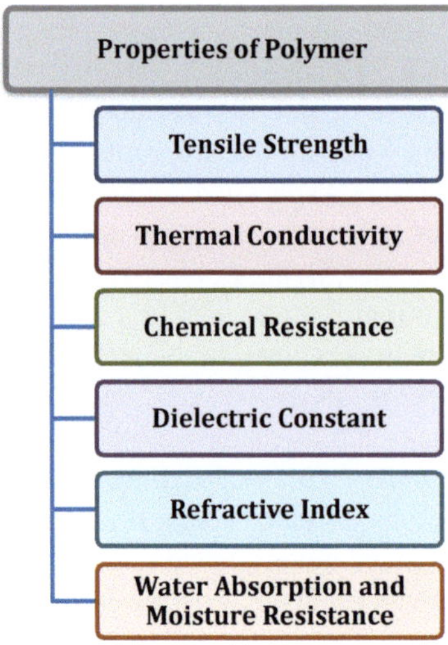


**Fig. 5.1** Various properties of polymers in biomedical applications

properties are relevant in bioactive polymers used in electrical stimulation therapies, biosensors, or implantable electronic devices. Polymers with tailored electrical conductivity or dielectric properties are employed to modulate cellular responses, transmit electrical signals, or interface with electronic components [7–9]. Bioactive polymers with tunable optical properties find applications in biosensing, imaging, and drug delivery, where optical signals are used for diagnostic or therapeutic purposes. By manipulating the optical transparency, refractive index, or fluorescent properties of polymers, researchers can develop novel bioactive materials with enhanced functionality and performance.

Understanding and optimizing polymer properties are essential for the development of bioactive polymers with tailored characteristics for specific biomedical applications. By considering mechanical, thermal, electrical, and optical properties in the design and synthesis of bioactive polymers, researchers can create innovative materials that exhibit optimal performance, biocompatibility, and functionality in physiological environments. This interdisciplinary approach holds great promise for advancing healthcare technologies and addressing complex medical challenges through the use of bioactive polymers.

### *5.1.3 Biological Interactions with Polymers*

Biological interactions with polymers constitute a fascinating area of study at the intersection of polymer science and biology, focusing on how polymers interact with living organisms at the molecular, cellular, and tissue levels. Understanding these interactions is essential for designing bioactive polymers with tailored properties and functionalities for biomedical applications. Biological responses to polymers can vary widely depending on factors such as polymer composition, structure, surface chemistry, and degradation behavior. For instance, polymers used in medical implants or tissue engineering scaffolds must elicit favorable responses from cells and tissues, promoting cell adhesion, proliferation, and tissue regeneration while minimizing adverse reactions such as inflammation or fibrosis [10]. By elucidating the mechanisms underlying biological interactions with polymers, researchers aim to develop innovative materials that enhance medical treatments, diagnostics, and therapies, ultimately improving patient outcomes and quality of life.

#### 5.1.3.1 Biological Response to Polymers

The biological response to polymers is a complex and multifaceted phenomenon that involves interactions between the polymer material and various components of the biological environment, including cells, proteins, and tissues. Upon implantation or contact with biological fluids, polymers can elicit a range of responses from the body, including inflammation, immune activation, and tissue regeneration. The initial response to a polymer material often involves the adsorption of proteins onto its surface, forming a biomolecular layer that mediates subsequent cellular interactions [9, 10]. These proteins can influence cell adhesion, migration, and differentiation, ultimately shaping the tissue response to the polymer implant. The biocompatibility of a polymer material is a critical factor in determining its suitability for biomedical applications, as it dictates the extent to which the material interacts with biological systems without causing adverse effects. Biocompatible polymers are designed to mimic the properties of natural tissues and extracellular matrices, promoting favorable interactions with cells and tissues while minimizing inflammatory and immune responses. Strategies for enhancing the biocompatibility of polymers include surface modification techniques, such as plasma treatment or biomolecule grafting [10, 11], which can alter the surface properties of the material to improve cell adhesion and integration.

In addition to biocompatibility, the degradation behavior of polymers plays a crucial role in determining their long-term performance in vivo. Biodegradable polymers undergo gradual degradation and resorption within the body, allowing for the gradual replacement of the polymer scaffold with newly formed tissue. This process is particularly important in tissue engineering applications [9, 10], where temporary scaffolds are used to support cell growth and tissue regeneration before being gradually replaced by native tissue. By understanding the biological response to polymers and tailoring their properties to match specific biomedical requirements, researchers can develop innovative materials that support tissue repair, regeneration, and therapeutic delivery with minimal adverse effects.

#### 5.1.3.2 Mechanisms of Bioactivity

The mechanisms underlying the bioactivity of polymers are diverse and multifaceted, reflecting the intricate interactions between these materials and biological systems.

One key mechanism involves the surface properties of polymers, including topography, charge, and chemistry, which can influence the adsorption of biomolecules and the subsequent cellular response. For example, polymers with hydrophilic surfaces tend to promote protein adsorption and cell adhesion, while polymers with specific functional groups can facilitate cell signaling or modulate immune responses. Moreover, the nanoscale architecture of polymer surfaces, such as roughness or patterning, can affect cellular behavior by mimicking the extracellular matrix or providing cues for cell alignment and differentiation. Another important mechanism of polymer bioactivity involves the controlled release of bioactive molecules, such as drugs, growth factors, and signaling molecules, from polymer matrices or carriers [12]. By incorporating bioactive agents into polymer scaffolds or nanoparticles, researchers can achieve targeted delivery to specific tissues or cells, thereby enhancing therapeutic efficacy while minimizing systemic side effects. This mechanism is particularly relevant in drug delivery systems and tissue engineering applications, where precise control over the spatiotemporal release of bioactive molecules is essential for achieving desired therapeutic outcomes.

The degradation behavior of polymers can influence their bioactivity by modulating the release of degradation byproducts or eliciting specific cellular responses. Biodegradable polymers, for instance, undergo controlled degradation in response to biological cues, leading to the release of degradation products that can either promote tissue regeneration or modulate immune responses. Conversely, non-degradable polymers may elicit chronic inflammatory responses or foreign body reactions due to their persistent presence in the body. Understanding and engineering the degradation behavior of polymers is therefore crucial for tailoring their bioactivity to specific biomedical applications while ensuring biocompatibility and safety.

Overall, the mechanisms of bioactivity of polymers are intricately linked to their physicochemical properties, degradation behavior, and interactions with biological molecules and cells [11, 12]. By elucidating these mechanisms, researchers can design bioactive polymers with tailored properties and functionalities for a wide range of biomedical applications, including drug delivery, tissue engineering, regenerative medicine, and medical devices. This interdisciplinary approach holds great promise for advancing healthcare technologies and addressing unmet clinical needs through the development of innovative polymer-based solutions.

#### 5.1.3.3 Factors Influencing Biocompatibility

- **Chemical Composition**: The chemical structure and composition of bioactive polymers influence their biocompatibility, with materials containing non-toxic and biocompatible monomers being preferred for biomedical applications.
- **Surface Properties**: Surface characteristics such as roughness, charge, and hydrophilicity/hydrophobicity can impact the interaction of bioactive polymers with biological entities, affecting cell adhesion, protein adsorption, and, ultimately, biocompatibility [12, 13].

- **Degradation Behavior**: The degradation kinetics and products of bioactive polymers play a crucial role in their biocompatibility. Biodegradable polymers that degrade into non-toxic byproducts at a rate compatible with tissue healing are generally preferred.
- **Immunogenicity**: The potential of bioactive polymers to induce an immune response, including inflammation or foreign body reactions, can significantly influence their biocompatibility and long-term performance in vivo.
- **Mechanical Properties**: Mechanical properties such as elasticity, stiffness, and strength are important considerations in biocompatibility, as polymers used in biomedical devices or implants must withstand physiological stresses without causing tissue damage.
- **Sterilization Compatibility**: The ability of bioactive polymers to maintain their properties and functionality after sterilization procedures [11, 13], such as autoclaving or gamma irradiation, is essential for ensuring biocompatibility and safety in clinical applications.
- **Interaction with Biological Molecules**: The ability of bioactive polymers to interact with proteins, cells, and tissues in a controlled manner can influence their biocompatibility, with materials that promote favorable cellular responses being preferred for tissue engineering and regenerative medicine applications.

### *5.1.4 Design and Synthesis of Bioactive Polymers*

The design and synthesis of bioactive polymers represent a dynamic and interdisciplinary field at the intersection of polymer chemistry, materials science, and biomedical engineering. This area of research focuses on developing innovative polymer materials with tailored properties and functionalities to interact with biological systems for various biomedical applications. By strategically designing polymer structures and incorporating specific chemical moieties or functional groups, researchers aim to impart bioactive properties to the materials, enabling them to elicit desired responses from cells, tissues, or biological molecules. Moreover, the synthesis of bioactive polymers involves employing a range of synthetic techniques, including polymerization methods, functionalization strategies [14], and biomolecule conjugation approaches, to precisely control polymer composition, architecture, and bioactivity. Through systematic design and synthesis efforts, bioactive polymers hold immense promise for advancing healthcare technologies, enabling targeted drug delivery, tissue engineering, regenerative medicine, and diagnostics, while addressing complex challenges in disease treatment and patient care.

#### 5.1.4.1 Rational Design Approaches

Rational design approaches are fundamental in the development of bioactive polymers, enabling researchers to engineer materials with specific properties and functionalities tailored to biomedical applications. One key strategy involves a bottom-up approach, where the molecular structure of the polymer is carefully designed based on an understanding of the desired biological interactions and target applications. By selecting appropriate monomers and functional groups, researchers can control the physicochemical properties of the polymer, such as surface chemistry, degradation behavior, and mechanical properties [15], to optimize its bioactivity and biocompatibility. Another rational design approach focuses on mimicking the structure and properties of natural biomolecules, such as proteins or polysaccharides, which play essential roles in biological processes. By incorporating bioinspired motifs or sequences into polymer chains, researchers can create synthetic polymers that interact with biological receptors or enzymes, modulate cell signaling pathways, or mimic the extracellular matrix. This biomimetic approach allows for the development of bioactive polymers that closely resemble the biological environment, promoting specific cellular responses and tissue regeneration.

Rational design approaches leverage computational modeling and simulation techniques to predict and optimize the behavior of bioactive polymers at the molecular level. Molecular dynamics simulations, quantum mechanical calculations, and structure-property relationship analyses enable researchers to understand the molecular interactions between polymers and biological molecules, predict material properties, and guide the design of novel polymer structures with enhanced bioactivity. By combining experimental data with computational insights, researchers can accelerate the discovery and optimization of bioactive polymers with tailored properties for biomedical applications.

Lastly, rational design approaches emphasize a systematic and iterative process of materials characterization, validation, and optimization to ensure the efficacy and safety of bioactive polymers. Through comprehensive physicochemical, biological, and in vivo evaluations, researchers can assess the performance of bioactive polymers in relevant biological contexts, identify structure-property relationships, and refine design parameters to meet specific application requirements [16]. This iterative approach allows for the continuous improvement and refinement of bioactive polymer materials, driving innovation in biomedicine and paving the way for the development of advanced therapies and diagnostics.

#### 5.1.4.2 Synthetic Strategies

Synthetic strategies for bioactive polymers encompass a diverse range of techniques aimed at controlling the chemical composition, structure, and functionality of polymer materials to achieve specific biological interactions. One common strategy involves the use of controlled polymerization techniques, such as living radical polymerization, ring-opening polymerization, and controlled/living polymerization, which allow for precise control over polymer chain length, architecture, and end-group functionality. These methods enable the synthesis of bioactive polymers with well-defined structures and tailored properties, facilitating their use in biomedical applications such as drug delivery, tissue engineering, and diagnostics. In addition to controlled polymerization techniques, modular or stepwise synthesis approaches are often employed to incorporate bioactive motifs or functional groups into polymer chains. This involves the sequential attachment of monomer units or pre-functionalized polymer segments through iterative chemical reactions, such as click chemistry, thiol-ene coupling [13, 15, 16], and peptide coupling reactions. By judiciously selecting and combining building blocks with specific biological functionalities, researchers can design bioactive polymers with precise control over the presentation and density of bioactive moieties, optimizing their interactions with biological systems.

Post-polymerization modification strategies offer flexibility and versatility in functionalizing polymer materials with bioactive groups or ligands. After polymerization, polymers can be chemically modified using a variety of reactions, such as esterification, amidation, and click chemistry, to introduce bioactive functionalities onto polymer backbones or side chains. This approach allows for the incorporation of bioactive molecules, such as peptides, proteins, and carbohydrates, onto polymer surfaces or within polymer matrices, enhancing their biological activity and specificity for targeted applications in drug delivery, tissue engineering, and biosensing. Lastly, supramolecular assembly techniques leverage non-covalent interactions, such as hydrogen bonding, $\pi$–$\pi$ stacking, and host-guest interactions, to construct bioactive polymer structures with hierarchical organization and responsiveness. By designing polymers with complementary self-assembly motifs or incorporating bioactive molecules into supramolecular aggregates [17], researchers can create materials with dynamic properties and stimuli-responsive behavior, enabling controlled release of bioactive agents or modulation of biological interactions in response to external stimuli. These synthetic strategies offer versatile and innovative approaches to designing bioactive polymers with tailored properties and functionalities for a wide range of biomedical applications.

#### 5.1.4.3 Functionalization Techniques

Functionalization techniques play a crucial role in tailoring the properties and functionalities of bioactive polymers for specific biomedical applications. One commonly employed strategy is surface functionalization, which involves modifying the surface of polymer materials with bioactive molecules or ligands to promote specific biological interactions. This can be achieved through techniques such as chemical grafting, physical adsorption, and layer-by-layer assembly, allowing for precise control over the presentation and density of bioactive moieties on the polymer surface. Surface functionalization enables the creation of biomimetic interfaces that mimic the extracellular matrix or cellular microenvironment [18], facilitating cell adhesion, proliferation, and tissue integration in applications such as tissue engineering scaffolds or implantable medical devices. Another important functionalization technique is the conjugation of bioactive molecules onto polymer backbones or side chains through chemical reactions, such as amidation, esterification, and click chemistry. This approach allows for the incorporation of specific biological functionalities, such as cell-targeting ligands, growth factors, and drug molecules, directly into the polymer structure. By covalently linking bioactive molecules to polymer chains, researchers can enhance the bioactivity and specificity of polymers for targeted applications in drug delivery, regenerative medicine, and biosensing.

Stimuli-responsive functionalization strategies enable the design of bioactive polymers that can dynamically respond to changes in their environment, such as pH, temperature, and enzyme activity. By incorporating stimuli-responsive moieties, such as pH-responsive polymers or thermo-responsive polymers, into polymer structures [17, 18], researchers can modulate the release of bioactive molecules or trigger specific biological responses in a controlled manner. These functionalization techniques offer opportunities for developing smart polymer materials with tunable properties and functionalities for applications in controlled drug delivery, tissue engineering, and diagnostics. Lastly, bioconjugation techniques enable the attachment of biomolecules, such as peptides, proteins, and nucleic acids, onto polymer surfaces or within polymer matrices through specific binding interactions, such as avidin–biotin, streptavidin-biotin, and antibody-antigen interactions. This approach allows for the functionalization of polymers with complex biological functionalities, such as cell-recognition motifs or enzymatic activity, enabling precise control over the interactions between polymers and biological systems. Bioconjugation techniques are particularly useful in applications such as biosensing, bioimaging, and targeted drug delivery [19], where the incorporation of biomolecules into polymer materials enhances their specificity, sensitivity, and biocompatibility.

**Exercise 1.**

**Fill Ups**

**(Topography, Polymerization, Biocompatibility, Polymerize, Interactions)**

**1.** In polymer chemistry, a monomer is a molecule that can ...........to form a larger polymer chain through chemical reactions such as polymerization.

**2.** The process by which monomers combine to form a polymer is called .................., which can occur through various mechanisms such as addition polymerization or condensation polymerization.

**3.** Biological interactions with polymers play a crucial role in determining their .................and compatibility when used in biomedical applications such as tissue engineering and drug delivery.

**4.** The surface properties of polymers, such as their ..............., charge, and functional groups, influence their interaction with biological molecules and cells.

**5.** The design and synthesis of bioactive polymers involve tailoring their chemical structure and properties to promote specific ................with biological systems, such as cell adhesion, proliferation, and tissue regeneration.

### *5.1.5 Bioactive Polymer Materials*

Bioactive polymer materials encompass a diverse array of substances that interact with biological systems, offering tailored solutions to various biomedical challenges. Derived from natural or synthetic sources, these materials exhibit unique properties such as biocompatibility, bioactivity, and stimuli responsiveness, making them versatile platforms for applications in tissue engineering, drug delivery, medical devices, and diagnostics. Examples of bioactive polymer materials include natural polymers like collagen and chitosan, which mimic the extracellular matrix and promote tissue regeneration, as well as synthetic polymers such as polyethylene glycol (PEG) and polylactic acid (PLA) [20], which offer tunable properties and controlled degradation kinetics. Stimuli-responsive polymers and composite materials combine the advantages of different polymer types, enabling targeted drug delivery and precise modulation of biological responses. Overall, bioactive polymer materials hold immense promise for advancing biomedicine and improving patient care through innovative materials and technologies.

#### 5.1.5.1 Natural Versus Synthetic Polymers

Natural and synthetic bioactive polymers represent two distinct classes of materials with unique properties and applications in biomedicine (Table 5.1). Natural bioactive polymers, derived from biological sources such as proteins, polysaccharides, and nucleic acids, offer inherent biocompatibility and bioactivity due to their resemblance to native extracellular matrix components. Examples include collagen, hyaluronic acid, and chitosan, which have been extensively utilized in tissue engineering, wound healing, and drug delivery applications. Natural polymers often exhibit superior biodegradability and cell affinity compared to synthetic counterparts, making them ideal for promoting tissue regeneration and minimizing adverse immune responses in vivo. In contrast, synthetic bioactive polymers are chemically engineered materials designed to mimic or enhance the properties of natural polymers while offering additional functionalities and tunability. Synthetic polymers offer advantages such as precise control over molecular structure, mechanical properties, and degradation kinetics, enabling customization for specific biomedical applications. Examples include polyethylene glycol (PEG), polylactic acid (PLA), and poly(caprolactone) (PCL), which are commonly used in drug delivery systems, scaffolds for tissue engineering, and medical implants [21]. Synthetic polymers can be tailored to exhibit controlled degradation, surface modification, or stimuli-responsive behavior, allowing for fine-tuning of bioactivity and therapeutic efficacy in biomedical applications.

While natural bioactive polymers offer inherent biocompatibility and bioactivity, they may also present limitations such as batch-to-batch variability, immunogenicity,

**Table 5.1** Major differences between natural and synthetic bioactive polymers

| Aspect | Natural bioactive polymers | Synthetic bioactive polymers |
|---|---|---|
| Source | Derived from biological sources such as proteins, polysaccharides, and nucleic acids | Chemically engineered materials designed to mimic or enhance properties of natural polymers |
| Biocompatibility | Generally exhibit inherent biocompatibility due to resemblance to native extracellular matrix components | Biocompatibility can be tailored through precise control over molecular structure |
| Bioactivity | Possess inherent bioactivity, often promoting cell adhesion, proliferation, and tissue regeneration | Bioactivity can be enhanced through incorporation of specific functional groups or bioactive molecules |
| Mechanical properties | May have variable mechanical properties depending on source and processing methods | Mechanical properties can be precisely controlled and tailored to match specific application requirements |
| Degradation behavior | Typically biodegradable, and degradation products often exhibiting biological activity | Degradation kinetics can be engineered to match tissue healing processes, offering controlled release of bioactive agents |

and limited mechanical strength. Synthetic bioactive polymers, on the other hand, provide greater control over material properties and functionality but may lack the complexity and biological cues present in natural materials [20, 21]. Thus, the choice between natural and synthetic bioactive polymers depends on factors such as the desired application, specific biological requirements, and regulatory considerations. In many cases, hybrid approaches that combine natural and synthetic components offer synergistic benefits, leveraging the strengths of both materials to achieve optimal biocompatibility, bioactivity, and performance in biomedical applications.

#### 5.1.5.2 Biodegradable Polymers

Biodegradable bioactive polymer materials represent a significant advancement in the field of biomedical engineering, offering tailored solutions for various therapeutic and regenerative applications (Fig. 5.2). These materials undergo controlled degradation in physiological environments, breaking down into non-toxic byproducts that are naturally metabolized or excreted by the body. Biodegradable bioactive polymers have garnered considerable attention for their potential to address challenges such as tissue regeneration, controlled drug delivery, and medical device design while minimizing long-term foreign body reactions and toxicity concerns associated with non-degradable materials [22]. One key advantage of biodegradable bioactive polymers is their ability to serve as temporary scaffolds or carriers for delivering therapeutic agents such as drugs, growth factors, and cells to target tissues or organs. By incorporating bioactive molecules into polymer matrices or coatings, researchers can achieve controlled release kinetics, ensuring sustained therapeutic efficacy while minimizing systemic side effects. This capability is particularly valuable in applications such as tissue engineering, where biodegradable polymer scaffolds provide temporary support for cell growth and tissue regeneration before being gradually replaced by newly formed tissue. The degradation behavior of biodegradable bioactive polymers can be tailored to match tissue healing processes, offering precise control over the release of bioactive agents and modulation of cellular responses. Degradation kinetics can be adjusted by selecting appropriate polymer chemistries, molecular weights, and structural architectures, allowing for customized degradation rates ranging from days to months. This tunability enables the design of polymer materials that degrade synchronously with tissue regeneration, promoting optimal integration and functionality of engineered tissues or implants.

Biodegradable bioactive polymers hold promise for applications requiring transient therapeutic interventions or temporary medical devices, such as surgical implants, wound dressings, and drug-eluting stents. These materials provide transient support or functionality while minimizing the need for additional surgical procedures for device removal. The use of biodegradable materials reduces the risk of long-term complications such as infection, inflammation, and tissue encapsulation commonly associated with non-degradable implants [22]. Overall, biodegradable bioactive polymer materials offer versatile and innovative solutions for addressing complex biomedical challenges and advancing the field of regenerative medicine.

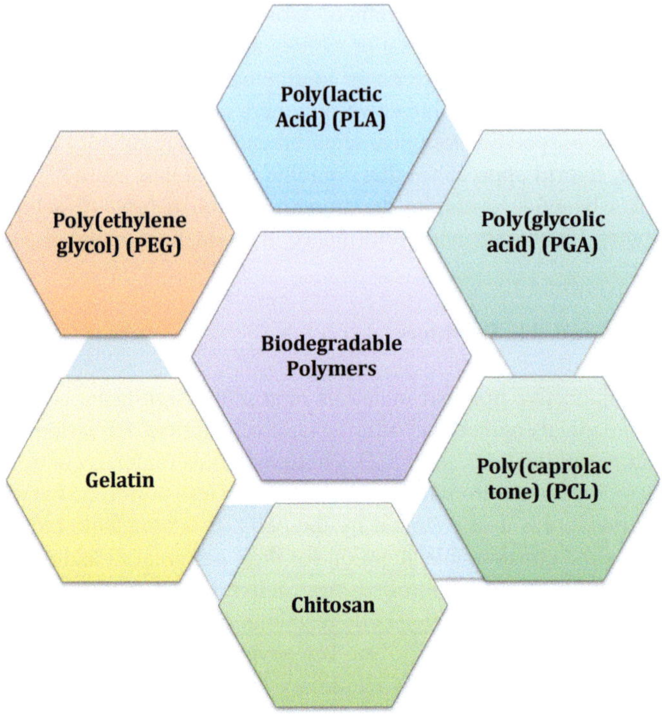


**Fig. 5.2** Examples of biodegradable polymers

#### 5.1.5.3 Smart Polymers

Smart polymers, also known as stimuli-responsive polymers, are a class of materials that exhibit reversible changes in their properties in response to external stimuli such as temperature, pH, light, and magnetic fields. These materials have garnered significant interest in biomedical research due to their ability to respond dynamically to changes in their environment, making them versatile platforms for various biomedical applications. Smart polymers can undergo conformational changes, solubility alterations, or structural transitions in response to specific stimuli, enabling precise control over drug release, cell adhesion, or tissue engineering processes. One of the key advantages of smart polymers in biomedical applications is their ability to enable controlled drug delivery [21, 22]. By incorporating drug molecules into polymer matrices or coatings, researchers can design smart polymer-based drug delivery systems that release therapeutics in response to specific physiological cues, such as changes in pH or enzyme activity at diseased tissue sites. This targeted drug release minimizes systemic side effects and improves therapeutic efficacy, making smart polymers promising candidates for precision medicine and personalized drug delivery strategies. Smart polymers offer opportunities for designing bioactive materials with tunable properties for tissue engineering and regenerative medicine applications. These materials can respond to cues from the surrounding biological environment, such as changes in temperature or mechanical stress, to promote cell adhesion, proliferation, and differentiation. For example, thermoresponsive polymers can undergo reversible phase transitions in response to changes in temperature, allowing for the fabrication of cell culture substrates or scaffolds with switchable adhesive properties that mimic the dynamic nature of native tissues. Smart polymers hold promise for the development of diagnostic and sensing platforms in healthcare [23]. By incorporating sensing elements or reporter molecules into polymer matrices, smart polymer-based sensors can detect specific biomarkers or analytes in biological samples, providing rapid and sensitive detection of disease biomarkers or environmental pollutants. Stimuli-responsive polymers can be integrated into wearable or implantable devices for real-time monitoring of physiological parameters, offering insights into disease progression or treatment efficacy. Overall, smart polymers represent a versatile and innovative class of bioactive materials with immense potential to revolutionize various aspects of biomedicine, from drug delivery and tissue engineering to diagnostics and personalized healthcare.

#### 5.1.5.4 Composite Materials

Composite polymer bioactive materials are innovative materials that combine the advantages of different types of polymers to achieve enhanced properties and functionalities for biomedical applications. These materials typically consist of a polymer matrix reinforced with bioactive components such as nanoparticles, fibers, and hydrogels, offering synergistic benefits that are not attainable with individual components alone. Composite polymer bioactive materials are designed to address specific biomedical challenges, such as promoting tissue regeneration, enhancing drug delivery, and improving medical device performance, by leveraging the unique properties of each constituent material. The advantage of composite polymer bioactive materials is their ability to tailor material properties to match the requirements of specific biomedical applications. By carefully selecting the type, composition, and morphology of the bioactive components [15, 16, 20], researchers can engineer composite materials with desired mechanical, biological, and degradation properties. For example, incorporating bioactive nanoparticles such as hydroxyapatite or calcium phosphate into polymer matrices can enhance the mechanical strength and bioactivity of scaffolds for bone tissue engineering, promoting osteogenesis and bone regeneration. Composite polymer bioactive materials offer opportunities for controlled drug delivery by incorporating drug-loaded nanoparticles or microparticles into polymer matrices. This enables sustained release of therapeutic agents over time, improving drug bioavailability, reducing dosing frequency, and minimizing systemic side effects. Composite drug delivery systems can be tailored to respond to specific physiological cues, such as changes in pH, temperature, and enzyme activity, allowing for targeted delivery to diseased tissue sites while sparing healthy tissues.

Composite polymer bioactive materials can be engineered to exhibit multifunctionality by incorporating different types of bioactive components with complementary properties. For example, combining antimicrobial agents with growth factors or cell-adhesive peptides in composite materials can promote tissue regeneration while preventing infection at implantation sites. Composite materials can be designed to incorporate imaging agents or contrast agents for real-time monitoring of tissue regeneration or drug delivery processes, offering insights into treatment efficacy and patient outcomes [24]. Overall, composite polymer bioactive materials represent a promising platform for developing advanced biomaterials with tailored properties and functionalities for a wide range of biomedical applications, from regenerative medicine and drug delivery to medical device design and diagnostics.

## 5.1.6 Applications of Bioactive Polymers

The applications of bioactive polymers span a wide spectrum of biomedical fields, offering innovative solutions to various healthcare challenges (Fig. 5.3). These versatile materials find extensive use in tissue engineering, where they serve as scaffolds to support cell growth and tissue regeneration. Bioactive polymers promote cell adhesion, proliferation, and differentiation, mimicking the natural extracellular matrix to facilitate the formation of functional tissues. Moreover, bioactive polymers are employed in drug delivery systems, enabling targeted and controlled release of therapeutic agents to specific tissue sites while minimizing systemic side effects. By encapsulating drugs within polymer matrices or conjugating them to polymer backbones, researchers can achieve sustained release kinetics, enhancing therapeutic efficacy and patient compliance [25]. In addition to tissue engineering and drug delivery, bioactive polymers play a crucial role in medical device design and fabrication. These materials are used to coat implantable devices such as stents, catheters, and prosthetics, promoting biocompatibility and reducing the risk of adverse reactions or complications. Surface modification techniques enable the incorporation of bioactive molecules or functional groups onto device surfaces, enhancing cell adhesion, tissue integration, and device performance in vivo. Bioactive polymers are employed in biosensing and diagnostic applications, where they serve as substrates for immobilizing biomolecules, detecting specific analytes, or imaging biological structures. By leveraging the unique properties of bioactive polymers, researchers continue to develop innovative materials and technologies to advance healthcare and improve patient outcomes.

### 5.1.6.1 Drug Delivery Systems

Bioactive polymers play a pivotal role in revolutionizing drug delivery systems, offering targeted and controlled release of therapeutic agents to specific sites within the body while minimizing systemic side effects. These polymers serve as versatile

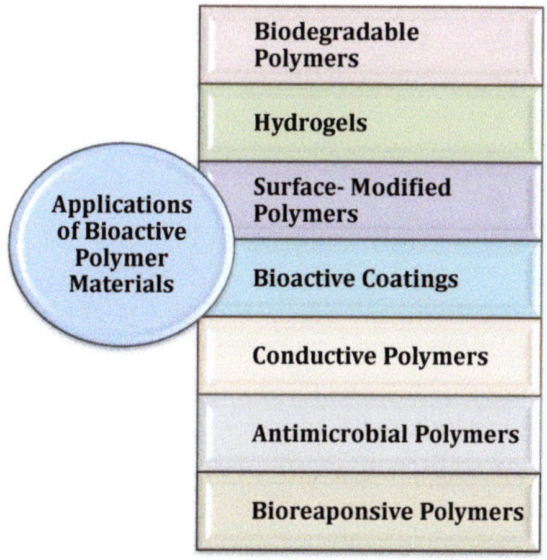


**Fig. 5.3** Various applications of bioactive polymer materials

carriers for drugs, allowing for the encapsulation, protection, and precise delivery of pharmaceutical compounds to target tissues or cells. One of the primary advantages of bioactive polymer-based drug delivery systems is their ability to tailor drug release kinetics to match therapeutic needs [23, 25], ranging from sustained release over extended periods to triggered release in response to specific physiological cues.

Bioactive polymers enable the design of drug delivery systems with enhanced stability, solubility, and bioavailability of encapsulated drugs, leading to improved therapeutic efficacy and patient compliance. By encapsulating drugs within polymer matrices or conjugating them to polymer backbones, researchers can protect drugs from enzymatic degradation or premature clearance in vivo, allowing for extended circulation times and enhanced drug delivery to target tissues or cells. Bioactive polymers can be functionalized with targeting ligands or cell-penetrating peptides to achieve site-specific drug delivery, enabling selective accumulation of drugs at diseased tissue sites while sparing healthy tissues from exposure to high drug concentrations.

Bioactive polymer-based drug delivery systems offer opportunities for multi-functionality and stimuli responsiveness, enabling precise control over drug release in response to changes in physiological conditions or external stimuli. Stimuli-responsive polymers, such as pH-sensitive or temperature-responsive polymers, can be designed to undergo conformational changes or structural transitions in response to changes in the local environment, triggering drug release at the desired site of action [26]. These smart drug delivery systems enable the on-demand release of therapeutic agents, offering precise control over drug dosage and timing to optimize therapeutic outcomes while minimizing side effects. Bioactive polymers have significantly advanced drug delivery technology, paving the way for personalized medicine and targeted therapies in various biomedical applications.

### 5.1.6.2 Tissue Engineering and Regenerative Medicine

Bioactive polymers have emerged as crucial components in tissue engineering, facilitating the development of engineered tissues and organs for regenerative medicine applications. These polymers serve as scaffolds that mimic the extracellular matrix (ECM) of native tissues, providing structural support and biochemical cues to promote cell adhesion, proliferation, and differentiation. By closely resembling the natural microenvironment of cells, bioactive polymer scaffolds create favorable conditions for tissue regeneration and repair. Bioactive polymers can be tailored to exhibit specific mechanical properties [2], degradation kinetics, and surface topographies to match the requirements of different tissues and organs. One of the key advantages of bioactive polymers in tissue engineering is their ability to promote vascularization and innervation within engineered tissues. Bioactive polymer scaffolds can be functionalized with angiogenic factors or cell-recognition motifs to stimulate the formation of blood vessels and nerve networks, facilitating nutrient and oxygen transport as well as integration with host tissues. Bioactive polymers enable the incorporation of growth factors, cytokines, and other bioactive molecules into scaffold matrices, providing spatiotemporal control over cellular responses and tissue development. This precise control over the biochemical microenvironment of cells enhances the functionality and maturity of engineered tissues, making them more closely resemble native tissues in terms of structure and function. Bioactive polymers offer opportunities for the development of composite scaffolds that combine the advantages of different polymer types or incorporate bioactive components such as nanoparticles, fibers, and hydrogels. Composite scaffolds can be engineered to exhibit tailored properties such as mechanical strength, porosity, and degradation behavior [16, 17, 22], enabling optimized tissue regeneration and integration. Bioactive polymers can be processed into three-dimensional (3D) scaffolds using advanced fabrication techniques such as electrospinning, 3D printing, and microfluidics, allowing for precise control over scaffold architecture and geometry. These advanced scaffolds offer opportunities for engineering complex tissues and organs with hierarchical structures and functional gradients, advancing the field of regenerative medicine toward the development of functional tissue substitutes for clinical applications.

Bioactive polymers play a crucial role in the development of bioinks for bioprinting applications, enabling the fabrication of complex tissue constructs with spatially defined cell distribution and organization. Bioactive polymer-based bioinks provide structural support and biochemical cues for encapsulated cells, promoting cell viability, proliferation, and differentiation within printed constructs. Bioinks can be functionalized with bioactive molecules or growth factors to induce specific cellular responses and tissue morphogenesis. By combining bioprinting technology with bioactive polymers, researchers can engineer tissues with controlled architectures and functionalities, offering new opportunities for personalized medicine, drug screening, and disease modeling. Bioactive polymers play a crucial role in the development of injectable hydrogels and cell delivery systems for minimally invasive therapies [20, 22, 23]. These materials offer opportunities for localized delivery of therapeutic agents, cells, or biomolecules to target tissues or organs, bypassing the need for traditional surgical procedures. Injectable hydrogels based on bioactive polymers provide temporary structural support and create favorable microenvironments for cell survival, proliferation, and differentiation. Bioactive polymers enable the encapsulation of cells within hydrogel matrices, protecting them from immune rejection and promoting their integration with host tissues, thereby enhancing the efficacy of cell-based therapies for tissue regeneration and repair.

Bioactive polymers are utilized in the development of bioresorbable implants and medical devices for regenerative medicine applications. These implants are designed to provide temporary support or functionality while facilitating tissue regeneration and remodeling. Bioactive polymers with controlled degradation kinetics are used to fabricate implants that gradually degrade over time, allowing for the gradual replacement of the implant with regenerated tissue [15, 16]. Bioactive polymers can be functionalized with bioactive molecules or growth factors to enhance tissue integration and vascularization, promoting the success of implantation procedures.

#### 5.1.6.3 Biomedical Devices

Bioactive polymers have become integral components in the design and development of various biomedical devices, offering unique properties and functionalities that enhance their performance and biocompatibility. These polymers are employed in a wide range of medical devices, including implants, prosthetics, sensors, and diagnostic tools, where they contribute to improved patient outcomes and healthcare delivery. One of the primary advantages of bioactive polymers in biomedical devices is their ability to promote biocompatibility and reduce the risk of adverse reactions or complications when in contact with biological tissues or fluids. Bioactive polymers play a crucial role in the fabrication of implantable medical devices such as stents [27], sutures, and orthopedic implants. These devices are designed to provide structural support or functionality while facilitating tissue regeneration and integration with host tissues. Bioactive polymers with controlled degradation kinetics are used to manufacture implants that gradually degrade over time, allowing for the replacement of the implant with regenerated tissue. Additionally, bioactive polymers can be surface-modified or functionalized with bioactive molecules to enhance tissue integration, reduce inflammation, and prevent infection at implantation sites, thereby improving the long-term performance and biocompatibility of the devices.

Bioactive polymers are utilized in the development of drug-eluting devices and controlled-release systems for targeted delivery of therapeutic agents. These devices, such as drug-eluting stents and patches, enable localized and sustained release of drugs to specific tissue sites, minimizing systemic side effects and improving therapeutic efficacy. Bioactive polymers serve as matrices for drug encapsulation or conjugation, allowing for precise control over drug release kinetics and dosage [2, 25, 26]. Bioactive polymers can be functionalized by targeting ligands or stimuli-responsive moieties to achieve site-specific drug delivery, offering opportunities for personalized medicine and precision therapies.

Furthermore, bioactive polymers are employed in the fabrication of biomedical sensors and diagnostic devices for monitoring physiological parameters and detecting biomarkers of disease. These sensors, which may be wearable, implantable, or handheld, utilize bioactive polymers as sensing elements, substrates, or coatings to enhance sensitivity, selectivity, and stability. Bioactive polymers enable the immobilization of biomolecules or receptors on sensor surfaces, allowing for specific recognition and detection of target analytes in biological samples. Bioactive polymer-based sensors can be integrated into diagnostic platforms for rapid and sensitive detection of disease biomarkers, offering opportunities for early diagnosis, disease monitoring, and personalized healthcare.

**Exercise 2.**

**Fill Ups**

**(Efficacy, Adhesion, Regenerative medicine, Functionality, Biomaterials)**

**1.** Bioactive polymer materials are designed to mimic the extracellular matrix, providing cues for cell ..................and tissue regeneration in biomedical applications.

**2.** Incorporating bioactive molecules such as peptides, growth factors, or drugs into polymer matrices enhances their ................., enabling targeted therapeutic delivery and tissue engineering applications.

**3.** Bioactive polymers find widespread application in ..........................., where they provide scaffolds for cell growth and tissue regeneration, aiding in wound healing and tissue engineering.

**4.** In drug delivery systems, bioactive polymers are utilized to encapsulate and release therapeutic agents at controlled rates, improving .....................and reducing side effects.

**5.** Bioactive polymers are also employed in the development of .................., where they can mimic the natural extracellular matrix to promote cell adhesion and proliferation for tissue repair and regeneration.

### *5.1.7 Characterization Techniques for Bioactive Polymers*

Characterization techniques for bioactive polymers are essential tools used to analyze and evaluate the properties and functionalities of these materials in biomedical applications (Fig. 5.4). These techniques encompass a diverse range of methods that provide insights into the chemical, physical, mechanical, and biological characteristics of bioactive polymers, enabling researchers to understand their structure-property relationships and optimize their performance for specific biomedical tasks. Common characterization techniques include spectroscopic methods such as Fourier-transform

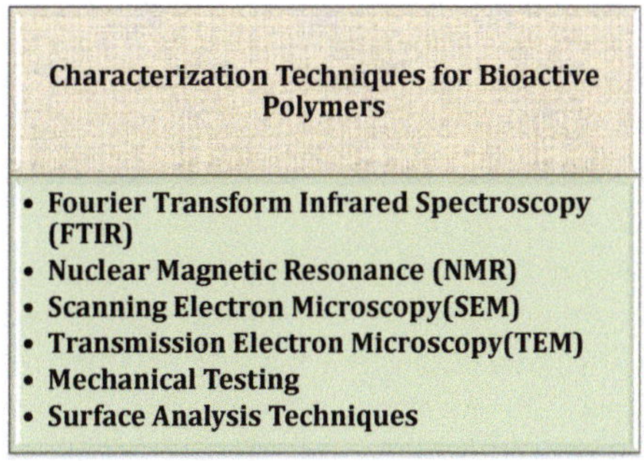


**Fig. 5.4** Different characterization techniques for bioactive polymers

infrared spectroscopy (FTIR) and nuclear magnetic resonance (NMR) spectroscopy, which are used to analyze the chemical composition, functional groups, and molecular structure of bioactive polymers [2]. Additionally, microscopy techniques such as scanning electron microscopy (SEM) and transmission electron microscopy (TEM) allow for the visualization of polymer morphology, surface topography, and interactions with cells or tissues at the micro- and nanoscale levels. Mechanical testing methods such as tensile testing and compression testing provide information on the mechanical properties, such as strength, elasticity, and stiffness, of bioactive polymers, which are crucial for designing implants, scaffolds, and medical devices. Moreover, biological characterization techniques such as cell culture assays, biocompatibility testing, and in vitro and in vivo studies are used to evaluate the biological responses, interactions, and biodegradation [25, 26] behavior of bioactive polymers in physiological environments, providing valuable insights into their suitability for biomedical applications. Overall, characterization techniques for bioactive polymers play a vital role in advancing the understanding and development of innovative materials and technologies in biomedicine, paving the way for improved healthcare outcomes and patient well-being.

#### 5.1.7.1 Morphological Analysis

Morphological analysis of bioactive polymers is a crucial aspect of their characterization, providing valuable insights into their structure, surface morphology, and internal architecture. This analysis involves various microscopy techniques that allow researchers to visualize and examine the physical characteristics of bioactive polymers at the micro- and nanoscale levels. Scanning electron microscopy (SEM) is one of the most commonly used techniques for morphological analysis, offering high-resolution imaging of polymer surfaces and providing information on surface roughness, porosity, and topographical features [2, 27]. Additionally, transmission electron microscopy (TEM) enables researchers to investigate the internal structure and nanoscale morphology of bioactive polymers, including the distribution and arrangement of polymer chains, nanoparticles, or other components within the material.

Atomic force microscopy (AFM) provides valuable information on the surface topography, mechanical properties, and interactions of bioactive polymers at the nanoscale level. AFM imaging allows for the visualization of surface features such as roughness, height variations, and molecular interactions, offering insights into the structure-property relationships of bioactive polymers. AFM-based techniques such as force spectroscopy enable the measurement of mechanical properties such as elasticity, adhesion, and stiffness, providing valuable information for the design and optimization of bioactive polymer materials for specific biomedical applications [28]. Morphological analysis of bioactive polymers plays a critical role in understanding their interactions with biological systems, including cells, tissues, and proteins. Techniques such as confocal microscopy and fluorescence microscopy allow researchers to visualize and study the interactions between bioactive polymers and cells or tissues in real time, providing insights into cell adhesion, proliferation, migration, and differentiation on polymer substrates. Techniques such as scanning probe microscopy (SPM) and electron microscopy combined with immunolabeling enable the visualization of protein adsorption, cell signaling, and other biological processes occurring on the surface of bioactive polymers, further elucidating their bioactivity and biocompatibility [2, 26, 27]. Overall, morphological analysis techniques are essential tools for characterizing bioactive polymers and understanding their structure-function relationships, guiding the development of advanced materials for biomedical applications such as tissue engineering, drug delivery, and medical devices.

#### 5.1.7.2 Mechanical Properties

Characterizing the mechanical properties of bioactive polymers is essential for understanding their suitability for various biomedical applications, including tissue engineering, drug delivery, and medical device design. Mechanical properties such as elasticity, strength, stiffness, and toughness play critical roles in determining the performance and functionality of bioactive polymers in physiological environments. One commonly used technique for characterizing mechanical properties is tensile testing, which involves applying a controlled force to a polymer sample and measuring its response in terms of stress and strain. Tensile testing provides valuable information on parameters such as Young's modulus, ultimate tensile strength, and elongation at break, allowing researchers to assess the material's mechanical behavior under tension. Compression testing is used to evaluate the compressive strength and stiffness of bioactive polymers, particularly those intended for load-bearing applications such as orthopedic implants and scaffolds [10, 12, 13]. This technique involves applying a compressive force to a polymer sample and measuring its response in terms of stress and strain. Compression testing provides insights into the material's ability to withstand compressive forces and maintain structural integrity under mechanical loading, which is crucial for ensuring the long-term performance and stability of biomedical devices.

Moreover, indentation testing techniques such as nanoindentation and microindentation are used to assess the hardness and modulus of bioactive polymers at small length scales. These techniques involve applying a controlled load to a polymer surface using a sharp indentation and measuring the resulting indentation depth or penetration force. Indentation testing provides information on the material's resistance to localized deformation and its ability to withstand mechanical wear and abrasion, which are important considerations for materials used in contact with biological tissues or fluids. Rheological analysis is employed to evaluate the viscoelastic properties of bioactive polymers, including their storage modulus, loss modulus, and complex viscosity. Rheological measurements involve subjecting the polymer to oscillatory or shear forces and analyzing its response in terms of deformation and stress relaxation [21–23]. Rheological analysis provides insights into the material's ability to flow, deform, and recover under applied stresses, which is crucial for applications such as injectable hydrogels, drug delivery systems, and tissue scaffolds.

#### 5.1.7.3 Chemical Analysis

Chemical analysis of bioactive polymers is essential for understanding their composition, structure, and functionality, providing valuable insights into their suitability for various biomedical applications. One commonly used technique for chemical analysis is Fourier-transform infrared spectroscopy (FTIR), which allows researchers to identify functional groups and chemical bonds present in bioactive polymers. FTIR spectra provide characteristic peaks corresponding to specific chemical bonds, enabling qualitative and quantitative analysis of polymer composition and molecular structure. This technique is particularly useful for confirming the presence of bioactive molecules [28], such as peptides, proteins, and drugs, within polymer matrices and evaluating their interactions with the polymer backbone. Nuclear magnetic resonance (NMR) spectroscopy is employed to analyze the chemical structure and molecular dynamics of bioactive polymers in solution or solid-state environments. NMR spectra provide information on the connectivity, stereochemistry, and conformational flexibility of polymer chains, allowing researchers to elucidate their macromolecular architecture and polymerization mechanisms. NMR spectroscopy can be used to monitor the degradation kinetics of bioactive polymers in physiological environments, providing insights into their stability and biodegradability over time [29]. This technique is invaluable for assessing the long-term performance and safety of bioactive polymer materials in biomedical applications.

Chromatographic techniques such as high-performance liquid chromatography (HPLC) and gas chromatography (GC) are utilized for the separation, identification, and quantification of individual components within bioactive polymer formulations. These techniques enable researchers to analyze the purity, molecular weight distribution, and release kinetics of bioactive molecules encapsulated within polymer matrices, providing critical information for drug delivery systems and pharmaceutical formulations [30]. The chromatographic analysis allows for the detection of degradation products and impurities in bioactive polymer materials, ensuring compliance with regulatory standards and quality control requirements for biomedical applications.

Furthermore, mass spectrometry (MS) is employed for the identification and structural characterization of bioactive polymers and their degradation products. MS spectra provide information on the molecular weight, fragmentation patterns, and chemical composition of polymer chains, allowing researchers to elucidate their primary structure and sequence. MS-based techniques such as matrix-assisted laser desorption/ionization (MALDI) and electrospray ionization (ESI) enable the analysis of biomolecules, such as peptides, proteins, and nucleic acids, immobilized on polymer surfaces or incorporated into polymer matrices. This information is critical for understanding the bioactivity, biocompatibility, and functionality of bioactive polymers in biomedical applications, guiding the development of innovative materials for tissue engineering, drug delivery, and medical devices.

#### 5.1.7.4 Biological Evaluation

Biological evaluation characterization of bioactive polymers is crucial for assessing their interactions with living organisms and biological systems, as well as their suitability for biomedical applications. These evaluations provide valuable insights into the biocompatibility, bioactivity, and safety of bioactive polymers in vivo and in vitro. One commonly used method for biological evaluation is cytotoxicity testing, which involves exposing cells to bioactive polymer extracts or surfaces and assessing their viability, proliferation, and morphology. Cytotoxicity assays such as MTT, AlamarBlue, and LDH release assays provide quantitative measures of cell response to polymer materials, helping to determine their biocompatibility and potential cytotoxic effects. Biological evaluation characterization includes studies on cell adhesion, spreading, and proliferation on bioactive polymer surfaces, which are essential for assessing their ability to support cell growth and tissue regeneration [25, 26, 31]. Techniques such as cell culture assays, immunofluorescence staining, and scanning electron microscopy (SEM) are employed to visualize and analyze cell behavior on polymer substrates, providing insights into cell-material interactions, cytoskeletal organization, and cell morphology. These studies help researchers evaluate the bioactivity and biocompatibility of bioactive polymers and optimize their surface properties to enhance cell adhesion and tissue integration.

Biological evaluation characterization encompasses studies on the immunological response to bioactive polymers, including inflammation, immune cell activation, and cytokine production. In vitro and in vivo experiments are conducted to assess the inflammatory potential of polymer materials and their ability to elicit immune responses in different biological systems. Techniques such as enzyme-linked immunosorbent assays (ELISA), flow cytometry, and histological analysis are used to quantify inflammatory markers, immune cell infiltration, and tissue reactions to polymer implants or scaffolds, providing valuable information for assessing their biocompatibility and immunogenicity [31, 32]. Biological evaluation characterization includes studies on the biodegradation behavior of bioactive polymers in physiological environments, such as enzymatic degradation, hydrolysis, and cellular metabolism. These studies involve monitoring changes in polymer mass, structure, and mechanical properties over time, as well as assessing the release of degradation byproducts and their effects on surrounding tissues. Techniques such as mass spectrometry, chromatography, and spectroscopic methods are employed to analyze degradation products and elucidate degradation mechanisms, providing insights into the long-term fate and safety of bioactive polymers in biological systems.

### *5.1.8 Case Studies*

The following are case studies showcasing successful applications of bioactive polymers.

#### 5.1.8.1 Drug Delivery System Using Bioactive Polymers

In this case study, bioactive polymers were utilized to develop a targeted drug delivery system for the treatment of cancer. Researchers engineered polymeric nanoparticles composed of biocompatible and biodegradable polymers, such as poly (lactic-co-glycolic acid) (PLGA) or chitosan, loaded with anticancer drugs. These nanoparticles were designed to passively accumulate in tumor tissues through the enhanced permeability and retention (EPR) effect and actively target cancer cells using ligands or antibodies attached to their surfaces. By encapsulating the drug within the polymer matrix, the system achieved controlled and sustained release of the drug at the tumor site, minimizing systemic toxicity and improving therapeutic efficacy [33]. Clinical trials demonstrated promising results, with reduced side effects and enhanced tumor regression compared to conventional chemotherapy, leading to the approval and commercialization of the bioactive polymer-based drug delivery system for cancer treatment.

#### 5.1.8.2 Tissue Engineering Scaffold for Bone Regeneration

In this case study, bioactive polymers were employed to fabricate a scaffold for bone regeneration in patients with critical-sized bone defects. Researchers developed a 3D scaffold using a combination of synthetic and natural polymers, such as poly (lactic acid) (PLA) or poly(caprolactone) (PCL) reinforced with hydroxyapatite nanoparticles or collagen fibers. The scaffold was designed to mimic the architecture and mechanical properties of natural bone tissue while providing a bioactive environment for cell adhesion, proliferation, and differentiation. Preclinical studies in animal models demonstrated successful bone regeneration, with the scaffold promoting new bone formation and vascularization within the defect site. Subsequent clinical trials confirmed the safety and efficacy of the bioactive polymer scaffold, leading to its approval by regulatory agencies for use in bone tissue engineering applications [34]. The scaffold has since been widely adopted in clinical practice, offering a minimally invasive and effective treatment option for patients with bone defects.

These case studies highlight the versatility and effectiveness of bioactive polymers in addressing critical healthcare challenges, ranging from targeted drug delivery to tissue regeneration. By harnessing the unique properties of bioactive polymers, researchers and clinicians continue to develop innovative solutions for improving patient outcomes and quality of life in diverse medical fields.

### *5.1.9 Current Challenges*

While bioactive polymers offer promising solutions for various biomedical applications [34], several challenges hinder their widespread adoption and optimization. One significant challenge is achieving precise control over the properties and functionalities of bioactive polymers to meet specific biomedical requirements. Designing polymers with tailored characteristics, such as degradation kinetics, mechanical strength, and bioactivity, remains a complex task, requiring a deep understanding of polymer chemistry, materials science, and biological interactions. Ensuring batch-to-batch consistency and reproducibility in polymer synthesis and processing presents challenges, particularly for large-scale manufacturing and clinical translation. Another challenge is the biocompatibility and long-term stability of bioactive polymers in physiological environments. While bioactive polymers are designed to interact favorably with biological systems, concerns remain regarding their potential immunogenicity, inflammatory response, and long-term biocompatibility. Ensuring that bioactive polymers do not elicit adverse reactions or trigger immune responses upon implantation or administration is critical for their safety and efficacy in clinical applications. Moreover, maintaining the stability and functionality of bioactive polymers over extended periods, particularly in dynamic physiological conditions, poses challenges, as degradation, hydrolysis, or other environmental factors may affect polymer performance and integrity over time. Achieving controlled and targeted delivery of bioactive molecules from polymer matrices remains a significant challenge in drug delivery applications. While bioactive polymers offer opportunities for encapsulating and releasing therapeutic agents with precise spatiotemporal control, challenges persist in optimizing drug loading, release kinetics, and stability within polymer matrices. Ensuring uniform distribution of drugs within the polymer matrix, achieving sustained release profiles, and overcoming barriers to drug diffusion and cellular uptake are ongoing challenges that require innovative formulation strategies and advanced drug delivery systems.

Regulatory and safety considerations present challenges for the clinical translation and commercialization of bioactive polymer-based products. Ensuring compliance with regulatory guidelines and demonstrating the safety, efficacy, and quality of bioactive polymer materials through rigorous preclinical and clinical testing is a time-consuming and resource-intensive process. Moreover, addressing concerns related to sterilization, shelf-life stability, and biodegradation products adds complexity to the development and regulatory approval of bioactive polymer-based medical devices and therapies.

**Exercise 3.**

**Choose The Correct Option**

1. Which of the following is a primary characteristic of polymers?
   a) They consist of monomers linked by covalent bonds
   b) They are exclusively made of inorganic materials
   c) They cannot undergo chemical reactions
   d) They are always crystalline in structure

2. What is the process called when monomers join together to form a polymer chain?
   a) Crystallization
   b) Polymerization
   c) Sublimation
   d) Condensation

3. What is the term used to describe the smallest repeating unit in a polymer chain?
   a) Monomer
   b) Polymer
   c) Oligomer
   d) Macromolecule

4. Bioactive polymers are designed to mimic which natural component in the body?
   a) Blood
   b) Bone
   c) Muscle
   d) Nerve tissue

5. What is one of the main purposes of incorporating bioactive molecules into polymer matrices?
   a) To decrease biocompatibility
   b) To reduce functionality
   c) To enhance therapeutic effects
   d) To increase degradation rate

6. Which application often utilizes bioactive polymer materials for promoting cell growth and tissue regeneration?
   a) Water treatment
   b) Food packaging
   c) Biomedical implants
   d) Automotive industry

7. Which technique is commonly used to analyze the surface morphology of bioactive polymer materials?
   a) Nuclear magnetic resonance (NMR)
   b) Fourier-transform infrared spectroscopy (FTIR)
   c) Scanning electron microscopy (SEM)
   d) Differential scanning calorimetry (DSC)

8. Which technique is employed to study the chemical composition and elemental analysis of bioactive polymer surfaces?
   a) X-ray diffraction (XRD)
   b) Atomic force microscopy (AFM)
   c) Energy-dispersive X-ray spectroscopy (EDS)
   d) Ultraviolet-visible spectroscopy (UV-Vis)

9. Which technique is useful for investigating the degradation behavior of bioactive polymers over time?
   a) Gel permeation chromatography (GPC)
   b) Dynamic light scattering (DLS)
   c) Thermogravimetric analysis (TGA)
   d) Mass spectrometry (MS)

10. Which technique is utilized to assess the mechanical properties of bioactive polymer materials?
   a) X-ray photoelectron spectroscopy (XPS)
   b) Raman spectroscopy
   c) Tensile testing
   d) Gas chromatography (GC)

## 5.2 Summary

The versatility of bioactive polymers is evident in their roles across different domains of medical science. In tissue engineering, these polymers serve as scaffolds that mimic the extracellular matrix, providing structural support and cues for cell adhesion, proliferation, and differentiation. This capability makes them invaluable in regenerative medicine, where they facilitate tissue repair and regeneration, offering new avenues for developing innovative therapies. Moreover, bioactive polymers play a crucial role in drug delivery by serving as carriers for controlled and targeted release of therapeutic agents. By encapsulating drugs within these polymers, researchers can enhance drug efficacy, minimize side effects, and enable precise delivery to specific sites within the body [5, 6]. Bioactive polymers find applications in wound healing, where they promote tissue regeneration and accelerate wound closure as dressings or coatings.

Furthermore, bioactive polymers contribute to the development of advanced medical devices and implants by providing functionalized surfaces that promote enhanced bio integration and reduce the risk of rejection. Their ability to be tailored and respond to environmental stimuli makes them ideal candidates for creating smart materials for controlled drug release and tissue regeneration [15, 17, 19]. These properties position bioactive polymers as invaluable tools for addressing emerging challenges in healthcare and biotechnology, inspiring ongoing research, and innovation in the field.

The chapter underscores the diverse capabilities and significant contributions of bioactive polymers to advancing biomedical science and improving patient care. From tissue engineering to drug delivery and beyond, bioactive polymers continue to drive innovation, offering promising prospects for addressing pressing healthcare challenges and paving the way for transformative medical interventions.

## Points to Remember

- **Some bioactive polymers possess self-healing properties, allowing them to repair damage autonomously when subjected to environmental stressors, extending their lifespan and durability.**
- **Bioactive polymer coatings are utilized in dentistry to promote remineralization of tooth enamel, aiding in the prevention and repair of dental cavities and enamel erosion.**
- **Hydrogels made from bioactive polymers can mimic the extracellular matrix of cartilage, providing a supportive environment for cartilage regeneration and repair in osteoarthritis and joint injuries.**
- **Some bioactive polymers exhibit inherent antimicrobial properties, making them effective against a wide range of pathogens, including bacteria, viruses, and fungi, and reducing the risk of infection in medical devices and wound dressings.**
- **Nano-sized particles made from bioactive polymers can selectively target cancer cells while sparing healthy tissues, offering a promising approach for targeted drug delivery and enhancing the efficacy of chemotherapy.**
- **Microneedle arrays made from bioactive polymers can painlessly penetrate the skin's outer layer to deliver drugs or vaccines, offering a convenient and efficient alternative to traditional injections.**
- **Composite materials combining bioactive polymers with bio ceramics or nanoparticles can mimic the structure and mechanical properties of natural bone, facilitating bone regeneration and fracture healing in orthopedic applications.**
- **Porous membranes made from bioactive polymers can support cell growth and differentiation, making them suitable scaffolds for tissue engineering applications such as skin grafts, organoids, and vascular grafts.**

## References

1. Jung K, Corrigan N, Wong EHH, Boyer C (2022) Bioactive synthetic polymers. Adv Mater (Deerfield Beach, Fla.) 35(2):e2105063. https://doi.org/10.1002/adma.202105063
2. Iovene A, Zhao Y, Wang S, Amoako K (2021) Bioactive polymeric materials for the advancement of regenerative medicine. J Funct Biomater 12(1):15. https://doi.org/10.3390/jfb12010015
3. Sharma S, Sudhakara P, Singh J, Ilyas RA, Asyraf MRM, Razman MR (2021) Critical review of biodegradable and bioactive polymer composites for bone tissue engineering and drug delivery applications. Polymers 13(16):2623. https://doi.org/10.3390/polym13162623
4. Muxika A, Etxabide A, Uranga J, Guerrero P, de la Caba K (2017) Chitosan as a bioactive polymer: Processing, properties and applications. Int J Biol Macromol 105(Pt 2):1358–1368. https://doi.org/10.1016/j.ijbiomac.2017.07.087

5. Drira M, Hentati F, Babich O, Sukhikh S, Larina V, Sharifian S, Homai A, Fendri I, Lemos MFL, Félix C, Félix R, Abdelkafi S, Michaud P (2021) Bioactive carbohydrate polymers-between myth and reality. Molecules (Basel, Switzerland) 26(23):7068. https://doi.org/10.3390/molecules26237068
6. Kaur G (2017) Polymers as bioactive materials II: synthetic/biodegradable polymers and composites. In: Series in bioengineering (pp. 53–76). https://doi.org/10.1007/978-3-319-45716-1_3
7. Shakeel A, Singh A, Das S et al (2017) Synthesis and morphological insight of new biocompatible smart hydrogels. J Polym Res 25:113. https://doi.org/10.1007/s10965-017-1267-7
8. Ahmed EM (2015) Hydrogel: preparation, characterization, and applications: a review. J Adv Res 6(2):105–121. https://doi.org/10.1016/j.jare.2013.07.006
9. Schmidt BVKJ (2021) Polymer chemistry: fundamentals and applications. Beilstein J Org Chem 17:2922–2923. https://doi.org/10.3762/bjoc.17.200
10. Ritsema van Eck GC, Chiappisi L, de Beer S (2022) Fundamentals and applications of polymer brushes in air. ACS Appl Polym Mater 5(5):3062–3087. https://doi.org/10.1021/acsapm.1c01615
11. Shahdeo D, Kesarwani V, Suhag D et al (2022) Corrigendum to "Self-assembled chitosan polymer intercalating peptide functionalized gold nanoparticles as nanoprobe for efficient imaging of urokinase plasminogen activator receptor in cancer diagnostics" [Carbohydrate Polymers, 266, (2021) 118138]. Carbohydr Polym. 285:119239. https://doi.org/10.1016/j.carbpol.2022.119239. PMID: 35287885
12. Williams DF (2022) Biocompatibility pathways and mechanisms for bioactive materials: the bioactivity zone. Bioact Mater 10:306–322. https://doi.org/10.1016/j.bioactmat.2021.08.014
13. Suhag D, Thakur P, Thakur A (2023) Introduction to nanotechnology. In: Suhag D, Thakur A, Thakur P (eds) Integrated nanomaterials and their applications. Springer, Singapore. https://doi.org/10.1007/978-981-99-6105-5_1
14. Chen CH, Woo EM (2018) Polymer science and technology. Chem Phys 15(7):2595–2506
15. Gong L, Xiang L, Zhang J, Chen J, Zeng H (2019) Fundamentals and advances in the adhesion of polymer surfaces and thin films. Langmuir: ACS J Surf Colloids 35(58):15915–15936. https://doi.org/10.1021/acs.langmuir.9b02123
16. Polymer Chemistry: From fundamentals to bio-based alternatives (2022). In: The royal society of chemistry eBooks, pp 105–120. https://doi.org/10.1039/9781837670895-00105
17. Wang M, Xu P, Lei B (2023) Engineering multifunctional bioactive citrate-based biomaterials for tissue engineering. Bioact Mater 19:511–537. https://doi.org/10.1016/j.bioactmat.2022.04.027
18. Rathore R, Suhag D, Wan F, Verma R, Thakur A, Thakur P (2023) Functionalization of biomaterials. In: Suhag D, Thakur A, Thakur P (eds) Integrated nanomaterials and their applications. Springer, Singapore. https://doi.org/10.1007/978-981-99-6105-4_11
19. Singh A, Suhag D (2023) Polymers and polymeric composites in nano/bio-medicine. In: Suhag D, Thakur A, Thakur P (eds) Integrated nanomaterials and their applications. Springer, Singapore. https://doi.org/10.1007/978-981-99-6105-5_8
20. Joyce K, Fabra GT, Bozkurt Y, Pandit A (2021) Bioactive potential of natural biomaterials: identification, retention and assessment of biological properties. Signal Transduct Target Ther 6(1). https://doi.org/10.1038/s41392-021-00512-8
21. Bolívar-Monsalve EJ, Alvarez MM, Hosseini S, Espinosa-Hernandez MA, Ceballos-González CF, Sanchez-Dominguez M, Shin SR, Cecen B, Hassan S, Di Maio E, Santiago GT (2021) Engineering bioactive synthetic polymers for biomedical applications: a review with emphasis on tissue engineering and controlled release. Mater Adv 2(14):4447–4478. https://doi.org/10.1039/d1ma00092f
22. Kurowiak J, Klekiel T, Będziński R (2023) Biodegradable polymers in biomedical applications: a review-developments, perspectives and future challenges. Int J Mol Sci 24(23):16952. https://doi.org/10.3390/ijms242316952

23. Suhag D, Bhatia R, Das S, Shakeel A, Ghosh A, Singh A, Sinha OP, Chakrabarti S, Mukherjee M (2015b) Physically cross-linked pH-responsive hydrogels with tunable formulations for controlled drug delivery. RSC Adv 5(66):53963–53972. https://doi.org/10.1039/c5ra07424j
24. Zagho MM, Hussein EA, Elzatahry AA (2018) Recent overviews in functional polymer composites for biomedical applications. Polymers 10(7):739. https://doi.org/10.3390/polym10070739
25. Iovene A, Zhao Y, Wang S, Amoako K (2021) Bioactive polymeric materials for the advancement of regenerative medicine. J Funct Biomater 12(1):14. https://doi.org/10.3390/jfb12010014
26. Hrochová M, Kotrchová L, Frejková M, Konefał R, Gao S, Fang J, Kostka L, Etrych T (2023) Adaptable polymerization platform for therapeutics with tunable biodegradability. Acta Biomater 171:517–527. https://doi.org/10.1016/j.actbio.2023.09.005
27. Teo AJ, Mishra A, Park I, Kim Y, Park W, Yoon Y (2016) Polymeric biomaterials for medical implants and devices. ACS Biomater Sci Eng 2(4):454–472. https://doi.org/10.1021/acsbiomaterials.5b00429
28. Disouza J, Kumbhar P, Manjappa A, Nadaf S, Kumbar V, Singh SK, Dua K, Patravale V (2024) Characterization of polymers. In: Elsevier eBooks (pp. 107–129). https://doi.org/10.1016/b978-0-323-95496-9.00003-x
29. Saalwächter K (2019) Applications of NMR in polymer characterization – an introduction. In: The royal society of Chemistry eBooks (pp. 1–22). https://doi.org/10.1039/9781788016483-00001
30. De Souza ALC, Pires ADR, Moraes CaF, De Matos CHC, Santos KIPD, Silva RCE, Acuña SPC, Araújo SDS (2023) Chromatographic methods for separation and identification of bioactive compounds. In: Drug Discovery and design using natural products (pp. 153–176). https://doi.org/10.1007/978-3-031-35205-8_6
31. Carbajal-De la Torre G, Zurita-Méndez NN, Ballesteros-Almanza ML, Ortiz-Ortiz J, Estévez M, Espinosa-Medina MA (2022) Characterization and evaluation of composite biomaterial bioactive glass-polylactic acid for bone tissue engineering applications. Polymers 15(15):3035. https://doi.org/10.3390/polym15153035
32. Li H, Chang J (2005) Preparation and characterization of bioactive and biodegradable wollastonite/poly(D,L-lactic acid) composite scaffolds. J Mater Sci Mater Med 15(10):1089–1095. https://doi.org/10.1023/B:JMSM.0000056390.09550.c2
33. Hunter AC, Moghimi SM (2017) Smart polymers in drug delivery: a biological perspective. Polym Chem 8(1):41–51. https://doi.org/10.1039/c6py00676k
34. Stratton S, Shelke NB, Hoshino K, Rudraiah S, Kumbar SG (2016) Bioactive polymeric scaffolds for tissue engineering. Bioact Mater 1(2):93–108. https://doi.org/10.1016/j.bioactmat.2016.11.001

# Chapter 6
# Scaffold Design in Tissue Engineering

## 6.1 Introduction

Tissue engineering and regenerative medicine represent innovative fields aiming to restore or replace damaged tissues and organs through the combination of biological principles, engineering techniques, and medical strategies. Tissue engineering focuses on the creation of functional biological substitutes that can restore, maintain, or improve tissue function. This interdisciplinary approach involves the design and fabrication of scaffolds, which serve as frameworks for cell attachment [1], proliferation, and differentiation, along with the integration of growth factors and signaling molecules to guide tissue regeneration. Regenerative medicine, on the other hand, encompasses a broader spectrum of therapeutic strategies aimed at harnessing the body's innate healing mechanisms to repair or regenerate damaged tissues and organs. Through the development of novel biomaterials, advanced fabrication techniques, and precise control over cellular behavior, tissue engineering and regenerative medicine offer promising solutions for addressing various medical challenges, from treating injuries and degenerative diseases to providing alternatives for organ transplantation [1, 2].

### *6.1.1 Basic Principles of Tissue Engineering*

Tissue engineering relies on a set of fundamental principles that guide the development of functional tissue substitutes. At its core, tissue engineering aims to mimic the complex microenvironment of native tissues by integrating cells, biomaterials, and biochemical cues. Central to this approach is the concept of cell-material interactions, where engineered scaffolds provide a supportive framework for cell attachment, proliferation, and differentiation. These scaffolds must possess suitable mechanical properties, porosity, and surface characteristics to mimic the native extracellular



D. Suhag et al., *Handbook of Biomaterials for Medical Applications, Volume 1*,
Biomedical Materials for Multi-functional Applications,
https://doi.org/10.1007/978-981-97-4818-1_6

matrix and facilitate cell adhesion and tissue formation. A key principle in tissue engineering is the selection of appropriate cell sources for seeding onto scaffolds [3]. Cells can be obtained from various tissue sources, including autologous, allogeneic, or stem cell-derived sources, each with its advantages and limitations. The choice of cell type depends on the desired tissue phenotype, functional requirements, and immunological considerations. The manipulation of cellular behavior through the incorporation of growth factors, cytokines, and other signaling molecules plays a crucial role in directing tissue development and regeneration. By modulating the biochemical cues presented to cells within the scaffold microenvironment, tissue engineers can promote specific cellular responses such as proliferation, differentiation, and extracellular matrix production. Another principle underlying tissue engineering is the concept of biomimicry, which involves the replication of key structural and functional features of native tissues. Biomimetic scaffolds are designed to recapitulate the hierarchical organization, mechanical properties, and biochemical composition of the target tissue, thereby promoting tissue-specific cell behavior and functionality [1–3]. This approach often involves the use of advanced fabrication techniques such as 3D printing, electrospinning, and microfluidics to create scaffolds with precise architecture and spatial distribution of biomaterials.

Tissue engineering embraces the principles of vascularization and innervation to address the challenges of tissue integration and long-term viability. The development of functional blood vessels and neural networks within engineered tissues is essential for ensuring adequate nutrient supply, waste removal, and communication with the host environment. Strategies such as pre-vascularization, co-culture techniques, and the incorporation of angiogenic and neurotrophic factors are employed to promote the formation of vascular and neural networks within engineered constructs. The basic principles of tissue engineering encompass the design of biomimetic scaffolds [2, 3], the selection and manipulation of appropriate cell sources, the modulation of cellular behavior through biochemical cues, and the promotion of vascularization and innervation for tissue integration and functionality. By leveraging these principles, tissue engineers strive to create innovative solutions for tissue repair, regeneration, and replacement, with the ultimate goal of restoring tissue function and improving patient outcomes.

### *6.1.2 Importance of Scaffold Design in Tissue Engineering*

The significance of scaffold design in tissue engineering cannot be overstated, as scaffolds serve as the architectural framework upon which tissue regeneration depends. These 3D structures provide a supportive environment for cells to attach, proliferate, and differentiate, mimicking the extracellular matrix found in native tissues. Scaffold design influences crucial aspects of tissue engineering, including cell behavior, tissue formation, and, ultimately, the functionality of the regenerated tissue. By carefully tailoring scaffold properties such as porosity, mechanical strength, and surface characteristics [4], tissue engineers can create microenvironments that promote specific cellular responses and guide tissue development. Furthermore, the choice of scaffold materials, fabrication techniques, and surface modifications play pivotal roles in determining the success of tissue regeneration efforts [3, 4]. Thus, a thorough understanding of scaffold design principles is essential for advancing the field of tissue engineering and developing effective therapies for tissue repair and regeneration.

#### 6.1.2.1 Role of Scaffolds in Tissue Regeneration

Scaffolds play a pivotal role in tissue regeneration by providing a structural framework that supports and guides the growth of new tissue. In essence, scaffolds act as temporary templates that facilitate the regeneration process, especially in situations where the body's natural healing mechanisms are insufficient. By mimicking the architecture and properties of the native extracellular matrix, scaffolds create a conducive microenvironment for cellular activities such as adhesion, proliferation, and differentiation. This ensures that the regenerated tissue closely resembles its natural counterpart in terms of structure and function. Scaffolds serve as delivery vehicles for bioactive molecules, growth factors, and other signaling cues that regulate cellular behavior and tissue development [5]. By incorporating these bioactive components into the scaffold matrix, tissue engineers can precisely control the spatial and temporal presentation of biochemical signals, thereby directing specific cellular responses. This targeted approach allows for the modulation of key processes such as cell migration, angiogenesis, and tissue remodeling, leading to enhanced tissue regeneration outcomes.

Scaffolds provide mechanical support to the newly formed tissue during the regeneration process, preventing mechanical stress and deformation that could impede proper tissue development. The mechanical properties of scaffolds, including stiffness, elasticity, and degradation kinetics, can be tailored to match those of the native tissue, ensuring optimal biomechanical compatibility and integration [4, 6]. This mechanical support not only promotes tissue formation and maturation but also facilitates the functional integration of the regenerated tissue with the surrounding host environment.

#### 6.1.2.2 Significance of Scaffold Materials and Structures

The significance of scaffold materials and structures in tissue engineering lies in their ability to dictate the success of tissue regeneration by providing a tailored microenvironment for cellular activities. The choice of scaffold materials is crucial as they serve as the foundation upon which cells adhere, proliferate, and differentiate. Natural polymers, such as collagen and hyaluronic acid, offer biocompatibility and bioactivity, mimicking the composition of the native extracellular matrix. Synthetic polymers, like polylactic acid and polyethylene glycol, provide tunable mechanical properties and degradation kinetics, allowing for precise control over scaffold design [6]. Composite materials, combining natural and synthetic components, offer synergistic advantages, such as improved mechanical strength and enhanced bioactivity, making them ideal candidates for tissue engineering applications. Scaffold structures play a critical role in regulating cellular behavior and tissue formation by influencing factors such as surface area, porosity, and interconnectivity. 3D scaffold architectures, including porous scaffolds, fibrous meshes, and micro/nanopatterned surfaces, facilitate nutrient and oxygen diffusion, waste removal, and cell–cell interactions, promoting homogeneous cell distribution and tissue ingrowth throughout the scaffold matrix [6]. The spatial arrangement of scaffold pores and channels also guides the alignment and organization of cells, essential for the development of functional tissues with desired mechanical properties and structural integrity.

Scaffold materials and structures influence the dynamic interactions between cells and their microenvironment, modulating cellular responses such as adhesion, migration, and differentiation. Surface modifications, such as chemical functionalization and incorporation of bioactive molecules, enable precise control over cell-scaffold interactions, enhancing cellular attachment and promoting specific lineage commitment. The mechanical properties of scaffold materials, including stiffness, elasticity, and viscoelasticity, play a crucial role in mechano transduction pathways, regulating cellular behavior and tissue development. By mimicking the mechanical cues present in native tissues, scaffold materials can promote tissue maturation and functional integration with the host environment. Scaffold materials and structures enable the design of patient-specific tissue engineering strategies tailored to individual anatomical and physiological requirements [7]. Advanced fabrication techniques, including 3D printing, electrospinning, and microfluidics, allow for precise control over scaffold architecture and composition, facilitating the development of customized tissue constructs with complex geometries and heterogeneous properties. This personalized approach not only enhances the compatibility and integration of engineered tissues but also reduces the risk of immune rejection and promotes long-term tissue functionality.

### *6.1.3 Principles of Scaffold Design*

The principles of scaffold design in tissue engineering encompass a multidisciplinary approach aimed at creating biomimetic structures that mimic the native extracellular matrix and promote tissue regeneration (Fig. 6.1). At the core of scaffold design principles lies the concept of providing a supportive microenvironment for cellular attachment, proliferation, and differentiation. Scaffold design involves careful consideration of factors such as material selection [6, 7], structural architecture, and surface properties to optimize cellular responses and tissue formation. By leveraging principles from materials science, biology, and engineering, tissue engineers can tailor scaffold properties to meet specific tissue regeneration requirements, ultimately leading to the development of functional tissue substitutes with enhanced biocompatibility and performance.

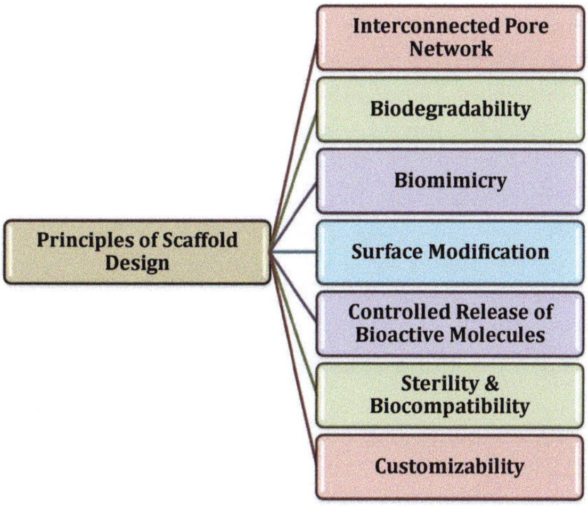


**Fig. 6.1** Principles to consider for scaffold designing

#### 6.1.3.1 Structural Requirements for Tissue Engineering

Structural requirements are paramount in tissue engineering, as they dictate the success of tissue regeneration by providing a framework that supports cellular activities and guides tissue formation. The structural integrity of scaffolds is crucial to withstand mechanical forces and maintain their shape during cell seeding, culture, and implantation. Scaffolds must possess sufficient mechanical strength and stability to withstand physiological loads and ensure long-term functionality [4, 6]. The architecture of scaffolds plays a vital role in facilitating nutrient and oxygen diffusion, waste removal, and cell–cell interactions throughout the scaffold matrix, promoting homogeneous cell distribution and tissue ingrowth. Scaffold porosity is a critical structural parameter that influences cellular behavior and tissue regeneration. Porous scaffolds provide a high surface area-to-volume ratio, allowing for increased cell attachment and infiltration into the scaffold matrix. The pore size and interconnectivity of scaffolds regulate cell migration, proliferation, and differentiation by influencing factors such as nutrient transport, cell–cell communication, and extracellular matrix deposition. Properly designed porous scaffolds promote tissue ingrowth and vascularization [7, 8], facilitating the formation of functional tissues with enhanced mechanical properties and biological functionality. The spatial arrangement of scaffold pores and channels guides the alignment and organization of cells, essential for the development of tissues with desired structural and functional properties. Scaffold architecture can be tailored to mimic the native tissue microenvironment, providing cues that promote tissue-specific cell behavior and phenotype. Strategies such as micro/nanopatterning [9], fiber alignment, and gradient scaffolds enable precise control over cellular responses and tissue development, facilitating the creation of biomimetic tissue constructs with enhanced regenerative potential.

The degradation kinetics of scaffold materials play a crucial role in tissue engineering, as they influence the temporal evolution of scaffold properties and the remodeling of the regenerated tissue. Biodegradable scaffolds undergo controlled degradation over time, allowing for the gradual replacement of the scaffold material with newly formed tissue. The rate of scaffold degradation must be carefully tuned to match the rate of tissue formation, ensuring proper integration of the regenerated tissue with the surrounding host environment. By considering these structural requirements [7–9], tissue engineers can design scaffolds that provide an optimal microenvironment for tissue regeneration, leading to the development of functional tissue substitutes for clinical applications.

#### 6.1.3.2 Cellular Interactions with Scaffolds

Cellular interactions with scaffolds are fundamental in tissue engineering, as they dictate the behavior and fate of seeded cells within the scaffold microenvironment. Upon seeding onto scaffolds, cells interact with the scaffold surface through a complex series of adhesion molecules and receptors, leading to cell attachment and spreading. Cell adhesion is mediated by integrin receptors, which bind to specific ligands present on the scaffold surface, initiating signaling pathways that regulate cellular responses such as migration, proliferation, and differentiation. The physical and chemical properties of the scaffold surface, including topography, chemistry, and stiffness, play crucial roles in modulating cellular adhesion and spreading, influencing cell behavior and tissue formation. Cellular interactions with scaffolds influence the process of cell migration within the scaffold matrix, essential for tissue ingrowth and colonization. Cells migrate through the scaffold pores and channels in response to chemotactic gradients [10], mechanical cues, and cell–cell interactions, navigating the 3D scaffold architecture to reach their desired destination. Proper scaffold design promotes uniform cell distribution and migration throughout the scaffold matrix, ensuring homogeneous tissue formation and integration. The spatial organization of scaffold pores and channels can guide cell migration and alignment, facilitating the development of tissues with desired structural and functional properties.

Cellular interactions with scaffolds regulate the process of cell proliferation and differentiation, essential for tissue regeneration and maturation. Scaffold properties such as porosity, surface chemistry, and mechanical stiffness influence cell proliferation rates and cell cycle progression, modulating tissue growth and development. Biochemical cues presented by the scaffold microenvironment, such as growth factors, cytokines, and extracellular matrix proteins, play pivotal roles in directing cellular fate decisions and lineage commitment. By providing a biomimetic microenvironment [10] that mimics the native tissue niche, scaffolds can promote specific cellular responses and tissue phenotypes, leading to the development of functional tissue substitutes for regenerative medicine applications. Cellular interactions with scaffolds play crucial roles in tissue remodeling and maturation, as cells actively remodel the scaffold matrix and deposit new extracellular matrix components over time. Scaffold degradation products and mechanical cues generated during tissue maturation can influence cellular behavior and tissue remodeling processes, promoting tissue maturation and functional integration with the surrounding host environment. The dynamic interactions between cells and scaffold materials can regulate the expression of matrix metalloproteinases, which play key roles in tissue remodeling and turnover.

#### 6.1.3.3 Scaffold Biocompatibility and Biofunctionality

Scaffold biocompatibility and biofunctionality are critical aspects of tissue engineering, ensuring that scaffolds provide a supportive environment for cell growth and tissue regeneration without eliciting adverse immune responses. Biocompatible scaffolds are designed to interact favorably with living tissues, minimizing cytotoxicity, inflammation [11], and immune rejection reactions. This requires careful selection of scaffold materials and fabrication techniques to ensure compatibility with biological systems. Natural polymers, such as collagen, hyaluronic acid, and chitosan, are widely used due to their inherent biocompatibility and similarity to the extracellular matrix, while synthetic polymers, like poly (lactic-co-glycolic acid) (PLGA) and polyethylene glycol (PEG), can be engineered to exhibit tailored biocompatibility profiles. In addition to biocompatibility, scaffold biofunctionality refers to the ability of scaffolds to actively interact with cells and regulate cellular behavior to promote tissue regeneration. Biofunctional scaffolds incorporate bioactive molecules, growth factors, and signaling cues that mimic the native tissue microenvironment and modulate cellular responses [9–11]. Surface modifications, such as chemical functionalization and immobilization of biomolecules, enable precise control over cell-scaffold interactions, enhancing cell adhesion, proliferation, and differentiation. Furthermore, scaffold degradation products can serve as bioactive cues that influence cellular behavior and tissue remodeling processes, promoting tissue maturation and functional integration with the surrounding host environment.

Scaffold biofunctionality extends beyond cellular interactions to include the ability to support tissue-specific functions and responses. Tissue-specific scaffolds are designed to replicate the biochemical, mechanical, and structural properties of native tissues, facilitating the development of functional tissue substitutes with enhanced biological activity and performance. For example, scaffolds for bone regeneration may incorporate osteoinductive and osteoconductive factors to promote bone formation [12], while scaffolds for cartilage repair may mimic the viscoelastic properties of cartilage tissue to support chondrogenic differentiation. By providing a biomimetic microenvironment that mimics the native tissue niche, biofunctional scaffolds can guide tissue development and regeneration, leading to the creation of functional tissue substitutes for clinical applications.

Moreover, scaffold biofunctionality encompasses the ability to integrate with the host environment and support long-term tissue viability and functionality. Scaffolds must be designed to degrade at a rate that matches tissue formation, ensuring proper integration of the regenerated tissue with the surrounding host tissue. Scaffolds should promote vascularization and innervation to facilitate nutrient and oxygen transport, waste removal, and communication with the host environment [9–11]. By supporting tissue integration and functionality, biofunctional scaffolds enable the development of durable tissue constructs with sustained therapeutic effects, advancing the field of tissue engineering and regenerative medicine.

**Exercise 1.**

**Fill Ups**

**(Tissue regeneration, Scaffold, Adhesion, 3D structure, Extracellular matrix)**

**1.** The principle of scaffold design involves creating a ..................to support cell growth and tissue regeneration.

**2.** ...............architecture influences cellular attachment, proliferation, and differentiation.

**3.** Surface properties of scaffolds dictate cell-scaffold interactions, including ................and signaling.

**4.** Scaffold degradation kinetics impact the timing of ..................and scaffold replacement with native tissue.

**5.** The principle of scaffold design aims to mimic the native ......................and provide a conducive environment for tissue regeneration.

## *6.1.4 Scaffold Materials and Structures*

In tissue engineering, scaffold materials and structures play a pivotal role in creating environments conducive to cell growth, proliferation, and tissue formation. Various types of scaffold materials and structures are utilized, ranging from natural polymers like collagen and fibrin to synthetic polymers such as poly (lactic-co-glycolic acid) (PLGA) and polyethylene glycol (PEG) [12]. These materials are often fabricated into porous structures, mimicking the extracellular matrix to provide mechanical support and facilitate nutrient and waste exchange. Scaffold designs may incorporate bioactive molecules or growth factors to further enhance cellular functions and tissue regeneration. The choice of scaffold material and structure is guided by factors such as biocompatibility, degradation kinetics, mechanical properties, and the specific tissue being targeted, highlighting the interdisciplinary nature of tissue engineering research and development.

### 6.1.4.1 Natural Biomaterials: Properties and Applications

Natural biomaterials, derived from biological sources such as plants, animals, and microorganisms, possess unique properties that make them invaluable for a wide range of applications in biomedical engineering and regenerative medicine (Fig. 6.2). One of the key advantages of natural biomaterials is their inherent biocompatibility, meaning they are well-tolerated by the body and elicit minimal immune response [13]. Examples include collagen, a fibrous protein found abundantly in connective tissues like skin, bone, and cartilage, and chitosan, derived from the exoskeletons of crustaceans. These biomaterials can be processed into scaffolds, hydrogels, or films, serving as platforms for cell attachment, proliferation, and tissue regeneration. Natural biomaterials often exhibit bioactive properties, meaning they can interact with biological systems to promote specific cellular responses. For instance, alginate, extracted from seaweed, forms gel-like structures in the presence of divalent cations, making it ideal for encapsulating cells in 3D environments for applications such as drug delivery and tissue engineering [11, 13]. Similarly, silk fibroin, obtained from silkworms, possesses remarkable mechanical strength and biodegradability, making it suitable for fabricating scaffolds for bone, cartilage, and vascular tissue regeneration. In addition to their biocompatibility and bioactivity, natural biomaterials offer versatility in terms of processing and modification. They can be easily functionalized with bioactive molecules, growth factors, or drugs to tailor their properties for specific applications [11]. For example, hyaluronic acid, a polysaccharide found in the extracellular matrix, can be modified to incorporate cell-adhesive peptides or degradation motifs, enhancing its suitability for tissue engineering applications such as wound healing and cartilage repair. Furthermore, natural biomaterials often exhibit tunable degradation profiles, allowing for controlled release of encapsulated bioactive agents and gradual replacement by newly formed tissue.

Beyond their use in tissue engineering, natural biomaterials find applications in drug delivery, wound dressing, and medical implants. For instance, keratin, a protein found in hair, wool, and feathers, has been utilized to develop dressings for chronic wounds and burns, thanks to its ability to promote wound healing and reduce inflammation [14]. Similarly, gelatin, derived from collagen, is commonly used in the fabrication of drug delivery systems and surgical sealants due to its biocompatibility, biodegradability, and ability to form thermally reversible hydrogels.

**Fig. 6.2** Properties of natural biomaterials

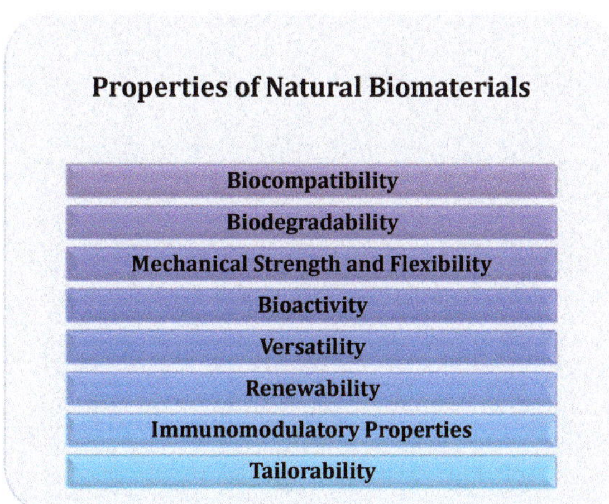

#### 6.1.4.2 Synthetic Biomaterials: Design and Characteristics

Synthetic biomaterials represent a diverse class of materials engineered for specific biomedical applications, offering precise control over their properties and functionalities (Table 6.1). These materials are typically synthesized from synthetic polymers, ceramics, or metals and are designed to mimic or surpass the properties of natural tissues. One key advantage of synthetic biomaterials is their tunable mechanical properties, allowing for the fabrication of scaffolds, implants, and drug delivery systems with tailored stiffness, elasticity, and strength. For example, poly (lactic-co-glycolic acid) (PLGA), a biodegradable copolymer, can be synthesized with varying ratios of lactic acid and glycolic acid to modulate its degradation rate and mechanical strength, making it suitable for applications ranging from tissue engineering to controlled drug delivery [15]. Synthetic biomaterials offer precise control over their chemical composition and surface properties, enabling the incorporation of bioactive molecules, functional groups, or targeting ligands to promote specific cellular responses or tissue regeneration. For instance, polyethylene glycol (PEG), a synthetic polymer widely used in biomaterials, can be modified with cell-adhesive peptides, growth factors, or nanoparticles to enhance cell adhesion, proliferation, and differentiation. Surface modification techniques such as plasma treatment or chemical grafting can be employed to impart desired properties such as hydrophilicity, protein adsorption resistance, and antimicrobial activity, further expanding the versatility of synthetic biomaterials in biomedical applications.

Furthermore, synthetic biomaterials offer excellent batch-to-batch consistency and reproducibility, crucial for the development and commercialization of medical devices and therapies. Unlike natural biomaterials, which may vary in composition and properties depending on their biological source, synthetic biomaterials can

**Table 6.1** Advantages and disadvantages of natural, synthetic, and composite polymers

| Material type | Advantage | Disadvantage |
|---|---|---|
| Natural polymer | Biocompatible and bioactive, closely mimicking the native extracellular matrix | Limited mechanical strength and stability compared to synthetic materials |
| | Promote cell adhesion, proliferation, and differentiation | Batch-to-batch variability in properties |
| Synthetic polymers | Tailorable mechanical properties, allowing for precise control over scaffold stiffness and degradation kinetics | May lack bioactivity, requiring surface modification for cell adhesion and functionality |
| | Consistent in properties and composition, enabling reproducible scaffold fabrication | Limited cellular recognition and interaction compared to natural polymer |
| Composite materials | Combine the advantages of natural and synthetic material, offering enhanced mechanical strength, bioactivity, and degradation properties | Complex fabrication process, requiring precise control over material composition and distribution |

be precisely engineered with controlled molecular weight, purity, and functionality. This consistency is essential for ensuring product performance, safety, and regulatory compliance in clinical applications. Synthetic biomaterials often exhibit enhanced stability and durability compared to their natural counterparts, making them suitable for long-term implantation and sustained drug release. Despite their numerous advantages, synthetic biomaterials also pose challenges related to biocompatibility, immunogenicity, and long-term degradation products [16]. The design and selection of synthetic biomaterials must consider factors such as cytotoxicity, inflammation, and foreign body response to minimize adverse effects on host tissues. Strategies such as surface modification, polymer blending, and the incorporation of bioactive agents can mitigate these challenges and improve the biocompatibility of synthetic biomaterials. Ongoing research efforts focus on developing next-generation synthetic biomaterials with enhanced biocompatibility, bioactivity, and therapeutic efficacy, driving innovation in the field of biomedical engineering and regenerative medicine.

#### 6.1.4.3 Composite Materials: Enhancing Scaffold Properties

Composite materials, composed of two or more distinct constituents, offer unique opportunities to enhance scaffold properties for biomedical applications (Table 6.1). By combining different materials with complementary characteristics, composite scaffolds can achieve a synergistic balance of mechanical strength [16], biocompatibility, and bioactivity. For example, composite scaffolds may consist of a polymer matrix reinforced with ceramic or metallic nanoparticles to improve mechanical properties such as stiffness and toughness. This reinforcement enhances scaffold stability and durability, making them suitable for load-bearing applications in bone tissue engineering and orthopedic implants. Composite scaffolds can be engineered to mimic the hierarchical structure and composition of natural tissues, thereby promoting cell adhesion, proliferation, and tissue regeneration. By incorporating biomimetic components such as extracellular matrix proteins, growth factors, and mineral phases, composite scaffolds can create microenvironments that closely resemble native tissue niches. For instance, composite scaffolds for cartilage repair may contain a combination of biodegradable polymers and chondroitin sulfate to provide mechanical support and stimulate chondrogenic [17] differentiation of mesenchymal stem cells, leading to the formation of functional cartilage tissue.

Composite scaffolds offer versatility in terms of fabrication techniques and processing methods, allowing for the creation of complex structures with controlled architecture and spatial distribution of components. Techniques such as electrospinning, 3D printing, and sol-gel encapsulation enable precise control over scaffold morphology, porosity, and interconnectivity, facilitating cellular infiltration and nutrient diffusion [14, 16]. Composite scaffolds can also be tailored to exhibit stimuli-responsive properties, such as pH-responsive drug release or piezoelectricity for mechano transduction, further expanding their potential applications in drug delivery, tissue engineering, and regenerative medicine.

In addition to enhancing scaffold properties, composite materials can address specific challenges associated with individual components, such as degradation kinetics, immunogenicity, and lack of bioactivity. By combining materials with complementary degradation profiles or immunomodulatory properties, composite scaffolds can achieve controlled degradation and minimize adverse host responses. The integration of bioactive molecules or signaling cues within composite scaffolds can promote tissue-specific regeneration and accelerate healing processes.

## 6.1.5 Fabrication Techniques for Scaffold Production

Fabrication techniques are crucial in scaffold production for tissue engineering, offering diverse methods to create structures that mimic the extracellular matrix and support tissue regeneration. Techniques like electrospinning, 3D printing, freeze drying, and solvent casting provide versatile approaches to tailor scaffold properties such as architecture, porosity, and mechanical strength. Each method offers unique advantages: electrospinning enables the creation of nanofibrous scaffolds with high surface area, 3D printing allows for precise control over scaffold geometry and internal architecture, freeze drying produces porous scaffolds with interconnected pores, and solvent casting offers simplicity and scalability in scaffold fabrication [17]. These techniques collectively contribute to the development of scaffolds with biomimetic properties, facilitating successful tissue engineering applications.

### 6.1.5.1 Electrospinning: Fabricating Nanostructured Scaffolds

Electrospinning is a versatile and widely used technique for fabricating nanostructured scaffolds in tissue engineering. It involves the extrusion of a polymer solution or melt through a charged spinneret toward a grounded collector, resulting in the formation of ultrafine fibers with diameters ranging from tens of nanometers to several micrometers. The electrospinning process offers precise control over scaffold morphology, allowing for the creation of nanofibrous structures that closely mimic the dimensions and architecture of the native extracellular matrix. By adjusting parameters such as polymer concentration [17, 18], solution viscosity, and applied voltage, researchers can tailor the properties of electrospun scaffolds to meet specific tissue engineering requirements. One of the key advantages of electrospinning is its ability to produce high surface area scaffolds with interconnected porosity, which enhances cell attachment, proliferation, and infiltration into the scaffold matrix. The small fiber diameter and porous structure of electrospun scaffolds facilitate nutrient and oxygen diffusion, waste removal, and cell–cell interactions, promoting homogeneous cell distribution and tissue ingrowth throughout the scaffold. The nanofibrous architecture of electrospun scaffolds can be modulated to mimic the hierarchical organization of native tissues, providing cues for directing cellular behavior and tissue development. Electrospun scaffolds offer tunable mechanical properties, making them suitable for a wide range of tissue engineering applications. By varying parameters such as polymer type, molecular weight, and fiber alignment, researchers can adjust the mechanical strength, stiffness [14, 16, 17], and elasticity of electrospun scaffolds to match those of the target tissue. This enables the development of scaffolds with tailored mechanical properties that can withstand physiological loads and support tissue regeneration. Moreover, electrospinning allows for the incorporation of bioactive molecules, growth factors, and nanoparticles into the scaffold matrix, further enhancing its biofunctionality and therapeutic potential. By exploiting the versatility and scalability of electrospinning, researchers can continue to advance the field of tissue engineering by creating nanostructured scaffolds with tailored properties for various biomedical applications.

#### 6.1.5.2 3D-Printing: Customizable Scaffold Manufacturing

3D printing, also known as additive manufacturing, has revolutionized scaffold manufacturing in tissue engineering by offering customizable and precise control over scaffold geometry, internal architecture, and material composition. This technique enables the layer-by-layer deposition of biomaterials, allowing for the fabrication of scaffolds with complex 3D structures tailored to specific patient needs. With 3D printing, researchers can create scaffolds with intricate geometries, including porous structures, interconnected channels, and patient-specific shapes, providing precise spatial control over scaffold properties [19]. This customization capability is particularly advantageous in applications such as bone regeneration, where scaffolds must closely match the anatomical features of the defect site for optimal tissue integration and functionality.

3D printing offers versatility in scaffold material selection, allowing for the incorporation of a wide range of biomaterials, including natural polymers, synthetic polymers, ceramics, and hydrogels. This versatility enables the fabrication of scaffolds with tailored mechanical properties, degradation kinetics, and bioactivity to match those of the native tissue. By combining multiple materials within a single scaffold or incorporating bioactive molecules and growth factors, researchers can create multifunctional scaffolds with enhanced therapeutic potential. 3D printing allows for precise control over scaffold porosity and pore size distribution, influencing factors such as nutrient transport, cell infiltration, and tissue ingrowth, thereby promoting optimal tissue regeneration outcomes [3, 6]. Furthermore, 3D printing enables rapid prototyping and iterative design optimization, accelerating the development of new scaffold designs and fabrication techniques. By rapidly fabricating and evaluating scaffold prototypes, researchers can iteratively refine scaffold designs to improve their performance and functionality. This iterative design process allows for the exploration of a wide range of scaffold parameters, including geometry, material composition, and surface properties, to optimize scaffold properties for specific tissue engineering applications. 3D printing offers scalability and cost-effectiveness in scaffold manufacturing, making it suitable for both research and clinical applications [20, 21]. As 3D printing technology continues to advance, it holds great promise for the development of customized tissue engineering solutions that address the diverse needs of patients suffering from tissue defects and degenerative diseases.

#### 6.1.5.3 Freeze Drying and Solvent Casting: Traditional Methods for Scaffold Fabrication

Freeze drying and solvent casting are traditional methods for scaffold fabrication in tissue engineering, offering simplicity, versatility, and scalability in scaffold production. Freeze drying, also known as lyophilization, involves freezing a polymer solution or suspension followed by sublimation of the solvent under vacuum, resulting in the formation of porous scaffolds with a highly interconnected pore structure. This method preserves the scaffold's 3D architecture and maintains the integrity of sensitive biomolecules [22], making it suitable for incorporating bioactive agents such as growth factors and cytokines into the scaffold matrix. Freeze-dried scaffolds exhibit low density and high porosity, facilitating cell infiltration, nutrient diffusion, and tissue ingrowth, making them suitable for a wide range of tissue engineering applications.

Solvent casting is another traditional method for scaffold fabrication that involves the dissolution of a polymer in a volatile solvent, casting the solution into a mold, and evaporating the solvent to form a solid scaffold. This technique allows for the fabrication of scaffolds with controlled geometry, surface topography, and mechanical properties, making it suitable for applications requiring precise control over scaffold architecture. Solvent-cast scaffolds can be fabricated from a variety of polymers, including natural polymers such as collagen and gelatin, as well as synthetic polymers such as polylactic acid (PLA) and poly (lactic-co-glycolic acid) (PLGA). Furthermore, solvent casting enables the incorporation of bioactive molecules and nanoparticles into the scaffold matrix, enhancing its bioactivity and therapeutic potential [18, 19]. Both freeze drying and solvent casting offer advantages in scaffold fabrication, including simplicity, scalability, and versatility, making them attractive options for researchers and clinicians alike. These traditional methods provide cost-effective solutions for producing scaffolds with tailored properties for specific tissue engineering applications. However, they also have limitations, such as limited control over scaffold porosity and pore size distribution, which may impact cellular behavior and tissue regeneration outcomes. Despite these limitations, freeze drying and solvent casting remain valuable techniques in scaffold fabrication, particularly in applications where simplicity, reproducibility, and cost-effectiveness are paramount. As tissue engineering continues to evolve, these traditional methods may be complemented by more advanced fabrication techniques to further enhance scaffold properties and performance.

### 6.1.6 *Surface Modification of Scaffolds*

Surface modification of scaffolds is a critical aspect of tissue engineering, aimed at enhancing the interaction between scaffolds and cells to promote tissue regeneration. By modifying the surface properties of scaffolds, such as topography, chemistry, and charge, researchers can tailor the scaffold microenvironment to better mimic the native extracellular matrix and guide cellular behavior. Surface modification techniques enable the incorporation of bioactive molecules, growth factors, and signaling cues onto the scaffold surface, providing spatially controlled cues that regulate cellular adhesion, proliferation, and differentiation [23]. These modifications can also influence the immune response, vascularization, and tissue integration of engineered constructs, ultimately improving the efficacy and functionality of tissue-engineered implants.

#### 6.1.6.1 Biofunctionalization for Enhanced Cellular Adhesion

Biofunctionalization of scaffolds is a powerful strategy in tissue engineering to enhance cellular adhesion, proliferation, and differentiation. By modifying scaffold surfaces with bioactive molecules, such as peptides, proteins, and growth factors, researchers can create microenvironments that mimic the native extracellular matrix and promote specific cellular responses. One common approach involves the immobilization of cell-adhesive peptides, such as RGD (arginine-glycine-aspartic acid), onto scaffold surfaces, which serve as recognition sites for integrin receptors on cell membranes, facilitating cellular adhesion and spreading. Biofunctionalization can involve the incorporation of extracellular matrix proteins, such as collagen or fibronectin, onto scaffold surfaces to provide cues for cell attachment and migration, further enhancing cellular interactions with the scaffold matrix [23]. Biofunctionalization enables the presentation of growth factors and cytokines on scaffold surfaces, which can regulate cellular behavior and tissue development. Growth factors such as transforming growth factor-beta (TGF-β), bone morphogenetic proteins (BMPs), and vascular endothelial growth factor (VEGF) play crucial roles in cell proliferation, differentiation, and tissue regeneration. By immobilizing these growth factors onto scaffold surfaces, researchers can spatially control their presentation and release, leading to localized effects that promote tissue-specific responses. This targeted approach allows for precise modulation of cellular behavior and tissue formation, enhancing the efficacy and functionality of tissue-engineered constructs.

Biofunctionalization can be used to create gradients of bioactive molecules within scaffold matrices, providing spatially controlled cues that guide cellular migration and tissue organization. By immobilizing gradients of cell-adhesive peptides or growth factors onto scaffold surfaces, researchers can create microenvironments that mimic the complex gradients found in native tissues, influencing cell behavior and tissue development in a spatially defined manner. This gradient-based approach enables the creation of tissue-engineered constructs with heterogeneous properties and structures, allowing for the development of complex tissue architectures and interfaces [23]. Biofunctionalization can be used to incorporate antibacterial or anti-inflammatory agents onto scaffold surfaces, reducing the risk of infection and inflammation in tissue-engineered implants, and promoting tissue regeneration in a controlled and sterile environment.

#### 6.1.6.2 Incorporating Growth Factors and Signaling Molecules

Incorporating growth factors and signaling molecules into scaffold matrices is a fundamental strategy in tissue engineering to regulate cellular behavior and promote tissue regeneration. Growth factors play crucial roles in cell proliferation, differentiation, and tissue development, making them key players in orchestrating the complex processes of tissue regeneration. By immobilizing growth factors within scaffold matrices, researchers can provide localized and controlled cues to cells, guiding their behavior and promoting tissue-specific responses. This targeted delivery approach ensures that growth factors exert their effects precisely where they are needed, enhancing the efficacy and functionality of tissue-engineered constructs [24]. One common method for incorporating growth factors into scaffold matrices is physical entrapment or encapsulation, where growth factors are physically trapped within the scaffold material during fabrication. This approach offers sustained release of growth factors over time, ensuring prolonged exposure of cells to bioactive cues and promoting tissue regeneration. Alternatively, growth factors can be chemically immobilized onto scaffold surfaces using cross-linking agents or bioconjugation techniques. This immobilization approach provides spatial control over-growth factor presentation and release, allowing for localized effects that promote tissue-specific responses and minimize off-target effects. Scaffold-based delivery systems enable the incorporation of multiple growth factors and signaling molecules into a single scaffold matrix, allowing for synergistic effects that enhance tissue regeneration. By combining growth factors with complementary functions, such as promoting cell proliferation, differentiation, and extracellular matrix synthesis, researchers can create multifunctional scaffolds with enhanced therapeutic potential. Scaffold-based delivery systems can be engineered to release growth factors in response to specific stimuli, such as changes in pH, temperature, and enzyme activity, enabling spatiotemporal control over-growth factor release and tissue regeneration processes.

Furthermore, incorporating growth factors and signaling molecules into scaffold matrices can be combined with other biofunctionalization strategies, such as surface modification with cell-adhesive peptides or extracellular matrix proteins, to further enhance cellular interactions and tissue regeneration. By tailoring scaffold surfaces and bulk properties with bioactive cues, researchers can create microenvironments that mimic the native tissue niche and promote specific cellular responses [23, 24], leading to improved efficacy and functionality of tissue-engineered constructs.

#### 6.1.6.3 Surface Patterning Techniques for Guiding Cell Behavior

Surface patterning techniques are invaluable tools in tissue engineering for guiding cell behavior and directing tissue development. These techniques involve the creation of spatially controlled patterns or features on scaffold surfaces, which can influence cellular adhesion, migration, proliferation, and differentiation. One commonly used surface patterning technique is microcontact printing, which involves the transfer of patterned molecules, such as cell-adhesive peptides or proteins, onto scaffold surfaces using elastomeric stamps. Microcontact printing enables precise control over pattern geometry, size, and spacing, allowing researchers to create well-defined microenvironments that dictate cellular responses and tissue organization. Photolithography is a powerful surface patterning technique that uses light to selectively expose and polymerize photosensitive materials on scaffold surfaces. Photolithography enables the creation of microscale and nanoscale features [24, 25], such as grooves, ridges, and pits, which can influence cell morphology, alignment, and behavior. By controlling the spatial arrangement and dimensions of surface features, researchers can create substrates that mimic the topographical cues present in native tissues, guiding cellular responses and tissue organization in a controlled manner.

Electron beam lithography (EBL) and focused ion beam (FIB) milling are advanced surface patterning techniques that offer high resolution and precision in creating nanostructured surfaces for tissue engineering applications. These techniques enable the fabrication of intricate patterns and structures with submicron resolution, allowing researchers to create surfaces with precise control over feature size, shape, and arrangement. Nanostructured surfaces produced by EBL and FIB milling can influence cell adhesion, spreading, and differentiation, providing cues that mimic the nanoscale architecture of the extracellular matrix and promote specific cellular responses [26].

Moreover, self-assembly techniques, such as layer-by-layer assembly and colloidal lithography, offer versatile approaches for creating hierarchical surface patterns on scaffold surfaces. These techniques involve the sequential deposition of alternating layers of functional molecules or nanoparticles onto scaffold surfaces, leading to the formation of ordered structures with tailored properties. Self-assembled surface patterns can influence cell adhesion, proliferation, and differentiation by presenting spatially controlled cues that mimic the organization of native tissues.

**Exercise 2.**

**Fill Ups**

**(Surface roughness, Scaffold biocompatibility, Functional, Tissue engineering, Surface modification)**

**1.** .................. of scaffold involves altering the chemical or physical properties of the scaffold's outer layer.

**2.** The process enhances.................., bioactivity, and interaction with cells or biological molecules.

**3.** .................. groups added during modification can promote cell adhesion, proliferation, and differentiation.

**4.** ....................influences cell attachment, spreading, and migration on the scaffold.

**5.** Surface modification of scaffold aims to improve its performance in ......................applications, promoting better integration with host tissues and enhancing regenerative outcomes.

### *6.1.7 Characterization Methods for Scaffold Evaluation*

Characterization methods for scaffold evaluation are essential tools in tissue engineering to assess the physical, chemical, and biological properties of scaffolds (Fig. 6.3). These methods provide valuable insights into scaffold structure, morphology, porosity, mechanical properties, degradation kinetics, and biocompatibility, guiding the design and optimization of tissue-engineered constructs. Common characterization techniques include scanning electron microscopy (SEM), Fourier-transform infrared spectroscopy (FTIR), X-ray diffraction (XRD), mechanical testing, and in vitro cell culture assays [27]. By utilizing a combination of these techniques, researchers can comprehensively evaluate scaffold performance, identify potential limitations, and refine scaffold designs to improve tissue regeneration outcomes. Overall, characterization methods play a critical role in advancing the field of tissue engineering by providing essential information for the development of functional and biocompatible scaffolds for various biomedical applications.

#### 6.1.7.1 Morphological Analysis of Scaffold Structures

Morphological analysis of scaffold structures is a crucial aspect of scaffold characterization in tissue engineering, providing insights into scaffold architecture, pore morphology, and surface topography. Scanning electron microscopy (SEM) is a

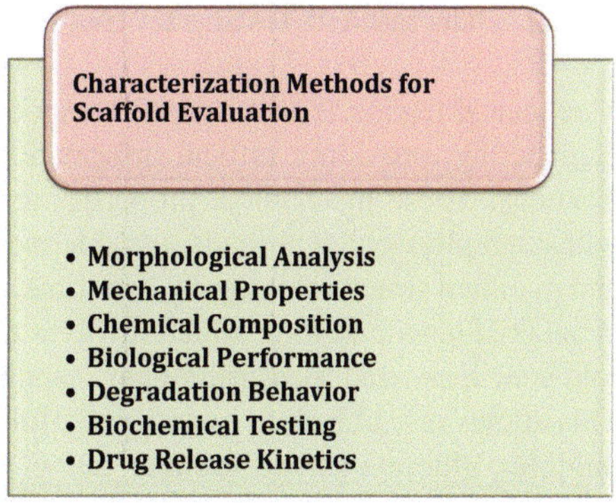


**Fig. 6.3** Characterization methods for scaffold evaluation

widely used technique for morphological analysis, allowing for high-resolution imaging of scaffold surfaces and internal structures at the micro- and nanoscale. SEM enables researchers to visualize scaffold morphology, including pore size, shape, distribution, and interconnectivity, which are critical factors influencing cell infiltration, nutrient diffusion, and tissue ingrowth [27]. Additionally, SEM can reveal surface roughness and texture, which play important roles in cellular adhesion, spreading, and proliferation on scaffold surfaces. Microcomputed tomography (micro-CT) is another valuable technique for morphological analysis of scaffold structures, providing 3D visualization and quantitative analysis of scaffold porosity, pore size distribution, and spatial arrangement. Micro-CT enables non-destructive imaging of scaffold internal structures with high resolution and accuracy, allowing researchers to assess scaffold architecture and pore interconnectivity in a non-invasive manner [12, 16]. This information is essential for understanding scaffold permeability, fluid flow, and cell migration within the scaffold matrix, aiding in the design and optimization of tissue-engineered constructs for specific applications.

Confocal laser scanning microscopy (CLSM) is a powerful technique for morphological analysis of scaffold-cell interactions in three dimensions. CLSM allows for visualization of cell distribution, morphology, and behavior within scaffold matrices, providing spatially resolved information on cell-scaffold interactions and tissue organization. By using fluorescently labeled cells or scaffold components, researchers can track cell migration, proliferation, and differentiation over time, elucidating dynamic cellular processes within tissue-engineered constructs. CLSM can be combined with advanced imaging modalities, such as fluorescence lifetime imaging microscopy (FLIM) and second harmonic generation (SHG) microscopy, to provide further insights into cellular metabolism, signaling, and matrix remodeling in 3D culture environments [27, 28]. By utilizing a combination of SEM, micro-CT, CLSM, and other imaging techniques, researchers can comprehensively evaluate scaffold morphology and optimize scaffold design parameters to enhance tissue regeneration outcomes.

#### 6.1.7.2 Mechanical Testing for Scaffold Strength and Elasticity

Mechanical testing is an essential component of scaffold characterization in tissue engineering, providing critical information about scaffold strength, stiffness, and elasticity. These mechanical properties are crucial for ensuring that scaffolds can withstand physiological loads, provide mechanical support to regenerating tissues, and maintain structural integrity over time. One common mechanical testing method is uniaxial tensile testing, which involves applying controlled tensile forces to scaffold specimens and measuring their response in terms of stress and strain [26, 27]. This allows researchers to quantify scaffold mechanical properties such as Young's modulus, ultimate tensile strength, and elongation at break, providing insights into scaffold behavior under tensile loading conditions.

Compressive testing is another widely used mechanical testing method for evaluating scaffold strength and stiffness. In compressive testing, scaffold specimens are subjected to compressive forces along a single axis, and their response in terms of stress and strain is measured. This allows researchers to assess scaffold compressive modulus, compressive strength, and deformation behavior under compressive loading conditions. Compressive testing provides valuable information about scaffold mechanical properties relevant to applications such as bone and cartilage tissue engineering, where scaffolds must withstand compressive forces and maintain structural integrity within load-bearing environments [28]. Scaffold mechanical properties can also be evaluated using techniques such as indentation testing and shear testing, which assess scaffold response to localized forces and shear stresses, respectively. Indentation testing involves applying controlled forces to scaffold surfaces using a sharp probe and measuring the resulting indentation depth, allowing researchers to quantify scaffold hardness, stiffness, and viscoelastic properties. Shear testing, on the other hand, involves applying parallel forces to scaffold specimens in opposite directions and measuring their response in terms of shear stress and strain [26]. These mechanical testing techniques provide additional insights into scaffold mechanical behavior under specific loading conditions, complementing tensile and compressive testing methods.

Mechanical testing plays a crucial role in scaffold characterization, providing essential information about scaffold strength, stiffness, and elasticity in tissue engineering applications. By assessing scaffold mechanical properties using techniques such as tensile testing, compressive testing, indentation testing, and shear testing, researchers can optimize scaffold designs, tailor material compositions, and enhance scaffold performance for specific tissue regeneration applications.

#### 6.1.7.3 Biocompatibility Assessment and Cytotoxicity Studies

Biocompatibility assessment and cytotoxicity studies are integral parts of scaffold evaluation in tissue engineering, ensuring that scaffolds are compatible with biological systems and do not elicit adverse effects on cells or tissues. Biocompatibility assessment involves evaluating the interactions between scaffolds and living organisms, including cells, tissues, and the immune system. Cytotoxicity studies specifically focus on assessing the toxic effects of scaffold materials or degradation products on cells in vitro. These studies are critical for predicting the safety and efficacy of tissue-engineered constructs before they are used in clinical applications. One commonly used method for biocompatibility assessment and cytotoxicity studies is the ISO 10993 standard, which provides guidelines for evaluating the biological safety of medical devices, including tissue-engineered scaffolds [7–10]. This standard outlines various tests, including cell viability assays, cell proliferation assays, and cell morphology analysis, to assess the effects of scaffold materials on cell behavior and health. These assays measure parameters such as cell viability, metabolic activity, and morphological changes in response to scaffold exposure, providing quantitative and qualitative data on scaffold cytotoxicity and biocompatibility.

In vitro cell culture assays are widely used for biocompatibility assessment and cytotoxicity studies, allowing researchers to culture cells on scaffold surfaces and evaluate their response over time. Cell culture assays enable researchers to assess parameters such as cell adhesion, proliferation, differentiation, and viability in the presence of scaffold materials or degradation products. By utilizing a variety of cell types, including primary cells, stem cells, and cell lines, researchers can evaluate the effects of scaffold materials on different cell populations and tissue types, providing comprehensive insights into scaffold biocompatibility and cytotoxicity.

Biocompatibility assessment and cytotoxicity studies often involve in vivo animal studies to evaluate scaffold performance and safety in living organisms. Animal studies allow researchers to assess scaffold integration, tissue regeneration [29], and immune responses in a physiological environment, providing valuable information on scaffold biocompatibility and efficacy.

#### 6.1.7.4 Degradation Kinetics and Long-Term Stability Evaluation

Degradation kinetics and long-term stability evaluation are critical aspects of scaffold characterization in tissue engineering, providing insights into scaffold degradation behavior and its impact on tissue regeneration outcomes. Scaffold degradation kinetics refer to the rate at which scaffold materials undergo degradation or breakdown over time, while long-term stability evaluation assesses the ability of scaffolds to maintain their structural integrity and mechanical properties during prolonged implantation in vivo. Understanding scaffold degradation kinetics and long-term stability is essential for predicting scaffold performance, optimizing scaffold design parameters, and ensuring the success of tissue engineering applications [30]. One common method for evaluating scaffold degradation kinetics is through in vitro degradation studies, where scaffold specimens are immersed in simulated physiological environments and monitored for changes in mass, morphology, and mechanical properties over time. These studies allow researchers to assess scaffold degradation rates, degradation mechanisms, and degradation products, providing insights into scaffold biocompatibility, tissue response, and degradation behavior. By adjusting scaffold composition, morphology, and processing parameters, researchers can modulate scaffold degradation kinetics to match tissue regeneration rates and optimize scaffold performance for specific applications [16, 17, 29]. Long-term stability evaluation involves assessing scaffold performance and functionality over extended periods, typically through in vivo animal studies or long-term in vitro culture experiments. In vivo animal studies allow researchers to evaluate scaffold integration, tissue regeneration, and degradation processes in a physiological environment over weeks, months, or even years. These studies provide valuable information on scaffold biocompatibility, host response, and long-term tissue outcomes, guiding the design of scaffolds with optimal degradation behavior and long-term stability. Long-term in vitro culture experiments enable researchers to assess scaffold performance and stability under controlled conditions over extended periods, providing insights into scaffold degradation kinetics and its impact on cellular behavior and tissue regeneration.

Degradation kinetics and long-term stability evaluation are essential for predicting scaffold performance, optimizing scaffold design parameters, and ensuring the success of tissue engineering applications [14, 30]. By comprehensively evaluating scaffold degradation behavior and long-term stability, researchers can develop scaffolds with tailored degradation profiles and optimal performance for specific tissue regeneration applications. These insights contribute to the advancement of tissue engineering technologies and the development of functional and biocompatible tissue-engineered constructs for clinical use.

### *6.1.8 Applications of Scaffold Design in Tissue Engineering*

Scaffold design plays a pivotal role in tissue engineering, offering versatile solutions for regenerating damaged or diseased tissues and organs. The applications of scaffold design in tissue engineering are diverse and encompass various biomedical fields, including orthopedics, cardiovascular, neural, and skin tissue regeneration (Fig. 6.4). In orthopedics, scaffolds are used for bone and cartilage repair and regeneration, providing mechanical support and guiding tissue growth in load-bearing environments [31]. In cardiovascular tissue engineering, scaffolds are utilized for vascular grafts and cardiac patches, aiming to repair damaged blood vessels and myocardial tissues and restore cardiac function. Moreover, in neural tissue engineering, scaffolds serve as substrates for nerve regeneration and neural tissue repair, facilitating axonal growth and functional recovery after spinal cord injury or peripheral nerve damage. In skin tissue engineering, scaffolds are employed for wound healing and skin regeneration, providing temporary coverage and promoting tissue remodeling in severe burns or chronic wounds.

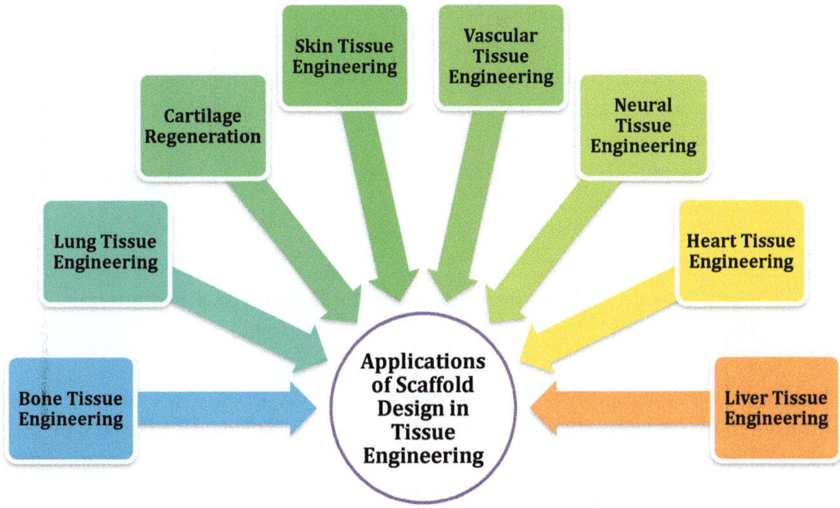


**Fig. 6.4** Various applications of scaffold designing in tissue engineering

#### 6.1.8.1 Bone Regeneration: Scaffold Strategies for Osteogenesis

Bone regeneration is a critical area in tissue engineering, with scaffold strategies playing a central role in promoting osteogenesis and restoring bone function. These strategies aim to address bone defects resulting from trauma, disease, or congenital abnormalities by providing structural support and creating a conducive environment for bone formation. Scaffold design for bone regeneration involves selecting appropriate biomaterials, such as natural polymers, synthetic polymers, ceramics, and composite materials, and optimizing scaffold architecture, porosity, and mechanical properties to mimic the native bone extracellular matrix [30, 31]. Scaffold strategies may incorporate bioactive molecules, growth factors, or cells to enhance osteogenic differentiation, angiogenesis, and tissue integration. By leveraging scaffold strategies for osteogenesis, tissue engineers aim to develop innovative solutions for bone repair and regeneration, addressing the clinical challenges associated with bone defects and improving patient outcomes.

#### 6.1.8.2 Cartilage Repair: Engineering Scaffolds for Chondrogenesis

Cartilage repair is a significant focus within tissue engineering, with scaffold design playing a crucial role in promoting chondrogenesis and restoring cartilage function. The development of effective scaffold strategies for cartilage repair aims to address the limitations of current treatment options for cartilage defects, such as osteoarthritis or traumatic injuries. These scaffold strategies involve the selection of biomaterials with suitable mechanical properties and biocompatibility to support chondrogenic differentiation and extracellular matrix deposition. Scaffold design may incorporate techniques to enhance cell adhesion, proliferation, and differentiation, as well as to mimic the biochemical and biomechanical cues present in native cartilage tissue [32]. By engineering scaffolds for chondrogenesis, tissue engineers aim to provide regenerative solutions that can repair damaged cartilage, alleviate pain, and restore joint function, ultimately improving the quality of life for patients suffering from cartilage-related conditions.

#### 6.1.8.3 Skin Tissue Engineering: Scaffolds for Dermal and Epidermal Regeneration

Skin tissue engineering holds immense promise for the treatment of severe burns, chronic wounds, and various dermatological conditions, with scaffold design playing a central role in facilitating dermal and epidermal regeneration. The development of scaffolds for dermal-epidermal regeneration aims to mimic the complex structure and function of native skin, providing mechanical support, promoting cell adhesion, and facilitating tissue integration. These scaffolds are engineered to support the growth and differentiation of dermal fibroblasts and epidermal keratinocytes, leading to the formation of organized skin tissue with proper barrier function and wound healing capacity. By leveraging scaffold-based approaches for dermal-epidermal regeneration, tissue engineers strive to develop advanced therapies that can improve wound healing outcomes, reduce scarring, and restore skin function for patients with extensive skin injuries or chronic wounds.

#### 6.1.8.4 Organ Transplantation: Scaffold Approaches for Organoid Development

Organ transplantation remains a critical medical intervention for patients with end-stage organ failure, but the shortage of donor organs presents significant challenges. Scaffold approaches for organoid development offer a promising alternative by enabling the engineering of functional tissue constructs that mimic the structure and function of native organs. These scaffolds serve as 3D templates for the growth and differentiation of stem cells or organ progenitor cells, facilitating the formation of organoids with complex architecture and functionality. By providing structural support, biochemical cues, and spatial organization, scaffolds guide the self-organization and maturation of cells into organoid structures resembling specific organs, such as the liver, kidney, pancreas, and heart [33]. The development of scaffold approaches for organoid development represents a paradigm shift in organ transplantation, offering opportunities to overcome the limitations of traditional organ transplantation, such as donor scarcity, immune rejection, and long-term organ dysfunction [34]. These scaffold-based strategies hold great promise for generating patient-specific organoids for transplantation, disease modeling, drug screening, and regenerative medicine applications. By leveraging scaffold approaches for organoid development, researchers aim to revolutionize the field of organ transplantation, providing personalized therapies that can improve patient outcomes and alleviate the burden of organ shortage on healthcare systems worldwide.

### 6.1.9 Challenges in Scaffold Design

Scaffold design in tissue engineering faces several challenges that must be addressed to achieve successful clinical translation. One significant challenge is scalability, as the fabrication of large-scale scaffolds with consistent properties remains technically challenging. Scaling up scaffold production while maintaining structural integrity, porosity, and bioactivity requires innovative manufacturing techniques and quality control measures to ensure reproducibility and uniformity across scaffolds. The scalability of scaffold fabrication is crucial for meeting the clinical demand for tissue-engineered constructs in regenerative medicine applications. Vascularization presents another critical challenge in scaffold design, as the lack of functional blood vessels within engineered tissues limits their viability, growth, and long-term functionality. Efficient vascularization strategies are essential for ensuring adequate nutrient and oxygen supply to cells within the scaffold matrix, promoting tissue integration and survival post implantation [35]. Various approaches, including biomimetic scaffold designs, co-culture systems, and angiogenic growth factor delivery, are being explored to promote vascular ingrowth and neovascularization within tissue-engineered constructs, but achieving robust and stable vascular networks remains a significant hurdle in the field.

Achieving proper integration of engineered tissues with the host tissue remains a challenge in scaffold design. Long-term functionality and durability of tissue-engineered constructs depend on their ability to seamlessly integrate with surrounding tissues, establish functional connections, and maintain physiological homeostasis [35]. Poor integration can lead to immune rejection, fibrosis, or graft failure, underscoring the importance of developing scaffolds with appropriate mechanical, biochemical, and structural properties to facilitate tissue integration and remodeling. Strategies such as surface modifications, cell seeding techniques, and scaffold degradation kinetics optimization are being investigated to improve tissue integration and long-term functionality of engineered tissues after implantation. In addition to integration, ensuring the long-term functionality of tissue-engineered constructs post integration poses a significant challenge in scaffold design [34, 36]. Engineered tissues must maintain their structural integrity, mechanical properties, and biological functionality over extended periods to provide durable therapeutic outcomes. Scaffold degradation kinetics, host tissue remodeling, and immune responses can all influence the long-term performance of tissue-engineered constructs, highlighting the need for comprehensive preclinical evaluation and long-term monitoring in translational studies [36]. Addressing these challenges in scaffold design is essential for advancing the field of tissue engineering and realizing the full potential of regenerative medicine for clinical applications.

**Exercise 3.**

**Choose The Correct Option**

1. Which of the following is a key principle of scaffold design in tissue engineering?
   a) Maximizing scaffold rigidity
   b) Minimizing scaffold porosity
   c) Mimicking the natural extracellular matrix
   d) Avoiding incorporation of bioactive molecules

2. What is a crucial consideration in scaffold design for tissue engineering?
   a) Increasing scaffold degradation rate
   b) Reducing scaffold biocompatibility
   c) Mimicking the mechanical properties of metal
   d) Tailoring scaffold architecture to guide tissue regeneration

3. Why is scaffold design important in tissue engineering?
   a) To inhibit cell growth
   b) To hinder tissue regeneration
   c) To provide structural support and guide cell growth
   d) To minimize biocompatibility

4. Which characteristic is essential for scaffolds in tissue engineering?
   a) Low porosity
   b) Slow degradation rate
   c) Lack of mechanical support
   d) Biocompatibility and porosity

5. Which characteristic is crucial for scaffold materials in tissue engineering?
   a) High toxicity
   b) Low biocompatibility
   c) High mechanical strength
   d) Biodegradability and biocompatibility

6. What is a key consideration in scaffold structure for tissue engineering?
   a) Uniform pore size
   b) Lack of interconnected pores
   c) Non-porous surface
   d) Limited surface area

7. What is a primary goal of surface modification of scaffolds in tissue engineering?
   a) Decreasing cell adhesion
   b) Increasing scaffold degradation rate
   c) Enhancing cell attachment and proliferation
   d) Minimizing scaffold porosity

8. Which technique is commonly used for surface modification of scaffolds?
   a) Increasing scaffold rigidity
   b) Decreasing scaffold biocompatibility
   c) Coating with bioactive molecules
   d) Avoiding scaffold porosity

9. How does scaffold design contribute to tissue engineering applications?
   a) By hindering cell growth
   b) By minimizing structural support
   c) By providing guidance for cell growth and tissue regeneration
   d) By avoiding biocompatibility requirements

10. What is a primary benefit of utilizing scaffold design in tissue engineering?
   a) Slowing down tissue regeneration
   b) Decreasing cell proliferation
   c) Enhancing tissue regeneration outcomes
   d) Neglecting scaffold architecture

## 6.2 Summary

This chapter on scaffold design in tissue engineering explores various aspects crucial for the successful development and implementation of tissue-engineered constructs. It begins by discussing the basic principles of tissue engineering, including scaffold materials, structures, and regenerative medicine approaches. The significance of scaffold design in tissue engineering is emphasized, highlighting its role in promoting tissue regeneration and addressing clinical needs in various biomedical fields. Key topics covered include scaffold materials and structures, incorporating natural polymers, synthetic polymers, and composite materials to tailor scaffold properties for specific applications [3, 4, 6]. The importance of scaffold biocompatibility, biofunctionality, and cellular interactions is also examined, along with advanced fabrication techniques such as electrospinning, 3D printing, and freeze drying for scaffold production. The chapter delves into challenges in scaffold design, including scalability, vascularization, integration of engineered tissues, and long-term functionality post integration. Strategies to overcome these challenges, such as innovative manufacturing techniques, vascularization strategies, and optimization of scaffold properties, are discussed to advance the field and realize the potential of tissue engineering for clinical translation [6, 8, 9].

Scaffold design plays a pivotal role in tissue engineering, offering tailored solutions for tissue repair, regeneration, and functional restoration in various clinical settings. By addressing challenges and leveraging innovative approaches, researchers aim to develop advanced therapies that improve patient outcomes and alleviate the burden of disease [13, 14]. Continued research and collaboration in scaffold design are essential for driving progress in tissue engineering and realizing the promise of regenerative medicine.

## Points to Remember

- **Some scaffolds are designed to change properties over time, such as degradation rate or mechanical strength, to accommodate the evolving needs of tissue regeneration.**
- **Scaffolds can be engineered to release growth factors in a controlled manner, promoting cell proliferation, differentiation, and tissue regeneration.**
- **Incorporating bioactive materials into scaffolds, such as hydrogels loaded with signaling molecules or extracellular matrix components, can enhance cellular response and tissue integration.**
- **Modern scaffold designs often aim to fulfill multiple functions, such as providing structural support, promoting vascularization, and delivering therapeutic agents simultaneously.**
- **Computational tools are increasingly used to optimize scaffold design, taking into account factors like mechanical properties, diffusion rates, and cell behavior to predict and enhance tissue regeneration outcomes.**
- **Advances in 3D printing technology allow for precise control over scaffold architecture and composition, enabling the fabrication of complex structures tailored to specific tissue engineering applications.**
- **Some scaffolds are designed to incorporate imaging agents or sensors, enabling non-invasive monitoring of tissue growth and integration in real-time.**
- **Decellularization techniques remove cellular components from natural tissues while preserving the extracellular matrix, providing a biomimetic scaffold with native tissue architecture and bioactive cues.**
- **Recent research explores the incorporation of immunomodulatory factors into scaffolds to regulate the immune response and promote tissue regeneration while minimizing inflammation and foreign body reactions.**

## References

1. Muschler GF, Nakamoto C, Griffith LG (2004) Engineering principles of clinical cell-based tissue engineering. J Bone Joint Surg Am 86(7):1641–1668. https://doi.org/10.2106/00004623-200407000-00029
2. Caplan AI (2003) Embryonic development and the principles of tissue engineering. Novartis Found Symp 249:17–241
3. de Isla N, Huseltein C, Jessel N, Pinzano A, Decot V, Magdalou J, Bensoussan D, Stoltz JF (2010) Introduction to tissue engineering and application for cartilage engineering. Bio-Med Mater Eng 20(3):127–133. https://doi.org/10.3233/BME-2010-0624
4. O'Brien FJ (2011) Biomaterials & scaffolds for tissue engineering. Mater Today 14(3):88–95. https://doi.org/10.1016/s1369-7021(11)70058-x
5. Okamoto M (2019) The role of scaffolds in tissue engineering. In: Elsevier eBooks (pp. 23–49). https://doi.org/10.1016/b978-0-08-102563-5.00002-2
6. Pattnaik A, Sanket AS, Pradhan S, Sahoo R, Das S, Pany S, Douglas TE, Dandela R, Liu Q, Rajadas J, Pati S, De Smedt SC, Braeckmans K, Samal SK (2023) Designing of gradient scaffolds and their applications in tissue regeneration. Biomaterials 296:122078. https://doi.org/10.1016/j.biomaterials.2023.122078

7. Krishani M, Shin WY, Suhaimi H, Sambudi NS (2023) Development of scaffolds from bio-based natural materials for tissue regeneration applications: a review. Gels (Basel, Switzerland) 9(2):100. https://doi.org/10.3390/gels9020100
8. Jang J, Min K, Kim C, Shin J, Lee J, Yi S (2023) Review: scaffold characteristics, fabrication methods, and biomaterials for the bone tissue engineering. Int J Precis Eng Manuf/Int J Korean Soc Precis Eng 24(3):511–529. https://doi.org/10.1007/s12541-022-00755-7
9. Handrea-Dragan IM, Botiz I, Tatar A, Boca S (2022) Patterning at the micro/nano-scale: polymeric scaffolds for medical diagnostic and cell-surface interaction applications. Colloids Surf B Biointerfaces 218:112730. https://doi.org/10.1016/j.colsurfb.2022.112730. PMID: 36287884
10. Wang F, Cai X, Shen Y, Meng L (2023) Cell–scaffold interactions in tissue engineering for oral and craniofacial reconstruction. Bioact Mater 23:16–44. https://doi.org/10.1016/j.bioactmat.2022.10.029
11. Carmagnola I, Ranzato E, Chiono V (2018) Scaffold functionalization to support a tissue biocompatibility. In: Elsevier eBooks (pp. 255–277). https://doi.org/10.1016/b978-0-08-100979-6.00011-2
12. Sharma P, Kumar P, Sharma R, Bhatt VD, Dhot PS (2019) Tissue engineering; current status & futuristic scope. J Med Life 12(3):226–229. https://doi.org/10.26122/jml-2019-0032
13. Brovold M, Almeida JI, Pla-Palacín I, Sainz-Arnal P, Sánchez-Romero N, Rivas JJ, Almeida H, Dachary PR, Serrano-Aulló T, Soker S, Baptista PM (2018) Naturally-derived biomaterials for tissue engineering applications. Advances in experimental medicine and biology 1077:421–449. https://doi.org/10.1007/978-981-13-0947-2_23
14. Chan BP, Leong KW (2008) Scaffolding in tissue engineering: general approaches and tissue-specific considerations. Eur Spine J: Off Publ Eur Spine Soc Eur Spinal Deform Soc Euro Sect Cerv Spine Res Soc 17(Suppl 4):467–479. https://doi.org/10.1007/s00686-008-0746-3
15. Yang S, Leong KF, Du Z, Chua CK (2001) The design of scaffolds for use in tissue engineering. Part I. Traditional factors. Tissue Eng 7(6):679–689. https://doi.org/10.1089/107632701763337646
16. Wei Q, Deng N, Guo J, Deng J (2018) Synthetic polymers for biomedical applications. Int J Biomater 2018:1–2. https://doi.org/10.1155/2018/7158621
17. Chen CH, Woo EM (2018) Polymer science and technology. Chem Phys 16(7):2496–2606
18. Lannutti J, Reneker D, Ma T, Tomasko D, Farson D (2007) Electrospinning for tissue engineering scaffolds. Mater Sci Eng C Biomimetic Mater Sens Syst 27(3):504–509. https://doi.org/10.1016/j.msec.2006.05.019
19. Ananth KP, Jayram ND (2024) A comprehensive review of 3D printing techniques for biomaterial-based scaffold fabrication in bone tissue engineering. Ann 3D Printed Med 13:100141. https://doi.org/10.1016/j.stlm.2023.100141
20. Wang P, Sun Y, Shi X, Shen H, Ning H, Liu H (2021) 3D printing of tissue engineering scaffolds: a focus on vascular regeneration. Bio-des Manuf 4(2):344–378. https://doi.org/10.1007/s42242-020-00109-0
21. Do AV, Khorsand B, Geary SM, Salem AK (2015) 3D printing of scaffolds for tissue regeneration applications. Adv Healthc Mater 4(12):1742–1762. https://doi.org/10.1002/adhm.201500168
22. Fereshteh Z (2018) Freeze-drying technologies for 3D scaffold engineering. In: Elsevier eBooks (pp. 151–174). https://doi.org/10.1016/b978-0-08-100979-6.00007-0
23. Rathore R, Suhag D, Wan F, Verma R, Thakur A, Thakur P (2023) Functionalization of biomaterials. In: Integrated nanomaterials and their applications (pp. 223–243). https://doi.org/10.1007/978-981-99-6105-4_11
24. De Witte T, Fratila-Apachitei LE, Zadpoor AA, Peppas NA (2018) Bone tissue engineering via growth factor delivery: from scaffolds to complex matrices. Regener Biomater 5(4):197–211. https://doi.org/10.1093/rb/rby013
25. Hasirci V, Kenar H (2006) Novel surface patterning approaches for tissue engineering and their effect on cell behavior. Nanomed (London, England) 1(1):73–90. https://doi.org/10.2217/17435889.1.1.73

26. Limongi T, Tirinato L, Pagliari F, Giugni A, Allione M, Perozziello G, Candeloro P, Di Fabrizio E (2016) Fabrication and applications of micro/nanostructured devices for tissue engineering. Nano-micro Lett 9(1). https://doi.org/10.1007/s40820-016-0103-7
27. Will J, Detsch R, Boccaccini AR (2013) Structural and biological characterization of scaffolds. In: Elsevier eBooks (pp. 299–310). https://doi.org/10.1016/b978-0-12-415800-9.00008-5
28. Haugen H, Bertoldi S (2017) Characterization of morphology—3D and porous structure. In: Elsevier eBooks (pp. 21–53). https://doi.org/10.1016/b978-0-08-100737-2.00002-9
29. Rathore R, Suhag D, Wan F, Thakur A, Thakur P (2023) Toxicology and toxicity studies of nano-biomaterials. In: Suhag D, Thakur A, Thakur P (eds) Integrated nanomaterials and their applications. Springer, Singapore. https://doi.org/10.1007/978-981-99-6105-4_20
30. Seguchi M, Seguchi M, Seguchi M, Seguchi M, Seguchi M, Seguchi M, Seguchi M (2023) Preclinical evaluation of the degradation kinetics of third-generation resorbable magnesium scaffolds. EuroIntervention 19(2):e167–e175. https://doi.org/10.4244/eij-d-22-00718
31. Abbasian M, Massoumi B, Mohammad-Rezaei R, Samadian H, Jaymand M (2019) Scaffolding polymeric biomaterials: are naturally occurring biological macromolecules more appropriate for tissue engineering? Int J Biol Macromol 134:673–694. https://doi.org/10.1016/j.ijbiomac.2019.04.197
32. Szustak M, Gendaszewska-Darmach E (2021) Nanocellulose-based scaffolds for chondrogenic differentiation and expansion. Front Bioeng Biotechnol 9:736213. https://doi.org/10.3389/fbioe.2021.736213
33. De Barros NR, Wang C, Maity S, Peirsman A, Nasiri R, Herland A, Ermis M, Kawakita S, Carvalho BG, Kouchehbaghi NH, Herculano RD, Tirpáková Z, Dabiri SMH, Tanaka JL, Falcone N, Choroomi A, Chen R, Huang S, Zisblatt E, … Khademhosseini A (2023) Engineered organoids for biomedical applications. Adv Drug Deliv Rev 203:115142. https://doi.org/10.1016/j.addr.2023.115142
34. Unagolla JM, Jayasuriya AC (2022) Recent advances in organoid engineering: a comprehensive review. Appl Mater Today 29:101582. https://doi.org/10.1016/j.apmt.2022.101582
35. Koyyada A, Orsu P (2020) Recent advancements and associated challenges of scaffold fabrication techniques in tissue engineering applications. Regener Eng Transl Med 7(2):147–159. https://doi.org/10.1007/s40883-020-00166-y
36. Echeverria Molina MI, Malollari KG, Komvopoulos K (2021) Design challenges in polymeric scaffolds for tissue engineering. Front Bioeng Biotechnol 9:617141. https://doi.org/10.3389/fbioe.2021.617141

# Chapter 7
# 3D Printing in Biomedicine

## 7.1 Introduction

3D bioprinting stands at the forefront of revolutionary technological advancements in the field of biomedicine, offering unprecedented opportunities for tissue engineering, regenerative medicine, and drug discovery. This cutting-edge technology enables the precise layer-by-layer deposition of biomaterials and living cells to fabricate intricate 3D structures mimicking native tissues and organs. Unlike traditional manufacturing methods, 3D bioprinting allows for the customization of constructs tailored to individual patient needs, promising personalized healthcare solutions and improved treatment outcomes. The genesis of 3D bioprinting can be traced back to the convergence of various disciplines, including engineering, biology, and materials science [1]. Over the years, significant strides have been made in refining bioprinting techniques, enhancing biomaterial formulations, and expanding the repertoire of printable cell types. This interdisciplinary synergy has propelled the field forward, unlocking new avenues for biomedical research and clinical applications. With ongoing advancements in printer technology and bioink development, the potential of 3D bioprinting to revolutionize healthcare continues to grow exponentially [2]. The current landscape of 3D bioprinting is marked by a diverse array of applications, ranging from the fabrication of tissue constructs for transplantation to the creation of organ-on-a-chip models for drug testing. Researchers and clinicians alike are harnessing the power of bioprinting to address critical challenges in regenerative medicine, such as organ shortage and tissue degeneration. Moreover, the versatility of bioprinting platforms enables the recreation of complex physiological environments, facilitating the study of disease mechanisms and the development of targeted therapeutics. Looking ahead, the future prospects of 3D bioprinting are boundless, with anticipated advancements in bio fabrication techniques, biomaterial design, and bioprinter capabilities. As technology continues to mature, it holds the potential to transform the landscape of healthcare delivery, ushering in an era of personalized medicine and regenerative therapies. However, along with its immense promise come ethical considerations,



D. Suhag et al., *Handbook of Biomaterials for Medical Applications, Volume 1*,
Biomedical Materials for Multi-functional Applications,
https://doi.org/10.1007/978-981-97-4818-1_7

regulatory challenges, and the need for continued collaboration across disciplines. By addressing these issues proactively, the field of 3D bioprinting is poised to make profound contributions to human health and well-being in the years to come.

## 7.1.1 *Historical Development*

The historical development of 3D bioprinting can be traced back to the early 2000s when researchers began exploring the potential of additive manufacturing techniques for fabricating biological constructs. While the concept of 3D printing itself dates back to the 1980s, it wasn't until the intersection of tissue engineering and rapid prototyping that the idea of bioprinting took shape. In 2003, Thomas Boland and his team at Clemson University demonstrated the feasibility of printing living cells using an inkjet-based approach, marking a pivotal moment in the field. Subsequent years saw rapid progress in bioprinting technology, with advancements in printer design, bioink formulation, and cell culture techniques [1, 2]. By the mid-2010s, bioprinting had emerged as a distinct discipline within regenerative medicine, attracting attention from researchers, clinicians, and industry stakeholders alike. Today, the field continues to evolve at a rapid pace, driven by ongoing innovations in biomaterials, bioprinter capabilities, and interdisciplinary collaboration.

## 7.1.2 *Fundamentals of 3D Bioprinting*

Understanding the fundamentals of 3D bioprinting is essential for appreciating the intricate interplay between technology, biology, and materials science in this groundbreaking field (Fig. 7.1). At its core, 3D bioprinting involves the precise deposition of living cells and biomaterials layer by layer to construct complex 3D structures that mimic native tissues and organs. This innovative approach to tissue engineering and regenerative medicine offers unprecedented opportunities for personalized healthcare solutions and novel therapeutic interventions [3]. By mastering the principles and techniques of bioprinting, researchers and clinicians can unlock the full potential of this transformative technology to address pressing challenges in healthcare and biomedical research.

### 7.1.2.1 Principles and Techniques of Bioprinting

Bioprinting relies on a combination of engineering principles and biological knowledge to fabricate intricate 3D structures composed of living cells and biomaterials. At its core, bioprinting follows three fundamental principles: computer-aided design (CAD), layer-by-layer deposition, and biomaterial selection. CAD software allows researchers to design precise digital models of the desired tissue or organ, specifying

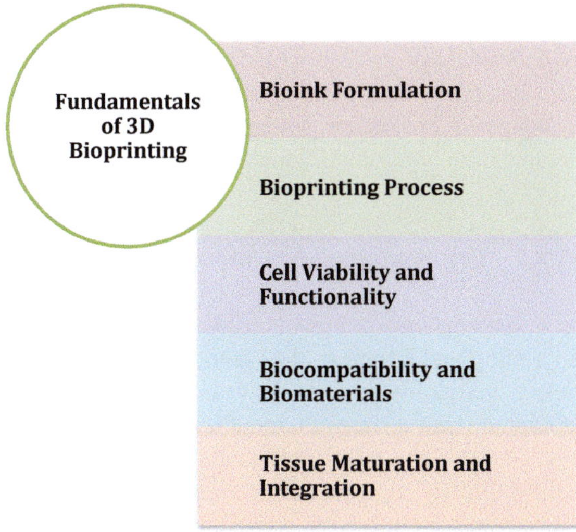


**Fig. 7.1** Fundamentals of 3D bioprinting

dimensions, geometries, and cellular arrangements [2, 3]. These digital blueprints serve as the basis for guiding the bioprinter's movements during the printing process. Layer-by-layer deposition involves the sequential deposition of bioinks, which are formulations containing living cells and biomaterials, onto a substrate or scaffold. This precise layering enables the creation of complex tissue architectures, ensuring spatial fidelity and cellular viability. Various bioprinting techniques have been developed to achieve different levels of resolution, speed, and scalability. Extrusion-based bioprinting, for example, utilizes pneumatic or mechanical forces to extrude bioinks through fine nozzles, allowing for high-throughput fabrication of large-scale tissue constructs [4]. Inkjet-based bioprinting, on the other hand, employs droplet ejection mechanisms to deposit tiny droplets of bioink onto a substrate, enabling precise patterning of cells and biomaterials with micron-scale resolution. Laser-assisted bioprinting utilizes focused laser beams to precisely pattern biomaterials and cells onto a substrate, offering excellent control over spatial distribution and cell viability. Each bioprinting technique has its advantages and limitations, making them suitable for different applications and research needs. Biomaterial selection plays a crucial role in bioprinting, influencing the mechanical properties, biocompatibility, and functionality of the resulting tissue constructs. Researchers must carefully choose biomaterials that mimic the native extracellular matrix of the target tissue, providing structural support and cues for cell adhesion, proliferation, and differentiation [5]. Commonly used biomaterials in bioprinting include hydrogels, polymers, and decellularized extracellular matrices, each offering unique advantages and considerations. Bioink formulation involves optimizing the rheological properties, crosslinking mechanisms, and cell encapsulation techniques to ensure proper printing fidelity and cellular viability throughout the bioprinting process.

The principles and techniques of bioprinting represent a convergence of engineering, biology, and materials science, offering unprecedented opportunities for tissue engineering, regenerative medicine, and drug discovery. By understanding the underlying principles of CAD design, layer-by-layer deposition, and biomaterial selection, researchers can harness the full potential of bioprinting to address complex challenges in healthcare and biomedical research, paving the way for personalized therapies and organ replacement strategies.

#### 7.1.2.2 Cell Sourcing and Culture Considerations

Cell sourcing and culture considerations play a pivotal role in the success of bioprinting endeavors, as the viability, functionality, and behavior of cells directly impact the quality and efficacy of bio printed constructs [3, 4]. The selection of appropriate cell sources is crucial, with factors such as cell type, phenotype, and source tissue influencing the suitability for bioprinting applications. Primary cells, derived directly from tissues or organs, offer physiological relevance and functionality but may be limited by availability and donor variability. Alternatively, immortalized cell lines provide a consistent and scalable cell source but may exhibit altered phenotypes or genetic stability over time. Stem cells, including embryonic stem cells (ESCs), induced pluripotent stem cells (iPSCs), and adult stem cells, hold immense potential for bioprinting due to their multipotency and self-renewal capabilities, enabling differentiation into various cell types for tissue regeneration and disease modeling [5]. Cell culture considerations encompass a multitude of factors that influence cell behavior and function in vitro, including culture media composition, substrate properties, and environmental conditions. Optimal cell culture conditions are essential for maintaining cell viability, proliferation, and differentiation throughout the bioprinting process. Culture media formulations must provide essential nutrients, growth factors, and signaling molecules to support cell growth and functionality. Furthermore, the choice of substrate materials and surface properties can influence cell adhesion, spreading, and morphology, affecting the overall architecture and mechanical properties of bio printed constructs [6]. Environmental factors such as temperature, pH, and oxygen tension must also be carefully controlled to mimic physiological conditions and promote cellular homeostasis.

The scalability and reproducibility of cell sourcing and culture techniques are critical considerations for translating bio printed constructs from the laboratory to clinical applications. Standardized protocols and quality control measures are essential for ensuring batch-to-batch consistency and reproducibility of bio printed tissues. Advancements in cell expansion techniques, such as bioreactor systems and microfluidic culture platforms, enable the large-scale production of cells for bioprinting applications, addressing the need for high cell densities and uniform distribution within tissue constructs [5, 7]. By optimizing cell sourcing and culture considerations, researchers can enhance the functionality, reliability, and clinical relevance of bio printed tissues, driving advancements in regenerative medicine, drug discovery, and personalized healthcare.

### 7.1.3 Bioprinting Technologies

Bioprinting technologies encompass a diverse array of approaches for fabricating 3D structures composed of living cells and biomaterials, each with its unique capabilities and applications (Fig. 7.2). Extrusion-based bioprinting utilizes pneumatic or mechanical forces to extrude bioinks, containing living cells and biomaterials, through fine nozzles onto a substrate or scaffold [7]. This approach offers high throughput and versatility, allowing for the fabrication of large-scale tissue constructs with precise control over cell distribution and spatial organization. Inkjet-based bioprinting, on the other hand, employs droplet ejection mechanisms to deposit tiny droplets of bioink onto a substrate, enabling precise patterning of cells and biomaterials with micron-scale resolution. Inkjet bioprinting is well-suited for applications requiring high-resolution printing, such as creating complex tissue architectures and microfluidic devices for drug screening and disease modeling.

Laser-assisted bioprinting utilizes focused laser beams to precisely pattern biomaterials and cells onto a substrate, offering excellent control over spatial distribution and cell viability. This approach enables the creation of intricate tissue structures with high resolution and fidelity, making it suitable for applications requiring precise control over cell positioning and alignment, such as engineering vascular networks and nerve regeneration scaffolds [8]. Stereolithography-based bioprinting utilizes photopolymerization techniques to solidify liquid bioinks layer by layer using ultraviolet (UV) light, allowing for the fabrication of complex 3D structures with high resolution and speed. This approach is particularly well-suited for creating biomimetic tissue constructs with precise control over microarchitecture and mechanical properties, making it ideal for applications in tissue engineering and regenerative medicine.

Magnetic bioprinting utilizes magnetic forces to manipulate and pattern cells within a magnetic field, offering a non-contact and biocompatible approach for bioprinting cells and biomaterials. By incorporating magnetic nanoparticles into bioinks, researchers can precisely control cell positioning and alignment within tissue constructs [7, 8], enabling the fabrication of complex structures with controlled cellular organization and functionality. Magnetic bioprinting holds promise for applications requiring dynamic cell manipulation, such as creating functional cardiac tissue patches and neural circuits for regenerative medicine and disease modelling.

### 7.1.4 Additive Manufacturing Techniques

Additive manufacturing techniques, also known as 3D printing, have revolutionized the manufacturing industry by enabling the fabrication of complex 3D objects layer by layer from digital designs. Unlike traditional subtractive manufacturing methods, which involve cutting away material from a solid block, additive manufacturing builds up objects from scratch, offering unparalleled design freedom

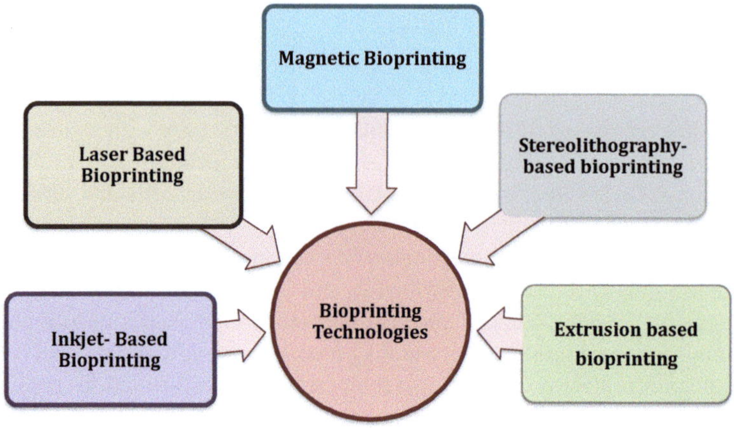


**Fig. 7.2** Different types of bioprinting technologies

and customization possibilities. These techniques encompass a diverse range of processes, including selective laser sintering (SLS), fused deposition modeling (FDM), digital light processing (DLP), electron beam melting (EBM), and direct metal laser sintering (DMLS), each with its unique advantages and applications (Fig. 7.3). Additive manufacturing techniques have found widespread use across various industries, including aerospace, automotive, healthcare, and consumer goods, driving innovation, efficiency, and cost-effectiveness in manufacturing processes [9].

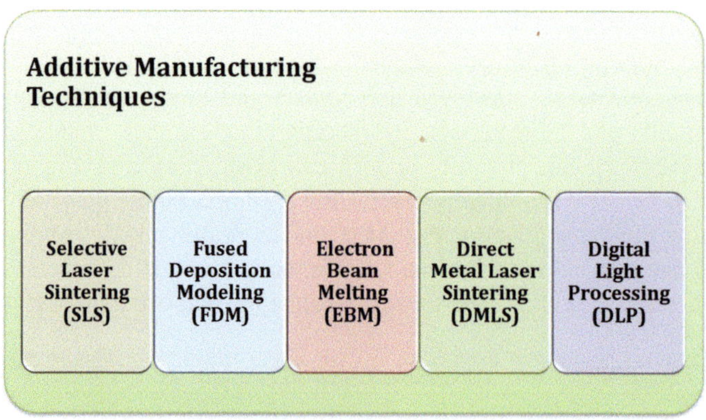


**Fig. 7.3** Additive manufacturing techniques

#### 7.1.4.1 Selective Laser Sintering (SLS)

Selective laser sintering (SLS) is an additive manufacturing technique that utilizes a high-power laser to selectively fuse powdered materials, typically polymers or metals, layer by layer to create 3D objects. The process begins with the deposition of a thin layer of powdered material onto a build platform [10]. A laser then scans the surface of the powder bed, selectively sintering the particles together based on the digital design of the object. Once a layer is completed, the build platform is lowered, and a new layer of powder is deposited on top. This layer-by-layer process is repeated until the entire object is formed. One of the key advantages of SLS is its ability to produce complex geometries with intricate internal structures without the need for support structures, as the surrounding powder acts as a natural support during printing. SLS offers several benefits over traditional manufacturing methods, including increased design freedom, rapid prototyping capabilities, and reduced material waste. Since SLS does not require molds or tooling, designers have greater flexibility to create complex shapes and geometries that would be difficult or impossible to achieve using conventional manufacturing techniques. Additionally, SLS allows for the rapid iteration of designs, enabling engineers and designers to quickly test and refine prototypes before moving into full-scale production [11]. Furthermore, because SLS is an additive process, it generates minimal material waste compared to subtractive manufacturing methods, making it a more sustainable option for manufacturing.

In addition to its applications in prototyping and product development, SLS is widely used in the production of end-use parts across various industries, including aerospace, automotive, and medical devices. In aerospace, SLS is used to fabricate lightweight and durable components for aircraft and spacecraft, such as air ducts, brackets, and engine components. In the automotive industry, SLS is employed to manufacture custom tooling, jigs, and fixtures for assembly lines, as well as functional prototypes for testing and validation. In the medical field, SLS is utilized to produce patient-specific implants [10, 11], prosthetics, and surgical guides, offering customized solutions for individual patients while reducing lead times and costs associated with traditional manufacturing methods.

With its ability to produce complex geometries with minimal material waste, SLS is well-suited for a wide range of applications across various industries, from aerospace and automotive to healthcare and consumer goods.

#### 7.1.4.2 Fused Deposition Modeling (FDM)

Fused Deposition Modeling (FDM) is an additive manufacturing technique that extrudes thermoplastic filament layer by layer to build 3D objects based on digital designs. In FDM, a filament of thermoplastic material is fed through a heated nozzle, where it is melted and extruded onto a build platform in a predetermined pattern. As each layer is deposited, it fuses with the previous layer, gradually building up the object from the bottom up. One of the distinguishing features of FDM is its simplicity and accessibility, making it one of the most widely used 3D printing technologies for both professional and hobbyist applications. FDM offers several advantages, including cost-effectiveness, speed, and versatility. Compared to other additive manufacturing techniques, FDM printers are relatively affordable and easy to use, making them accessible to a wide range of users [12]. Additionally, FDM is a fast and efficient process, capable of producing functional prototypes and end-use parts with quick turnaround times. FDM is compatible with a wide range of thermoplastic materials, including PLA, ABS, PETG, and nylon, allowing users to choose materials that best suit their specific application requirements. The versatility of FDM makes it suitable for a wide range of applications across various industries, including product design, prototyping, and manufacturing. In product design and development, FDM is used to create prototypes for testing form, fit, and function before moving into production. In manufacturing, FDM is employed to produce custom tooling, jigs, and fixtures for assembly lines, as well as end-use parts for consumer products, automotive components, and medical devices [12]. FDM is utilized in education and research settings to teach students about design principles, engineering concepts, and manufacturing processes, fostering innovation and creativity.

Fused Deposition Modeling (FDM) is a versatile and accessible additive manufacturing technique that offers numerous benefits, including cost-effectiveness, speed, and versatility. With its ability to produce functional prototypes and end-use parts with quick turnaround times, FDM is widely used across various industries for product design, prototyping, and manufacturing applications.

#### 7.1.4.3 Digital Light Processing (DLP)

Digital Light Processing (DLP) is an additive manufacturing technology that utilizes photopolymerization to create 3D objects layer by layer. In DLP, a digital light projector shines light onto a vat of liquid photopolymer resin, causing it to selectively solidify and form the desired shape. The digital light projector projects a 2D image of each layer onto the surface of the resin, curing the resin wherever the light strikes. As each layer is cured, the build platform moves incrementally downward [13], allowing the next layer of resin to be exposed to light and solidified. This layer-by-layer process continues until the entire object is formed. One of the key advantages of DLP is its ability to produce highly detailed and accurate parts with smooth surface finishes. The high-resolution projector used in DLP allows for precise control over the curing process, resulting in intricate features and fine details in the printed parts. DLP is a fast and efficient process, capable of producing parts at relatively high speeds compared to other additive manufacturing technologies. This makes DLP well-suited for applications that require high-resolution, high-precision parts, such as jewelry, dental prosthetics, and microfluidic devices.

DLP offers versatility in terms of material selection, as it can work with a wide range of photopolymer resins with varying properties and characteristics. These resins can be tailored to specific application requirements, such as mechanical strength, flexibility, and biocompatibility. DLP allows for the use of composite resins, which contain additives such as nanoparticles or fibers to enhance material properties, such as strength, conductivity, or thermal resistance [13]. This flexibility in material selection makes DLP suitable for a wide range of applications across various industries, from prototyping and product development to manufacturing and biomedical engineering. Digital Light Processing (DLP) is a powerful additive manufacturing technology that offers high resolution, accuracy, and speed in producing intricate 3D objects. With its ability to create detailed parts with smooth surface finishes and a wide range of material options, DLP is widely used in industries such as jewelry, dentistry, microfluidics, and prototyping.

#### 7.1.4.4 Electron Beam Melting (EBM)

Electron Beam Melting (EBM) is an additive manufacturing technique that utilizes an electron beam to selectively melt and fuse metal powder particles [14], layer by layer, to create 3D objects. In EBM, a high-energy electron beam is directed onto a bed of metal powder, causing the powder particles to melt and solidify to form a solid layer. The build platform is then lowered, and a new layer of powder is deposited on top of the previous layer. This process is repeated until the entire object is formed. One of the key advantages of EBM is its ability to produce fully dense metal parts with excellent mechanical properties, making it particularly well-suited for applications in aerospace, automotive, and medical industries where high-strength and high-performance materials are required. EBM offers several benefits over traditional manufacturing methods, including the ability to produce complex geometries with minimal material waste and excellent material properties. Since EBM is an additive process, it allows for the fabrication of parts with intricate internal structures and features that would be difficult or impossible to achieve using conventional manufacturing techniques. Additionally, EBM produces fully dense metal parts without the need for post-processing, resulting in parts with uniform mechanical properties and superior performance. EBM is a fast and efficient process, capable of producing parts at relatively high speeds compared to other additive manufacturing technologies, making it suitable for both prototyping and production applications.

The versatility of EBM makes it well-suited for a wide range of applications across various industries, including aerospace, automotive, and medical devices [14]. In aerospace, EBM is used to fabricate lightweight and high-strength components for aircraft and spacecraft, such as turbine blades, brackets, and structural components. In the automotive industry, EBM is employed to produce custom tooling, molds, and dies for manufacturing processes, as well as end-use parts for engines, transmissions, and chassis components. In the medical field, EBM is utilized to manufacture patient-specific implants, prosthetics, and surgical instruments, offering customized solutions for individual patients while reducing lead times and costs associated with traditional manufacturing methods [14].

With its versatility and efficiency, EBM is widely used in industries such as aerospace, automotive, and medical devices to produce high-performance parts for a wide range of applications.

#### 7.1.4.5 Direct Metal Laser Sintering (DMLS)

Direct Metal Laser Sintering (DMLS) is an additive manufacturing technique that utilizes a high-powered laser to selectively fuse metal powder particles together, layer by layer, to create fully dense 3D objects. In DMLS, a thin layer of metal powder is spread onto a build platform, and a laser beam is directed onto the powder bed, selectively melting and solidifying the particles according to a digital design. Once a layer is completed, the build platform is lowered, and a new layer of powder is spread on top [15]. This process is repeated until the entire object is formed. DMLS is particularly well-suited for producing complex metal parts with intricate geometries and fine details. One of the key advantages of DMLS is its ability to produce fully dense metal parts with excellent mechanical properties and material integrity. Since DMLS involves the melting and solidifying of metal powder particles, it results in parts that are free from voids and defects, offering uniform mechanical properties throughout the entire structure. DMLS allows for the fabrication of parts with complex geometries and internal features that would be difficult or impossible to achieve using traditional manufacturing methods. This makes DMLS ideal for producing lightweight, high-strength components for aerospace, automotive, and medical applications.

DMLS offers versatility in terms of material selection, as it can work with a wide range of metal powders, including stainless steel, titanium, aluminum, and nickel alloys. These materials can be tailored to specific application requirements, such as strength, corrosion resistance, and thermal conductivity, making DMLS suitable for a wide range of industrial and commercial applications. Furthermore, DMLS enables the production of functional prototypes and end-use parts with quick turnaround times, allowing for rapid iteration and validation of designs before moving into full-scale production [16]. Direct Metal Laser Sintering (DMLS) is a powerful additive manufacturing technology that offers numerous benefits, including the ability to produce fully dense metal parts with excellent mechanical properties, complex geometries, and material versatility. With its ability to fabricate parts with intricate details and high precision, DMLS is widely used in industries such as aerospace, automotive, and medical devices to produce high-performance components for a wide range of applications.

**Exercise 1.**

**Fill Ups**

**(Photopolymer material, Electron Beam Melting, Biomaterials, Binder Jetting, 3D bioprinting)**

**1.** In............., precise layer-by-layer deposition of biological materials such as cells and biomaterials are utilized to create complex three-dimensional structures.

**2.** The selection and preparation of living cells, .....................and other bioactive agents are crucial steps in the 3D bioprinting process, tailored to the specific tissue or organ being printed and the intended application.

**3.** .....................is an additive manufacturing technique that uses a high-energy electron beam to selectively melt and fuse layers of metal powder, enabling the production of complex metal parts with excellent mechanical properties.

**4.** ............ is an additive manufacturing technique where a liquid binding agent is selectively deposited onto a powder bed, binding together particles layer by layer to form a solid object.

5. Material Jetting is an additive manufacturing technique similar to inkjet printing, where droplets of ...................... are selectively deposited and cured layer by layer to create precise, high-resolution parts.

### *7.1.5 3D Printing in Medical Device Fabrication*

3D printing has emerged as a disruptive technology in the fabrication of medical devices, offering numerous advantages in terms of customization, efficiency, and cost-effectiveness. By leveraging additive manufacturing techniques, medical device manufacturers can produce complex geometries with high precision and accuracy, tailored to the specific needs of patients. This level of customization is particularly beneficial in orthopedics, where implants can be designed to match the unique anatomy of individual patients, improving fit and functionality [17]. 3D printing enables rapid prototyping, allowing designers to iterate and refine their designs more quickly than traditional manufacturing methods. This accelerated design process not only reduces time-to-market, but also facilitates innovation, as designers can experiment with novel concepts and materials with relative ease. As a result, the landscape of medical device development is evolving rapidly, with new and innovative solutions emerging to address a wide range of clinical needs. In addition to customization and rapid prototyping, 3D printing offers cost advantages for medical device fabrication. By eliminating the need for specialized tooling and reducing material waste, additive manufacturing can significantly lower production costs, particularly for low-volume or highly complex devices. This cost-effectiveness makes 3D printing an attractive option for medical device companies looking to bring innovative products to market without breaking the bank. 3D printing allows for the production of patient-specific surgical guides and instrumentation, enhancing the precision and accuracy of surgical procedures [17]. By leveraging preoperative imaging data, surgeons can plan and execute procedures with greater confidence, reducing the risk of complications and improving patient outcomes. This level of customization is particularly valuable in complex surgeries, where anatomical variations present challenges for traditional surgical techniques.

Despite these advantages, challenges remain in the widespread adoption of 3D printing for medical device fabrication. Regulatory considerations, including validation and quality assurance requirements, must be addressed to ensure the safety and efficacy of printed devices. Ongoing advancements in materials science and printing technologies are needed to further expand the capabilities of 3D printing in medical device manufacturing [17]. Nonetheless, the potential of this technology to revolutionize the healthcare industry is undeniable, with continued innovation driving progress toward more personalized and effective medical solutions.

### *7.1.6 Applications of 3D Bioprinting in Biomedicine*

The applications of 3D bioprinting in biomedicine are vast and transformative, offering innovative solutions to address complex medical challenges. One prominent application lies in tissue engineering and regenerative medicine, where bio printed constructs hold the potential to regenerate damaged tissues and organs. By precisely depositing living cells and biomaterials layer by layer, bioprinting enables the creation of tissue scaffolds with intricate architectures that mimic the native tissue microenvironment, promoting cell adhesion, proliferation, and differentiation. The bio printed tissues have the potential to revolutionize transplantation procedures by providing patient-specific implants that reduce the risk of rejection and improve long-term outcomes. Bio printed tissues can serve as models for studying disease mechanisms, drug screening, and personalized medicine, offering insights into complex biological processes and facilitating the development of targeted therapeutics.

#### 7.1.6.1 Tissue Engineering and Regenerative Medicine

The application of 3D bioprinting in tissue engineering and regenerative medicine represents a groundbreaking approach to addressing the critical need for functional replacements for damaged or diseased tissues and organs. With its ability to precisely deposit living cells and biomaterials in predefined patterns and structures, 3D bioprinting offers unparalleled opportunities for fabricating complex tissue constructs that closely resemble native tissues. In tissue engineering, bio printed scaffolds serve as templates to support cell attachment [17, 18], proliferation, and differentiation, providing a framework for the regeneration of functional tissues. These scaffolds can be customized to match the size, shape, and mechanical properties of the target tissue, offering patient-specific solutions for tissue repair and replacement. One of the key advantages of 3D bioprinting in tissue engineering is its ability to create vascularized tissues, which are essential for delivering oxygen and nutrients to cells and supporting tissue growth and function. Bioprinting enables the precise patterning of blood vessels within tissue constructs, mimicking the intricate vascular networks found in native tissues. This vascularization facilitates rapid integration with the host tissue and promotes the formation of functional blood vessels, enhancing the viability and functionality of bio printed tissues. Vascularized tissue constructs hold immense promise for applications in regenerative medicine, including the repair and regeneration of complex organs such as the liver, kidney, and heart.

3D bioprinting allows for the incorporation of multiple cell types within tissue constructs, enabling the recreation of complex tissue microenvironments and cell–cell interactions. By co-printing different cell types, such as stem cells, progenitor cells, and support cells, researchers can create biomimetic tissues with enhanced functionality and regenerative potential. These multicellular constructs can recapitulate the structure and function of native tissues more accurately, making them valuable tools for studying tissue development, disease progression, and therapeutic interventions [19]. Bioprinting offers the flexibility to incorporate bioactive molecules, growth factors, and extracellular matrix components into tissue constructs, further enhancing their regenerative properties and therapeutic efficacy.

The application of 3D bioprinting in tissue engineering and regenerative medicine holds immense promise for revolutionizing healthcare by providing patient-specific solutions for tissue repair and regeneration. With its ability to create vascularized, multicellular tissue constructs with biomimetic properties, bioprinting offers unprecedented opportunities to address complex medical challenges and improve patient outcomes.

#### 7.1.6.2 Organ-on-a-Chip Models

The application of 3D bioprinting in organ-on-a-chip models represents a significant advancement in biomedical research, offering a sophisticated platform to mimic the structure and function of human organs in vitro. Organ-on-a-chip models consist of microfluidic devices lined with living cells that replicate the physiological environment and functionality of specific organs or tissues. By integrating 3D bioprinting techniques into the fabrication process, researchers can create complex multicellular constructs with precise control over cellular composition, spatial organization, and microarchitecture, mimicking the complexity of native tissues and organs. The bio printed organ-on-a-chip models serve as powerful tools for studying disease mechanisms, drug responses, and toxicity testing, offering insights into human biology that cannot be achieved using traditional cell culture models or animal studies [20]. One of the key advantages of 3D bioprinting in organ-on-a-chip models is its ability to recreate the intricate structure and cellular organization of human organs, providing a more physiologically relevant model for drug discovery and development. Bio printed organ-on-a-chip models can replicate the complex cellular interactions and tissue interfaces found in native organs, allowing researchers to study how drugs interact with different cell types and tissues within the body. This enables more accurate prediction of drug efficacy, safety, and toxicity, reducing the reliance on animal models and accelerating the drug development process.

3D bioprinting allows for the creation of customized organ-on-a-chip models tailored to specific research needs and disease conditions. By incorporating patient-derived cells, researchers can create personalized organ-on-a-chip models that capture the genetic and phenotypic variability observed in human populations. These personalized models offer a valuable platform for studying individualized drug responses [18, 19], disease mechanisms, and treatment strategies, paving the way for precision medicine approaches tailored to the needs of individual patients. Bio printed organ-on-a-chip models can be used to model rare diseases and genetic disorders that are difficult to replicate in traditional cell culture systems, providing new insights into disease pathogenesis and potential therapeutic interventions. The application of 3D bioprinting in organ-on-a-chip models holds immense promise for advancing biomedical research and drug discovery by providing more physiologically relevant and customizable models of human organs and tissues. With their ability to mimic the complexity of native tissues and organs, bio printed organ-on-a-chip models offer a powerful platform for studying disease mechanisms, drug responses, and personalized medicine approaches, ultimately leading to improved treatments and better outcomes for patients [20]. As technology continues to advance, bio printed organ-on-a-chip models may become indispensable tools for drug development, disease modeling, and preclinical testing, revolutionizing the field of biomedical research.

#### 7.1.6.3 Drug Screening and Development

The application of 3D bioprinting in drug screening and development represents a revolutionary approach to evaluating drug efficacy, toxicity, and safety in a more physiologically relevant context. Traditional drug screening methods often rely on 2D cell cultures or animal models, which may not accurately recapitulate the complex 3D architecture and microenvironment of human tissues and organs. By contrast, 3D bioprinting allows researchers to create multicellular tissue constructs that mimic the structure and function of specific organs, providing a more accurate representation of human physiology. The bio printed tissues can be used as platforms for screening potential drug candidates, evaluating their effects on tissue function, and predicting their response in vivo [21]. One of the key advantages of 3D bioprinting in drug screening is its ability to create high-throughput and reproducible models of human tissues and organs. Bioprinting enables the precise deposition of cells and biomaterials to create complex tissue constructs with consistent architecture and cellular composition, allowing for standardized testing conditions across experiments. This standardization facilitates the comparison of drug responses between different tissues and experimental conditions, improving the reliability and reproducibility of drug screening assays. Bio printed tissue models can be customized to match the specific requirements of different drug screening assays, such as incorporating patient-derived cells or disease-specific biomarkers.

Furthermore, 3D bioprinting enables the creation of personalized tissue models for drug screening, tailoring the tissue constructs to match the genetic and phenotypic characteristics of individual patients. By incorporating patient-derived cells, researchers can create tissue models that capture the variability observed in human populations [21], allowing for more accurate prediction of drug responses and personalized treatment strategies. These personalized tissue models offer valuable insights into inter-individual variability in drug metabolism, efficacy, and toxicity, facilitating the development of precision medicine approaches tailored to the needs of individual patients. The application of 3D bioprinting in drug screening and development holds immense promise for improving the efficiency and effectiveness of the drug discovery process. By providing more physiologically relevant tissue models, bioprinting allows for more accurate prediction of drug responses, reducing the need for costly and time-consuming animal studies and clinical trials. With their ability to create standardized, reproducible, and personalized tissue models, bio printed tissues offer a powerful platform for accelerating drug discovery, optimizing treatment strategies, and ultimately improving patient outcomes.

#### 7.1.6.4 Personalized Medicine and Patient-Specific Implants

The application of 3D bioprinting in personalized medicine and patient-specific implants holds significant promise for revolutionizing healthcare by providing tailored solutions to individual patients' needs. Traditional medical treatments often adopt a one-size-fits-all approach, but each patient's anatomy and physiology are unique [19, 22], leading to variations in treatment outcomes. 3D bioprinting offers a solution by allowing for the fabrication of personalized tissue constructs and implants that match the specific anatomical features and biological characteristics of individual patients. These patient-specific implants can improve treatment efficacy, reduce complications, and enhance patient satisfaction. One of the primary applications of 3D bioprinting in personalized medicine is the fabrication of patient-specific implants for reconstructive surgery and tissue regeneration. By utilizing medical imaging techniques such as CT scans or MRIs, clinicians can generate precise digital models of a patient's anatomy, including bones, joints, and soft tissues [22]. These digital models serve as the basis for designing customized implants that perfectly match the patient's unique anatomy. With 3D bioprinting, these implants can be fabricated from biocompatible materials and seeded with the patient's own cells, promoting integration with the surrounding tissues and enhancing long-term stability and functionality.

Moreover, 3D bioprinting enables the creation of personalized tissue constructs for regenerative medicine applications, such as bone grafts, cartilage implants, and skin substitutes. By combining patient-specific anatomical data with biomaterials and living cells, researchers can fabricate tissue constructs that closely resemble the native tissue, promoting faster healing and better outcomes. These personalized tissue constructs can be tailored to match the patient's specific requirements, such as size, shape, and mechanical properties, offering customized solutions for tissue repair and regeneration [22]. Bio printed tissues can incorporate bioactive molecules and growth factors to enhance tissue regeneration and accelerate the healing process further. The application of 3D bioprinting in personalized medicine and patient-specific implants holds immense potential for improving patient care and treatment outcomes. By providing customized solutions that match the individual patient's anatomy and biology, bioprinting offers new possibilities for precision medicine and personalized healthcare.

### *7.1.7 Bio Printed Tissue Constructs*

Bio printed tissue constructs represent a groundbreaking approach in regenerative medicine, offering a promising avenue for the fabrication of functional tissues and organs for transplantation and therapeutic applications. Unlike traditional tissue engineering methods, which rely on scaffold-based approaches, bio-printing enables the precise deposition of living cells and biomaterials layer by layer to create complex 3D structures that closely mimic the architecture and functionality of native tissues [19, 22]. These bio printed constructs hold immense potential for addressing the critical need for tissue replacements in patients suffering from injuries, diseases, or congenital defects. By recapitulating the microenvironment and cellular composition of native tissues, bio printed constructs offer a promising solution for promoting tissue regeneration, enhancing patient outcomes, and advancing personalized medicine approaches.

#### 7.1.7.1 Vascularized Tissue Constructs

Vascularized tissue constructs represent a significant advancement in tissue engineering, aiming to recreate the intricate network of blood vessels found in native tissues. These constructs play a crucial role in promoting tissue viability, functionality, and integration upon transplantation. The vascular network serves as a conduit for delivering oxygen, nutrients, and signaling molecules to cells throughout the tissue, as well as facilitating the removal of metabolic waste products. By incorporating vascularization into tissue constructs, researchers aim to overcome the limitations of traditional tissue engineering approaches, which often struggle to maintain cell viability and function over large distances from nutrient and oxygen sources [23]. One of the primary challenges in engineering vascularized tissue constructs lies in recreating the complex architecture and functionality of native blood vessels. Traditional approaches typically involve seeding endothelial cells onto pre-formed scaffolds or embedding microchannels within hydrogel matrices to promote angiogenesis. However, these methods often result in limited vascularization and poor integration with the host tissue. In contrast, bio-printing offers a promising solution by enabling the precise deposition of cells and biomaterials to create perfusable vascular networks within tissue constructs. By layering endothelial cells, pericytes, and supporting biomaterials in a spatially controlled manner, researchers can recreate the hierarchical structure of blood vessels, including capillaries, arterioles, and venules, to support tissue perfusion and function.

Moreover, vascularized tissue constructs hold immense promise for a wide range of applications in regenerative medicine [24], including the repair and replacement of damaged or diseased tissues and organs. These constructs can be tailored to match the specific requirements of different tissues and organs, such as the heart, liver, kidney, and skin, offering patient-specific solutions for tissue repair and regeneration. Vascularized tissue constructs can serve as platforms for studying disease mechanisms, drug responses, and personalized medicine approaches. By recreating the complex interactions between cells and blood vessels within tissues, researchers can gain valuable insights into disease progression and develop targeted therapeutics for a wide range of conditions, including cancer, cardiovascular disease, and diabetes. Vascularized tissue constructs represent a promising approach in tissue engineering, offering new possibilities for tissue repair, regeneration, and disease modeling. By incorporating perfusable vascular networks into bio printed tissue constructs, researchers aim to overcome the limitations of traditional tissue engineering methods and create functional tissues and organs for transplantation and therapeutic applications.

#### 7.1.7.2 Nerve Regeneration Scaffolds

Nerve regeneration scaffolds represent a critical component in the field of regenerative medicine, offering a promising approach to repairing and restoring damaged nerves. Nerve injuries, whether due to trauma, disease, or surgical intervention, can result in debilitating symptoms and loss of function. Traditional treatments often have limited success in promoting nerve regeneration and functional recovery. Nerve regeneration scaffolds aim to address this challenge by providing a supportive framework for guiding the growth of regenerating nerve fibers and facilitating the restoration of neural function. These scaffolds are designed to mimic the extracellular matrix (ECM) of native nerve tissue, providing physical and biochemical cues that promote cell adhesion, proliferation, and differentiation. One of the key considerations in designing nerve regeneration scaffolds is the selection of biomaterials with suitable properties for supporting nerve regeneration. Ideally, these biomaterials should be biocompatible, biodegradable, and possess mechanical properties that match those of native nerve tissue. The scaffolds should have an interconnected porous structure that allows for the infiltration of cells and the diffusion of nutrients and growth factors throughout the scaffold. Various biomaterials, such as natural polymers (Fig. 7.4) (e.g., collagen, fibrin, and chitosan) and synthetic polymers (e.g., poly (lactic-co-glycolic acid) (PLGA) and poly(caprolactone) (PCL)), have been investigated for

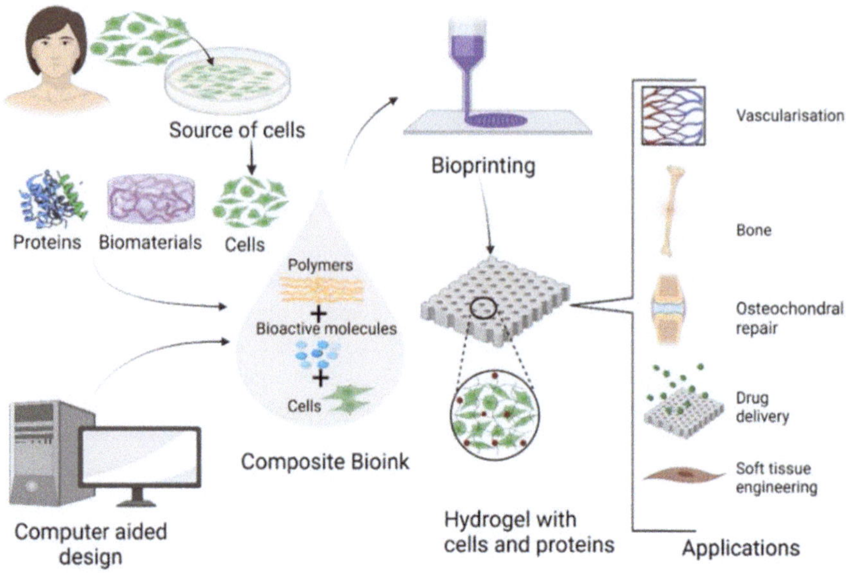


**Fig. 7.4** Bioprinting using natural polymers. https://doi.org/10.3390/polym13193321 [25]

use in nerve regeneration scaffolds, each with its unique advantages and limitations [25]. Nerve regeneration scaffolds can be functionalized with bioactive molecules and growth factors to enhance their regenerative properties further. These molecules can promote cell survival, axonal outgrowth, and myelination, facilitating the regeneration of functional nerve tissue. Nerve regeneration scaffolds can be designed to provide physical guidance cues, such as aligned microfibers or topographical features, to direct the growth of regenerating nerve fibers along desired pathways [26]. By incorporating these design elements into nerve regeneration scaffolds, researchers aim to improve the success rate of nerve regeneration and enhance functional recovery following nerve injury.

Nerve regeneration scaffolds hold immense promise for promoting nerve regeneration and functional recovery in patients with nerve injuries. By providing a supportive environment for guiding the growth of regenerating nerve fibers, these scaffolds offer new possibilities for restoring neural function and improving patient outcomes.

#### 7.1.7.3 Bone and Cartilage Constructs

Bone and cartilage constructs are pivotal applications of 3D bioprinting, representing cutting-edge approaches in tissue engineering and regenerative medicine. These constructs are developed to address the complex challenges associated with bone and cartilage injuries, such as fractures, degeneration, and defects resulting from trauma or disease. Traditional treatment methods often have limitations in providing long-term functional restoration and can lead to complications, underscoring the need for innovative solutions. Bio printed bone and cartilage constructs offer promising alternatives by providing customized scaffolds that mimic the native tissue microenvironment and facilitate tissue regeneration [25]. One of the primary advantages of bio printed bone and cartilage constructs lies in their ability to precisely control the composition, structure, and mechanical properties of the scaffolds. By combining biomaterials, cells, and growth factors in a spatially controlled manner, researchers can tailor the constructs to match the specific requirements of the target tissue. This customization allows for the fabrication of patient-specific implants that closely mimic the anatomical and physiological characteristics of individual patients, offering personalized solutions for bone and cartilage repair and regeneration.

Moreover, bio printed bone and cartilage constructs can replicate the hierarchical organization and intricate architecture of native tissues, facilitating tissue integration and functional restoration upon transplantation [27]. These constructs can be designed to emulate the microenvironmental cues present in the native tissue, such as the presence of growth factors, extracellular matrix proteins, and mechanical stimuli. By providing a biomimetic environment that closely resembles the native tissue, bio printed constructs enhance the regenerative capacity of implanted tissues and improve long-term outcomes for patients.

Bio printed bone and cartilage constructs hold promise for advancing our understanding of tissue biology, disease pathology, and therapeutic interventions [25, 27]. These constructs serve as valuable platforms for studying tissue development, disease progression, and drug responses in vitro. By providing researchers with controlled and reproducible experimental systems, bio printed constructs accelerate scientific discovery and facilitate the development of novel therapeutics for bone and cartilage-related disorders.

#### 7.1.7.4 Skin Substitutes and Wound Healing

Skin substitutes and wound healing are critical areas of application for bio printed tissue constructs, offering innovative solutions for treating acute and chronic wounds. Traditional wound care methods often involve the use of skin grafts or synthetic dressings, which may have limitations in promoting optimal wound healing and tissue regeneration. Bio printed skin substitutes, however, provide a more biomimetic approach by replicating the complex structure and functionality of native skin tissue [28]. These constructs offer a promising alternative for promoting wound closure, reducing scarring, and restoring skin function following injury or surgery. One of the key advantages of bio printed skin substitutes lies in their ability to recreate the multilayered structure of native skin, including the epidermis, dermis, and subcutaneous tissue. By precisely depositing cells and biomaterials layer by layer, researchers can mimic the architecture and composition of native skin, promoting cell proliferation, migration, and differentiation. Bio printed skin substitutes can incorporate growth factors, cytokines, and other bioactive molecules to enhance wound healing processes, such as angiogenesis, collagen deposition, and epithelialization [28]. These constructs offer a conducive microenvironment for promoting tissue regeneration and functional restoration, leading to improved outcomes for patients with acute and chronic wounds. Bio printed skin substitutes hold promise for personalized medicine approaches, allowing for the fabrication of patient-specific constructs tailored to match the individual requirements of each patient. By utilizing patient-derived cells and biomaterials, researchers can create customized skin substitutes that closely mimic the genetic and physiological characteristics of the recipient's skin [1]. This personalized approach enhances the compatibility and effectiveness of the skin substitutes, leading to improved wound healing outcomes and reduced risk of rejection. Bio printed skin substitutes offer opportunities for studying disease mechanisms, drug responses, and tissue interactions in vitro, facilitating the development of novel therapeutic strategies for wound healing and skin regeneration.

**Exercise 2.**

**Fill Ups**

**(Bioprinted tissue, Clinical translation, Adoption, Medical devices, Efficacy, Commercialization)**

**1.** Additive manufacturing techniques like Selective Laser Sintering (SLS) and Fused Deposition Modeling (FDM) are commonly used in the fabrication of............

**2.** ..................and ............involve the process of transitioning innovative medical technologies from laboratory research to real-world applications, including testing in human clinical trials.

**3.** Successful clinical translation and commercialization require rigorous testing through preclinical studies, regulatory approval, and ultimately, market launch and..................

**4.** Key factors influencing the success of clinical translation and commercialization include funding availability, regulatory compliance, market demand, and the ability to demonstrate safety and................

**5.** .................. constructs are engineered using living cells and biomaterials, allowing for the creation of complex three-dimensional structures that closely mimic the native architecture and function of human tissues, offering promising solutions for regenerative medicine and tissue engineering.

### *7.1.8 Clinical Translation and Commercialization*

The clinical translation and commercialization of 3D bioprinting in biomedicine represents a pivotal step toward bringing innovative tissue engineering solutions from the laboratory to the bedside. As 3D bioprinting technologies continue to advance, there is a growing interest in leveraging these techniques to develop clinically relevant therapies for a wide range of medical applications. From regenerative medicine to drug screening and personalized healthcare, 3D bioprinting holds immense promise for improving patient outcomes and transforming the practice of medicine [28]. However, the successful translation of bioprinting technologies into clinical practice requires overcoming various challenges, including regulatory approval, scalability, and cost-effectiveness, as well as ensuring the safety, efficacy, and reproducibility of bio printed products. Despite these challenges, the commercialization of 3D bioprinting technologies has already begun, with several companies and research institutions actively developing bio printed products for clinical use. As the field continues to evolve, collaborations between academia, industry, and regulatory agencies will be essential for driving forward the clinical translation and commercialization of 3D bioprinting technologies [2, 3], ultimately improving patient care and advancing the field of biomedicine.

#### 7.1.8.1 Regulatory Considerations and Approval Processes

Regulatory considerations and approval processes play a crucial role in the clinical translation and commercialization of 3D bioprinting technologies in biomedicine. As these technologies move from the laboratory to the clinic, they must adhere to stringent regulatory requirements to ensure their safety, efficacy, and quality. Regulatory agencies, such as the Food and Drug Administration (FDA) in the United States and the European Medicines Agency (EMA) in Europe, oversee the approval process for bio printed products, including tissue-engineered constructs, implants, and medical devices [29]. These agencies evaluate bio printed products based on preclinical and clinical data to assess their safety, effectiveness, and performance in treating specific medical conditions. The regulatory approval process for bio printed products involves several stages, including preclinical testing, clinical trials, and post-market surveillance. During preclinical testing, researchers conduct rigorous studies to evaluate the biocompatibility, biodegradability, and functionality of the bio printed products in vitro and in animal models. These studies provide crucial data on the safety and performance of the products, helping to inform the design and conduct of clinical trials [30]. Clinical trials are then conducted to assess the safety and efficacy of the bio printed products in human patients, typically in phases I, II, and III, involving increasing numbers of participants and evaluating different aspects of product performance.

In addition to demonstrating safety and efficacy, bio printed products must also meet quality standards and manufacturing requirements set forth by regulatory agencies. Good Manufacturing Practice (GMP) guidelines outline the quality control and quality assurance measures that must be implemented throughout the manufacturing process to ensure product consistency, purity, and sterility [30, 31]. Manufacturers of bio printed products must adhere to these guidelines to obtain regulatory approval and commercialize their products. Furthermore, post-market surveillance mechanisms are in place to monitor the safety and performance of bio printed products once they are on the market, allowing for timely identification and mitigation of any potential risks or adverse events.

Regulatory considerations and approval processes are critical factors in the successful clinical translation and commercialization of 3D bioprinting technologies in biomedicine [31]. By ensuring the safety, efficacy, and quality of bio printed products, regulatory agencies play a vital role in protecting patient safety and public health while facilitating the development of innovative treatments and therapies.

#### 7.1.8.2 Intellectual Property and Licensing Strategies

Intellectual property (IP) and licensing strategies are paramount in the development and commercialization of 3D bioprinting technologies in biomedicine. As the field continues to expand, researchers and companies invest significant resources in innovating new bioprinting techniques, materials, and applications [32]. Protecting these innovations through patents, trademarks, and copyrights is essential for safeguarding their value and ensuring a competitive advantage in the market. Intellectual property rights enable inventors to control the use, distribution, and commercialization of their creations, thereby incentivizing investment in research and development.

Licensing strategies play a crucial role in leveraging intellectual property assets to facilitate technology transfer and commercialization. Through licensing agreements, inventors can grant third parties the right to use their patented technologies in exchange for royalties or other financial considerations. Licensing agreements may encompass various forms of IP, including patents, trade secrets, know-how, and software. These agreements allow companies to access cutting-edge bioprinting technologies without the need for significant upfront investment in research and development, accelerating the development and commercialization process. Licensing strategies enable collaborations between academic institutions, research organizations, and industry partners, fostering innovation and knowledge exchange in the field of bioprinting. Universities and research institutions often license their patented technologies to companies for further development and commercialization, generating revenue streams to support ongoing research and education efforts. Industry partnerships facilitate the translation of academic discoveries into practical applications, leading to the development of new bio printed products and therapies for healthcare [4].

In addition to licensing agreements, intellectual property strategies may also involve cross-licensing arrangements, joint ventures, and strategic alliances to maximize the value of IP portfolios and foster innovation. By pooling resources and expertise, companies can overcome technological barriers, mitigate risks, and capitalize on market opportunities more effectively.

### 7.1.9 Challenges in 3D Printing in Biomedicine

- **Biocompatibility**: Ensuring that 3D printed materials and constructs are biocompatible and non-toxic to human tissues is a significant challenge. Materials used in printing must not elicit adverse immune responses or cause tissue rejection.
- **Resolution and Accuracy**: Achieving high resolution and accuracy in 3D printing is crucial, especially for printing intricate anatomical structures or microscale features. Improving printing resolution can enhance the functionality and performance of bio printed tissues and implants.
- **Vascularization**: Incorporating vascular networks into bio printed tissues to mimic the complex vasculature of native tissues remains challenging. Ensuring adequate perfusion and nutrient supply to cells throughout the construct is essential for tissue viability and functionality.
- **Cell Viability and Functionality**: Maintaining cell viability and functionality during the printing process and post-printing is critical for the success of bio printed tissues. Optimizing printing parameters, biomaterial formulations, and culture conditions is necessary to preserve cell viability and promote tissue maturation.
- **Regulatory Approval**: Navigating the regulatory pathways for approval of bio printed products, including tissue-engineered constructs, implants, and medical devices, can be complex and time-consuming. Compliance with regulatory requirements for safety, efficacy, and quality is essential for clinical translation and commercialization.
- **Scalability and Production Cost**: Scaling up bioprinting processes to meet clinical demand while maintaining cost-effectiveness is a challenge. Developing scalable and cost-effective manufacturing methods for bio printed products is essential for widespread adoption in healthcare.
- **Term Stability and Integration**: Ensuring the long-term stability and integration of bio printed tissues and implants within the host environment is crucial for successful clinical outcomes. Promoting tissue integration, remodeling, and functionality over time remains a challenge in bioprinting.
- **Ethical and Social Implications**: Addressing ethical considerations surrounding the use of bioprinting technologies, such as patient consent, privacy, and equitable access to healthcare, is essential. Balancing technological advancements with ethical and societal concerns is critical for responsible innovation in biomedicine.
- **Education and Training**: Developing standardized protocols, guidelines, and training programs for researchers, clinicians, and regulatory professionals is necessary to ensure safe and effective use of bioprinting technologies in biomedicine. Enhancing education and training opportunities can facilitate broader adoption and implementation of bioprinting in healthcare.

**Exercise 3.**

**Choose The Correct Option**

1. Which of the following best describes 3D printing in biomedicine?
   a) Printing of 2D images onto paper
   b) Creating three-dimensional objects using biological materials
   c) Manufacturing electronic devices
   d) Producing synthetic fabrics

2. What is the primary advantage of using 3D printing in biomedicine?
   a) Reduced cost
   b) Increased manufacturing time
   c) Decreased complexity
   d) Customization and personalization

3. Which of the following is NOT a potential application of 3D printing in biomedicine?
   a) Tissue engineering
   b) Prosthetic limb fabrication
   c) Space exploration
   d) Drug delivery system development

4. What type of materials can be used in 3D bioprinting?
   a) Only metals
   b) Only ceramics
   c) Biological materials, such as living cells and biomaterials
   d) Only plastics

5. Which 3D printing technique is commonly used for the fabrication of patient-specific implants?
   a) Selective Laser Sintering (SLS)
   b) Stereolithography (SLA)
   c) Fused Deposition Modeling (FDM)
   d) Inkjet bioprinting

6. What role does computer-aided design (CAD) play in 3D printing in biomedicine?
   a) It is not used in 3D printing
   b) It converts 2D images into 3D objects
   c) It designs digital models of the desired object
   d) It provides medical diagnoses

7. Which of the following medical specialties benefits the most from 3D printing technology?
   a) Neurology
   b) Dermatology
   c) Ophthalmology
   d) Orthopedics

8. What is the primary limitation of current 3D bioprinting technology?
   a) Lack of regulatory approval
   b) Limited material availability
   c) Inability to produce complex structures
   d) Challenges in vascularization and tissue integration

9. Which term refers to the process of converting medical imaging data into a 3D-printable format?
   a) Bio fabrication
   b) Medical imaging
   c) Digitalization
   d) Medical segmentation

10. What is the future outlook for 3D printing in biomedicine?
    a) It is expected to remain stagnant
    b) Limited to research applications only
    c) Continued growth with advancements in technology and expanded applications
    d) Likely to be replaced by traditional manufacturing methods

## 7.2 Summary

The chapter on 3D bioprinting in biomedicine explores the transformative potential of this technology in revolutionizing healthcare. It begins with an introduction to 3D bioprinting, covering its historical development, current landscape, and future prospects. Fundamentals of 3D bioprinting, including principles, techniques, and bioprinting technologies, are discussed in detail, emphasizing their role in fabricating complex tissue constructs. The applications of 3D bioprinting in biomedicine are then explored, highlighting its diverse range of uses in tissue engineering, regenerative medicine, organ-on-a-chip models, drug screening, and personalized medicine. Specific applications, such as vascularized tissue constructs, nerve regeneration scaffolds, and bone and cartilage constructs, demonstrate the potential of bioprinting to address critical healthcare needs. The chapter also delves into clinical translation and commercialization, addressing regulatory considerations, approval processes, and intellectual property strategies. Challenges in 3D printing in biomedicine, including biocompatibility, resolution, vascularization, and regulatory approval, are identified, alongside potential solutions and future directions.

Overall, the chapter underscores the transformative impact of 3D bioprinting on healthcare, offering personalized solutions for tissue repair, regeneration, and disease modeling. As the field continues to advance, collaboration between academia, industry, and regulatory agencies will be essential for driving forward the clinical translation and commercialization of bioprinting technologies, ultimately improving patient care and advancing the field of biomedicine.

## Points to Remember

- During the COVID-19 pandemic, 3D printing technology was utilized to produce various medical supplies, including face shields, ventilator parts, and nasal swabs. This rapid manufacturing approach helped address shortages and supported healthcare workers on the front lines.
- Dental laboratories are increasingly using 3D printing to create custom dental implants, crowns, and orthodontic devices. This technology allows for precise fitting and aesthetic customization, leading to better patient outcomes in dental care.
- Surgeons are incorporating 3D-printed models into their preoperative planning process, especially for complex surgeries like spinal fusion and joint replacements. These patient-specific models help surgeons visualize anatomical structures and practice surgical approaches, leading to improved outcomes and reduced operating times.
- Pharmaceutical companies are utilizing 3D printing to prototype and manufacture drug formulations with controlled release profiles. This technology enables the production of personalized dosage forms tailored to individual patient needs, potentially improving drug efficacy and compliance.
- 3D-printed anatomical models are becoming essential tools in medical education and training programs. Medical students and healthcare professionals can interact with lifelike replicas of organs and anatomical structures, enhancing their understanding of complex medical concepts and procedures.
- Several 3D-printed medical devices and implants have received approval from the U.S. Food and Drug Administration (FDA), demonstrating the growing acceptance and adoption of this technology in mainstream healthcare.

## References

1. Papaioannou TG, Manolesou D, Dimakakos E, Tsoucalas G, Vavuranakis M, Tousoulis D (2019) 3D Bioprinting methods and techniques: applications on artificial blood vessel fabrication. Acta Cardiol Sinica 35(3):284–289. https://doi.org/10.7515/ACS.201905_35(3).20181115A
2. Kačarević ŽP, Rider PM, Alkildani S, Retnasingh S, Smeets R, Jung O, Ivanišević Z, Barbeck M (2018) An introduction to 3D bioprinting: possibilities, challenges and future aspects. Materials (Basel, Switzerland) 11(11):2199. https://doi.org/10.3390/ma11112199
3. Chae S, Cho DW (2023) Biomaterial-based 3D bioprinting strategy for orthopedic tissue engineering. Acta Biomater 157:4–20. https://doi.org/10.1017/j.actbio.2022.08.004
4. Assad H, Assad A, Kumar A (2023) Recent developments in 3D bio-printing and its biomedical applications. Pharmaceutics 15(1):255. https://doi.org/10.3390/pharmaceutics15010255
5. Mandrycky C, Wang Z, Kim K, Kim DH (2017) 3D bioprinting for engineering complex tissues. Biotechnol Adv 34(4):422–434. https://doi.org/10.1017/j.biotechadv.2015.12.011
6. Dey M, Ozbolat IT (2020) 3D bioprinting of cells, tissues and organs. Sci Rep 10(1):14023. https://doi.org/10.1038/s41598-020-70087-y
7. Ozbolat IT (2016) 3D bioprinting: fundamentals, principles and applications. Academic Press

8. Guillotin B, Ali M, Ducom A, Catros S, Keriquel V, Souquet A, Remy M, Fricain J, Guillemot F (2013) Laser-assisted bioprinting for tissue engineering. In: Elsevier eBooks (pp. 95–118). https://doi.org/10.1016/b978-1-4557-2852-7.00006-8
9. Rothbauer M, Eilenberger C, Spitz S, Bachmann BEM, Kratz SRA, Reihs EI, Windhager R, Toegel S, Ertl P (2022) Recent advances in additive manufacturing and 3D bioprinting for organs-on-a-chip and microphysiological systems. Front Bioeng Biotechnol 10:837087. https://doi.org/10.3389/fbioe.2022.837087
10. Song Y, Ghafari Y, Asefnejad A, Toghraie D (2024) An overview of selective laser sintering 3D printing technology for biomedical and sports device applications: processes, materials, and applications. Opt Laser Technol/Opt Laser Technol 171:110459. https://doi.org/10.1016/j.optlastec.2023.110459
11. Gueche YA, Sanchez-Ballester NM, Cailleaux S, Bataille B, Soulairol I (2021) Selective laser sintering (sLS), a new chapter in the production of solid oral forms (SOFs) by 3D printing. Pharmaceutics 13(8):1212. https://doi.org/10.3390/pharmaceutics13081212
12. Fused deposition modeling based 3D printing (2021). In: Materials forming, machining and tribology. https://doi.org/10.1007/978-3-030-68024-4
13. Amini A, Guijt RM, Themelis T, De Vos J, Eeltink S (2023) Recent developments in digital light processing 3D-printing techniques for microfluidic analytical devices. J Chromatogr A/J Chromatogr 1692:463842. https://doi.org/10.1016/j.chroma.2023.463842. PMID: 35287884
14. Kovalchuk D, Ivasishin O (2019) Profile electron beam 3D metal printing. In: Elsevier eBooks (pp. 213–233). https://doi.org/10.1016/b978-0-12-814062-8.00012-1
15. Anand M, Das AK (2021) Issues in fabrication of 3D components through DMLS technique: a review. Opt Laser Technol/Opt Laser Technol 139:106914. https://doi.org/10.1016/j.optlastec.2021.106914
16. Humnabad PS, Tarun R, Das I (2022) An overview of direct Metal Laser sintering (DMLS) technology for metal 3D printing. J Min Met Fuels 70(3A):127. https://doi.org/10.18311/jmmf/2022/30681
17. Ng WL, An J, Chua CK (2024) Process, material, and regulatory considerations for 3D printed medical devices and tissue constructs. Engineering. https://doi.org/10.1016/j.eng.2024.01.028
18. Gupta S, Bit A (2021) 3D bioprinting in tissue engineering and regenerative medicine. Cell Tissue Bank 23(2):199–212. https://doi.org/10.1007/s10561-021-09936-6
19. Tripathi S, Mandal SS, Bauri S, Maiti P (2022) 3D bioprinting and its innovative approach for biomedical applications. MedComm 4(1):e194. https://doi.org/10.1002/mco2.194
20. Park JY, Jang J, Kang H (2018) 3D bioprinting and its application to organ-on-a-chip. Microelectron Eng 200:1–11. https://doi.org/10.1016/j.mee.2018.08.004
21. Peng W, Datta P, Ayan B, Ozbolat V, Sosnoski D, Ozbolat IT (2017) 3D bioprinting for drug discovery and development in pharmaceutics. Acta Biomater 57:26–46. https://doi.org/10.1016/j.actbio.2017.05.025
22. Bg PK, Mehrotra S, Marques SM, Kumar L, Verma R (2023) 3D printing in personalized medicines: a focus on applications of the technology. Mat Today Commun 35:105875. https://doi.org/10.1016/j.mtcomm.2023.105875
23. Chae S, Ha DH, Lee H (2023) 3D bioprinting strategy for engineering vascularized tissue models. Int J Bioprinting 9(5):748. https://doi.org/10.18063/ijb.748
24. Zhang Y, Kumar P, Lv S, Xiong D, Zhao H, Cai Z, Zhao X (2021) Recent advances in 3D bioprinting of vascularized tissues. Mater Des 199:109398. https://doi.org/10.1016/j.matdes.2020.109398
25. Troy E, Tilbury MA, Power AM, Wall JG (2021) Nature-based biomaterials and their application in biomedicine. Polymers 13(19):3321. https://doi.org/10.3390/polym13193321
26. Liu X, Hao M, Chen Z, Zhang T, Huang J, Dai J, Zhang Z (2021) 3D bioprinted neural tissue constructs for spinal cord injury repair. Biomaterials 272:120771. https://doi.org/10.1016/j.biomaterials.2021.120771
27. Zhou J, Li Q, Tian Z, Yao Q, Zhang M (2023) Recent advances in 3D bioprinted cartilage-mimicking constructs for applications in tissue engineering. Mater Today Bio 23: 100870. https://doi.org/10.1016/j.mtbio.2023.100870

28. Lian Q, Jiao T, Zhao T, Wang H, Yang S, Li D (2021) 3D bioprinted skin substitutes for accelerated wound healing and reduced scar. J Bionic Eng/J Bionics Eng 18(4):900–914. https://doi.org/10.1007/s42235-021-0053-8
29. Mladenovska T, Choong PF, Wallace GG, O'Connell CD (2023) The regulatory challenge of 3D bioprinting. Regener Med 18(8):659–674. https://doi.org/10.2217/rme-2022-0194
30. Kirillova A, Bushev S, Abubakirov A, Sukikh G (2020) Bioethical and legal issues in 3D bioprinting. Int J Bioprinting 6(3):272. https://doi.org/10.18063/ijb.v6i3.272
31. Pountos I, Tellisi N, Ashammakhi N (2019) Three-dimensional bioprinting: safety, ethical, and regulatory considerations. In: Springer eBooks (pp. 191–203). https://doi.org/10.1007/978-3-030-23906-0_7
32. Ferrill ED, Rajan K (2022) 3D bioprinting, nanotechnology, and intellectual property. In: Elsevier eBooks (pp. 505–522). https://doi.org/10.1016/b978-0-12-824552-1.00006-2

# Chapter 8
# Hydrogels in Drug Delivery

## 8.1 Introduction

Hydrogels have emerged as versatile materials with a wide range of applications in various fields, with one of the most promising domains being drug delivery (Fig. 8.1). These 3D networks of hydrophilic polymers have garnered substantial attention from researchers and pharmaceutical industries alike due to their unique properties that make them ideal candidates for controlled and targeted drug release systems. In this chapter, we delve into the captivating world of hydrogels in drug delivery, exploring their fundamental characteristics, synthesis methods, drug loading mechanisms, and diverse applications that span oral, transdermal, injectable, ocular, and implantable drug delivery systems [1]. The foundation of understanding hydrogels in drug delivery begins with grasping the essential characteristics that set them apart. Hydrogels are hydrophilic polymers capable of absorbing and retaining large quantities of water or biological fluids while maintaining their structural integrity. This exceptional water-absorbing capacity forms the basis for their use in drug delivery, as it allows for the incorporation and controlled release of drugs in a hydrated environment. Their biocompatibility, tunable mechanical properties, and responsiveness to environmental cues further contribute to their appeal in drug delivery applications. Effective drug delivery systems require not only the controlled release of therapeutic agents, but also the ability to encapsulate and protect these agents from degradation [2]. This chapter will explore the various methods and techniques employed to synthesize and fabricate hydrogels, including natural and synthetic hydrogel materials, cross-linking strategies, and encapsulation methods. Understanding these techniques is crucial for tailoring hydrogel properties to meet specific drug delivery requirements. Once hydrogels are synthesized, the next critical step is loading drugs within their networks and achieving controlled release profiles [3]. We will investigate drug loading strategies, release kinetics, and the factors influencing drug release from hydrogels. This knowledge is essential for designing drug delivery systems that can effectively deliver therapeutic agents to target sites in the body. Hydrogels'

 
D. Suhag et al., *Handbook of Biomaterials for Medical Applications, Volume 1*,
Biomedical Materials for Multi-functional Applications,
https://doi.org/10.1007/978-981-97-4818-1_8

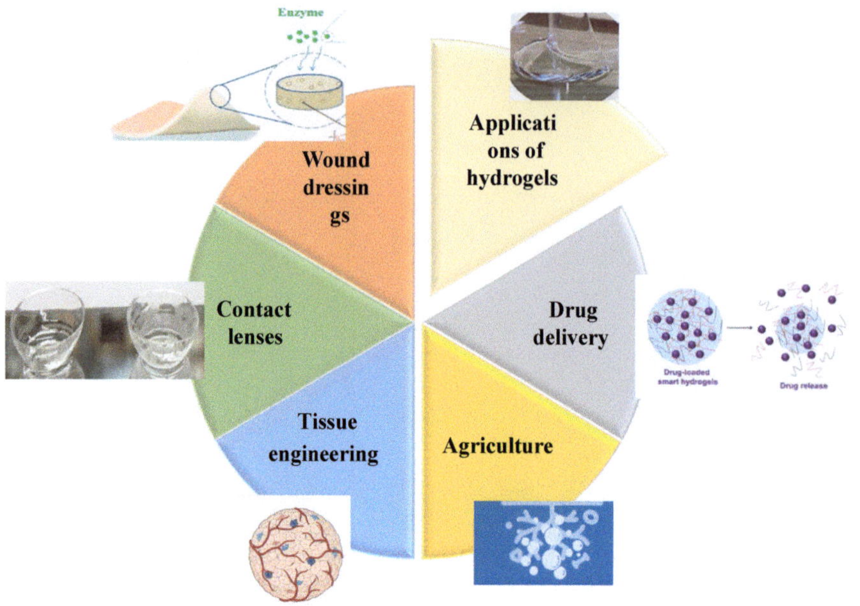


**Fig. 8.1** Hydrogels for various applications [3]

adaptability extends to a diverse range of drug delivery applications, each with its unique set of challenges and opportunities. As the field of drug delivery continues to evolve, hydrogels are at the forefront of innovation. Recent advances in smart hydrogels, personalized medicine, nanotechnology integration, and regulatory considerations are paving the way for more precise and effective drug delivery solutions. This chapter will discuss these developments and provide insights into the future prospects of hydrogels in drug delivery, emphasizing the potential for improved patient outcomes and the realization of more tailored therapies.

However, challenges and limitations must be acknowledged and addressed. Biocompatibility issues, stability concerns, and hurdles in clinical translation are among the challenges that researchers and industry professionals encounter when working with hydrogels in drug delivery. These obstacles will be examined in detail, offering valuable insights for overcoming them and advancing the field. Hydrogels represent a promising avenue for revolutionizing drug delivery by providing tailored, controlled, and targeted release of therapeutic agents. This chapter will serve as a comprehensive guide to understanding the fundamentals, applications, advances, and challenges associated with hydrogels in drug delivery, offering readers a thorough exploration of this exciting field at the intersection of materials science, pharmaceuticals, and healthcare.

**History of Hydrogels in Drug Delivery**

The history of hydrogels in drug delivery is marked by remarkable progress and innovation, driven by the need for more effective and controlled drug administration methods. Hydrogels, initially developed in the early twentieth century, found their way into drug delivery applications due to their unique properties. The concept of hydrophilic polymers that can absorb and retain large amounts of water or biological fluids dates back to the late nineteenth century, but it wasn't until the 1930s that the term "hydrogel" was coined by Wichterle and Lim, who pioneered the development of hydrogel contact lenses [4].

The 1960s marked a significant milestone with the introduction of the first hydrogel-based drug delivery system. These early hydrogel drug delivery devices primarily focused on delivering ocular medications, showcasing the hydrogels' ability to provide sustained and controlled release of drugs to the eye. This breakthrough laid the foundation for further exploration of hydrogel materials in drug delivery, especially in the field of ophthalmology. In the subsequent decades, hydrogels gained prominence in drug delivery due to their biocompatibility, tunable mechanical properties, and ease of modification to achieve desired drug release kinetics. Researchers began investigating various hydrogel materials, including natural polymers like alginate and synthetic polymers such as polyethylene glycol (PEG), to design drug delivery systems tailored to specific applications. The twenty-first century witnessed a surge in hydrogel-based drug delivery research and development [5]. Hydrogels expanded their presence from ocular drug delivery to diverse applications, including oral, transdermal, injectable, and implantable drug delivery systems. Researchers explored innovative methods for drug encapsulation and release from hydrogels, resulting in improved bioavailability and targeted therapies.

Today, hydrogels in drug delivery continue to evolve, with ongoing advancements in smart hydrogels that respond to physiological cues, personalized medicine approaches, and integration with nanotechnology for precise drug delivery. This rich history underscores the versatility and potential of hydrogels in revolutionizing drug delivery, offering patients more effective and patient-friendly treatment options.

## *8.1.1 Basics of Hydrogels*

Hydrogels are a class of materials with unique characteristics that distinguish them from conventional solids and liquids. These 3D networks of hydrophilic polymers have a defining characteristic—high water content—that plays a pivotal role in their properties and applications. Hydrogels are composed of water-absorbing polymer chains, creating a matrix structure capable of absorbing and retaining significant amounts of water or biological fluids while maintaining their structural integrity [6]. This property allows hydrogels to resemble biological tissues, making them highly biocompatible and suitable for various biomedical applications. One of the most distinguishing features of hydrogels is their ability to swell in response to changes in environmental conditions, particularly changes in the surrounding water content. This characteristic arises from the hydrophilic nature of the polymer chains within the hydrogel network. When placed in a hydrated environment, hydrogels can absorb water and swell, increasing in volume. Conversely, in a dehydrated state, hydrogels can deswell or shrink, demonstrating their responsiveness to moisture levels.

Hydrogels also exhibit tunable mechanical properties, which can be adjusted to suit specific applications. By varying the polymer composition, cross-linking density, and other parameters during synthesis, researchers can fine-tune the stiffness, elasticity, and overall mechanical behavior of hydrogels [7]. This versatility makes hydrogels adaptable for use in soft and hard tissue engineering, where matching the mechanical properties of native tissues is crucial. Another key feature of hydrogels is their high permeability to water and small molecules. This property allows for the efficient diffusion of solutes within the hydrogel network. It is a fundamental characteristic for drug delivery applications, as it enables the controlled release of therapeutic agents while protecting them from external factors that could lead to degradation.

Furthermore, hydrogels can exhibit excellent biocompatibility, making them suitable for use in contact with biological systems such as living cells and tissues. The water content and physical properties of hydrogels mimic those of natural tissues, reducing the risk of immune responses or adverse reactions when hydrogels are employed in biomedical applications [8]. This biocompatibility is a crucial factor in their success in various medical fields. Hydrogels are a class of materials characterized by their high water content, responsiveness to environmental moisture levels, tunable mechanical properties, permeability to solutes, and biocompatibility. These defining characteristics make hydrogels uniquely suited for a wide range of applications in areas such as drug delivery, tissue engineering, wound healing, and more, where their ability to mimic biological tissues and provide controlled hydration and drug release is highly advantageous.

### *8.1.2 Types of Hydrogels*

Hydrogels are a versatile class of materials, and they come in various types based on their composition, origin, and intended applications (Fig. 8.2). Here are some common types of hydrogels:

- **Physical Hydrogels**: Physical hydrogels form through physical interactions, such as hydrogen bonding or ionic interactions, between polymer chains. They do not require chemical cross-linking and can undergo reversible gel-sol transitions with changes in environmental conditions, like temperature or pH. Physical hydrogels are often used in drug delivery systems that respond to stimuli [9–11].

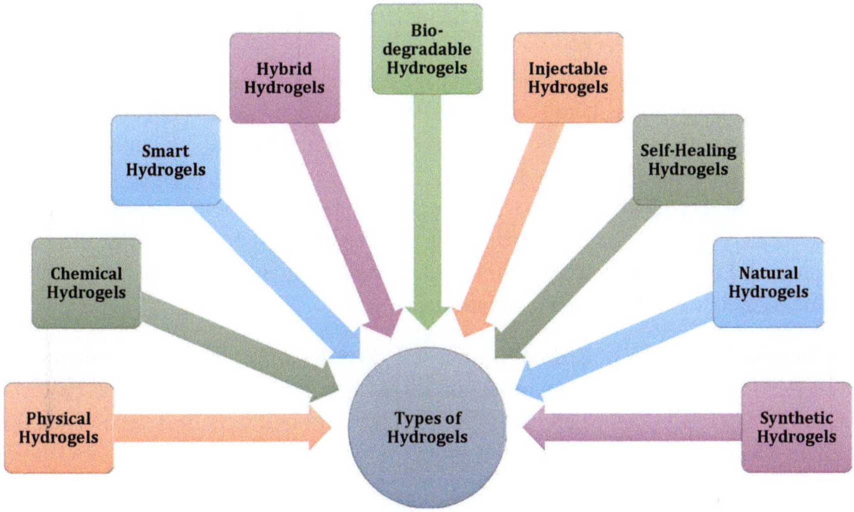


**Fig. 8.2** Types of hydrogels used for drug delivery

- **Responsive or Smart Hydrogels**: These hydrogels can undergo reversible volume changes in response to specific external stimuli, such as temperature, pH, or the presence of ions. This responsiveness is harnessed in applications like controlled drug release and tissue engineering [12].
- **Hybrid Hydrogels**: Hybrid hydrogels combine natural and synthetic components to leverage the strengths of both. For instance, a natural polymer like alginate can be combined with synthetic polymers to create a hydrogel with improved mechanical properties and stability [13].
- **Biodegradable Hydrogels**: These hydrogels are designed to degrade over time into non-toxic byproducts. They are commonly used in drug delivery systems where the hydrogel scaffold gradually breaks down, releasing the encapsulated drug [14].
- **Injectable Hydrogels**: Injectable hydrogels are in a liquid state at low temperatures but solidify at body temperature. They are useful for minimally invasive procedures and localized drug delivery [15].
- **Self-healing Hydrogels**: These hydrogels have the ability to repair themselves when damaged. They are being explored for applications in regenerative medicine and artificial organs [16].
- **Electroconductive Hydrogels**: These specialized hydrogels contain conductive materials, such as graphene or carbon nanotubes, and are used in bioelectronics, neural interfaces, and tissue engineering applications that require electrical conductivity [17].
- **Synthetic Hydrogels**: Synthetic hydrogels are chemically engineered from synthetic polymers like polyethylene glycol (PEG), polyacrylamide, and polyvinyl alcohol (PVA). They offer precise control over their properties,

including mechanical strength and degradation rates, making them adaptable for various applications [18].
- **Natural Hydrogels**: These hydrogels are derived from natural polymers, such as agar, alginate, chitosan, and collagen [19]. They often have excellent biocompatibility, making them suitable for biomedical applications like wound dressings, tissue engineering, and drug delivery.
- **Chemical Hydrogels**: These hydrogels are formed through covalent bonds between polymer chains, providing greater stability and long-term structural integrity [20]. They are commonly used in tissue engineering and as drug carriers due to their sustained release capabilities.

These various types of hydrogels offer a wide range of properties and functionalities, making them valuable materials in fields ranging from medicine and biotechnology to materials science and beyond. The choice of hydrogel type depends on the specific requirements of the intended application.

## *8.1.3 Hydrogel Properties Relevant to Drug Delivery*

Hydrogels have emerged as versatile materials with unique properties that hold immense promise in the field of drug delivery. These 3D networks of hydrophilic polymers possess several key characteristics that make them particularly well-suited for controlled drug release. With a high water content, tunable porosity, and biocompatibility, hydrogels offer a platform for precise and sustained drug administration. Their ability to respond to environmental cues, such as changes in pH or temperature, further enhances their utility in targeted therapy [2]. In this context, the below discussion explores the various properties of hydrogels that play a pivotal role in optimizing drug delivery systems, offering insights into their potential applications in modern medicine (Fig. 8.3).

### 8.1.3.1 High Water Content

Hydrogels are composed of a significant amount of water, typically ranging from 90 to 99% of their total weight. This high water content closely resembles the natural environment of living tissues, making hydrogels biocompatible and well-tolerated within the body. This property is crucial for drug delivery because it allows drugs to be dissolved or dispersed uniformly within the hydrogel matrix, facilitating efficient drug loading, and minimizing potential adverse reactions [3].

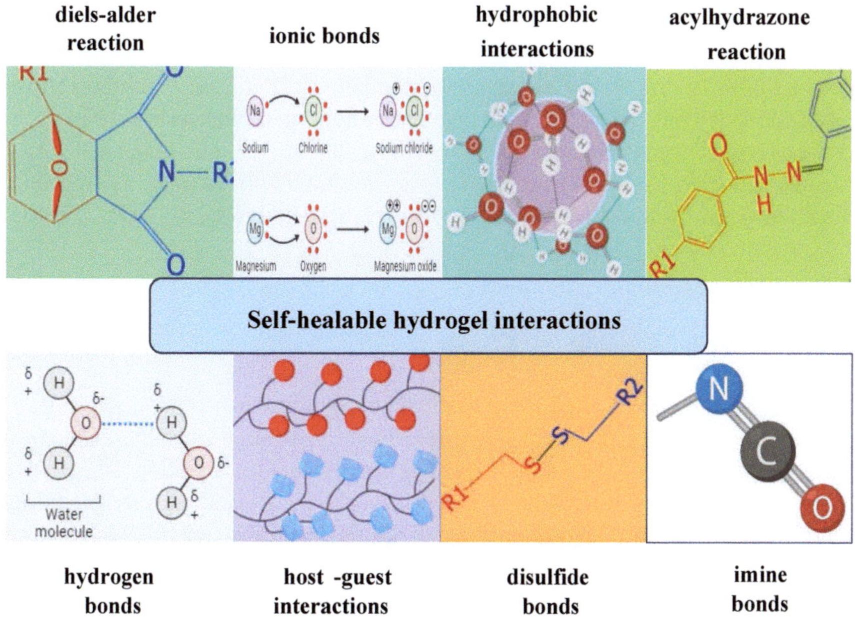


**Fig. 8.3** Different types of physical and chemical interactions behind self-healable hydrogels [16]

#### 8.1.3.2 Swelling Capacity

Hydrogels possess a remarkable ability to absorb and retain large volumes of water while maintaining their structural integrity. This property enables hydrogels to swell in response to changes in environmental conditions, such as pH, temperature, or ionic strength [5]. In drug delivery, this swelling capacity is harnessed to control the release of drugs. As the hydrogel swells, it can release the incorporated drug in a sustained and controlled manner, ensuring a prolonged therapeutic effect.

#### 8.1.3.3 Tunable Porosity

The porosity of hydrogels, which refers to the size and distribution of pores within the matrix, can be tailored to control the diffusion rate of drugs. By adjusting the cross-linking density or the polymer composition, researchers can modify the hydrogel's porosity to achieve specific drug release kinetics. This tunability is vital for designing drug delivery systems that match the therapeutic needs of different medications, ranging from rapid release to long-term sustained delivery.

#### 8.1.3.4 Biodegradability

Biodegradable hydrogels are designed to gradually degrade and break down over time. In drug delivery applications, biodegradable hydrogels offer several advantages. As the hydrogel degrades, it releases the encapsulated drug, eliminating the need for surgical removal or extraction of the delivery system. This property is especially valuable in cases where sustained drug release is required over an extended period of treatment.

#### 8.1.3.5 Functionalization

Hydrogel surfaces can be modified or functionalized with various chemical groups to enhance drug loading and release properties. Functionalization allows researchers to attach specific ligands or targeting molecules to the hydrogel, enabling the selective delivery of drugs to specific cells or tissues. This property facilitates targeted drug therapy, minimizing side effects and improving treatment efficacy [6].

#### 8.1.3.6 PH Responsiveness

Some hydrogels exhibit pH-responsive behavior, meaning they can swell or contract in response to changes in the pH of their surroundings. pH-responsive hydrogels are particularly advantageous for drug delivery in environments with varying acidity levels, such as the gastrointestinal tract or tumor microenvironments. By designing hydrogels that respond to pH changes, researchers can achieve on-demand drug release at specific locations in the body.

#### 8.1.3.7 Thermosensitivity

Thermosensitive hydrogels undergo reversible phase transitions in response to temperature variations. This property is exploited in drug delivery systems that release drugs in response to temperature changes. When exposed to the target temperature, such as body heat or external heating devices, thermosensitive hydrogels can trigger drug release, offering precise control over drug administration.

#### 8.1.3.8 Biocompatibility

Hydrogels are generally well-tolerated by living tissues, minimizing the risk of immune responses or toxicity when used for drug delivery. This biocompatibility ensures that the hydrogel-based drug delivery system is safe for patients and can be used for various therapeutic applications.

#### 8.1.3.9 Adhesive Properties

Certain hydrogels have adhesive properties, allowing them to adhere to specific tissues or surfaces. In drug delivery, adhesive hydrogels are valuable for targeted drug delivery to localized sites within the body. They can adhere to mucosal surfaces or surgical sites, ensuring that the drug is delivered precisely where it is needed.

#### 8.1.3.10 Sterility

Hydrogels can be sterilized through various methods, such as autoclaving or gamma irradiation. This property is crucial for drug delivery to prevent infections or complications associated with microbial contamination during administration.

Incorporating these diverse properties of hydrogels into drug delivery systems offers a wide range of options for tailoring drug release profiles, enhancing therapeutic efficacy, and minimizing adverse effects. Researchers continue to explore and innovate hydrogel-based drug delivery platforms to address various medical challenges.

### *8.1.4 Fundamentals of Drug Delivery Using Hydrogels*

Drug delivery is a crucial aspect of modern healthcare, aiming to transport pharmaceutical compounds to specific locations within the body in a controlled manner to achieve therapeutic effects. Drug delivery systems encompass a wide array of techniques and technologies designed to optimize drug administration, improve drug bioavailability, and minimize side effects. These systems can range from conventional oral medications and injections to advanced nanoparticles, liposomes, and hydrogels [21]. The primary goal is to enhance the therapeutic efficiency of drugs. Despite remarkable progress in the field, several challenges persist in drug delivery. One of the major hurdles is achieving precise drug targeting. Many drugs distribute throughout the body, potentially causing off-target effects and reducing therapeutic efficacy. Additionally, the inherent biopharmaceutical properties of drugs, such as poor solubility, instability, or low bioavailability, can limit their effective delivery [22]. Overcoming these challenges necessitates innovative drug formulation and delivery approaches, as well as a deeper understanding of drug pharmacokinetics.

Hydrogels, 3D networks of hydrophilic polymers, have emerged as promising candidates in drug delivery due to their unique properties. They are primarily composed of water, making them biocompatible and well-tolerated within the body. One key attribute is their high water content, which allows for efficient drug loading by dissolving or dispersing drugs within the hydrogel matrix. This property ensures a uniform distribution of drugs and minimizes adverse reactions when administered. The swelling capacity of hydrogels is another pivotal feature for drug delivery. Hydrogels can absorb and retain substantial amounts of water, leading to controlled drug release over extended periods. By altering the cross-linking density or polymer composition, researchers can customize the porosity of hydrogels. This tunability enables precise control of drug diffusion rates, a critical factor for tailoring drug release kinetics to specific therapeutic needs.

Biodegradable hydrogels, capable of gradual degradation within the body, offer significant advantages for drug delivery systems [23]. As the hydrogel breaks down, it releases the encapsulated drug, eliminating the need for invasive removal procedures. This property is particularly valuable for long-term drug delivery applications, such as in the treatment of chronic diseases, where sustained drug release is essential.

Hydrogels also excel in targeted drug delivery. By functionalizing the hydrogel surface with specific ligands or targeting molecules, researchers can direct drugs to specific cells or tissues. This approach minimizes off-target effects and enhances the therapeutic efficacy of drugs, making it particularly relevant in the context of cancer therapy and localized diseases. Some hydrogels exhibit pH-responsive behavior, meaning they can swell or contract in response to changes in pH levels. pH-responsive hydrogels are advantageous for oral drug delivery, as they can respond to the acidic environment of the stomach to control drug release. This property enhances drug bioavailability and enables targeted release to specific gastrointestinal regions, optimizing therapeutic outcomes.

**Exercise 1.**

**Fill Ups**

**(Systemic, Blood, Tissue regeneration, Hydrogels, Retention)**

1. Hydrogels aid in wound healing by providing a moist environment and controlled drug delivery, enhancing.........................
2. Hydrogels improve drug .................. on the ocular surface, reducing the need for frequent administration.
3. Hydrogels target cancer cells, delivering chemotherapy directly to tumors while minimizing ................ toxicity.
4. ...................... provide sustained release of anti-inflammatory drugs, reducing pain and inflammation in affected joints.
5. Hydrogels deliver insulin, mimicking physiological release patterns to regulate .................. glucose levels effectively.

Drug delivery is a fundamental aspect of modern medicine, encompassing various systems and technologies to ensure the effective and precise administration of pharmaceutical compounds. While challenges persist, innovative strategies continue to evolve, including the use of hydrogels. Hydrogels offer a versatile platform for drug delivery, with their high water content, tunable porosity, biodegradability, and potential for targeted release. As research in this field advances, hydrogel-based drug delivery systems hold great promise in revolutionizing healthcare and improving patient outcomes.

### *8.1.5 Synthesis and Fabrication of Hydrogels*

The synthesis and fabrication of hydrogels is a fascinating field of research that involves creating unique materials with a wide range of applications in biomedicine, drug delivery, tissue engineering, and more. Hydrogels are 3D networks of cross-linked hydrophilic polymers capable of absorbing and retaining large amounts of water or biological fluids. The synthesis of hydrogels typically begins with the selection of suitable polymers, such as polyethylene glycol (PEG), polyvinyl alcohol (PVA), or natural polymers like alginate or collagen [24]. These polymers are then chemically cross-linked to form a hydrogel structure. Cross-linking can be achieved through various methods, including chemical cross-linking with agents like glutaraldehyde or physical cross-linking through temperature or pH changes.

Fabrication techniques for hydrogels are diverse and depend on the intended application. Common methods include solution casting, photopolymerization, and electrospinning [25]. Solution casting involves dissolving the polymer in a solvent, casting it into a desired shape, and allowing it to solidify. Photopolymerization uses light to initiate cross-linking in a photosensitive polymer solution, enabling precise control over hydrogel shape and properties. Electrospinning creates nanofibrous hydrogels with high surface area for applications like wound dressings. Hydrogels' tunable properties, such as swelling behavior, mechanical strength, and degradation rates, make them incredibly versatile for tailored applications. They can be loaded with drugs, growth factors, or cells, making them valuable in controlled drug release systems and tissue engineering scaffolds [26, 27].

The synthesis and fabrication of hydrogels involve designing and creating innovative materials with unique properties; holding great promise in various fields. Researchers continue to explore novel techniques and applications, contributing to the advancement of science and technology in a diverse range of disciplines.

#### 8.1.5.1 Natural Versus Synthetic Hydrogels

Natural and synthetic hydrogels are distinct in terms of their composition, synthesis, and fabrication processes (Table 8.1).

#### 8.1.5.2 Crosslinking Techniques

Cross-linking techniques are crucial in the synthesis and fabrication of hydrogels, providing structural integrity and controlling various properties. Here are some common cross-linking techniques:

**Table 8.1** Synthesis and fabrication of natural vs. synthetic hydrogels

| | Natural | Synthetic |
|---|---|---|
| Composition | • These are derived from biocompatible materials found in nature, such as agar, alginate, or collagen<br>• They often have inherent biocompatibility and can be well-suited for biomedical applications | • These are created from synthetic polymers, like polyacrylamide or polyethylene glycol (PEG)<br>• Their composition can be precisely controlled for specific properties |
| Synthesis | • Typically, natural hydrogels are extracted from natural sources like seaweed<br>• They are synthesized through chemical modifications of natural polymers | • These hydrogels are synthesized through chemical reactions<br>• Cross-linking agents, such as N, N′-methylenebisacrylamide, are commonly used to link polymer chains together to form the gel structure |
| Fabrication | • Fabrication of natural hydrogels often involves casting or molding methods<br>• For example, agar hydrogels can be formed by cooling a heated agar solution | • Fabrication methods for synthetic hydrogels include photopolymerization, thermal gelation, or chemical cross-linking<br>• These methods provide control over gel properties, such as stiffness and porosity |

- **Chemical Cross-Linking**

Description: Chemical cross-linking involves the formation of covalent bonds between polymer chains. Cross-linking agents are used to connect polymer molecules.

Techniques: Common cross-linking agents include glutaraldehyde, ethylene glycol dimethacrylate (EGDMA), and N, N'-methylenebisacrylamide (MBAA). These agents are introduced into the hydrogel precursor solution and initiate cross-linking during polymerization.

- **Physical Cross-Linking:**

Description: Physical cross-linking relies on non-covalent interactions, such as hydrogen bonds, van der Waals forces, or hydrophobic interactions, to hold polymer chains together.

Techniques: This can involve processes like freeze-thaw cycling, where the hydrogel solution is repeatedly frozen and thawed to induce physical cross-linking through crystalline or gel-like structures.

- **Photopolymerization**

Description: Photopolymerization uses UV or visible light to initiate the cross-linking reaction in the presence of photoinitiators. It offers precise control and rapid gel formation.

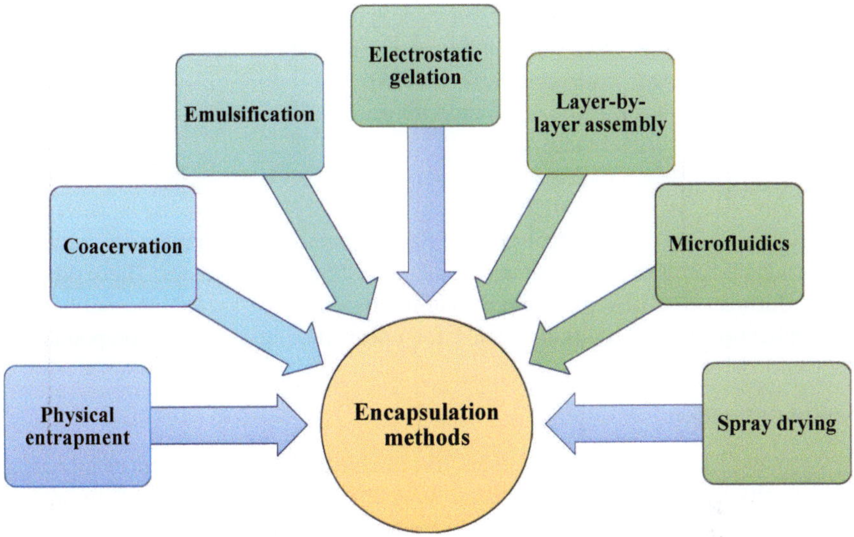


**Fig. 8.4** Different encapsulation methods for synthesis and fabrication of hydrogels

Techniques: A photoinitiator is mixed with the hydrogel precursor solution, and exposure to light triggers cross-linking. Common photoinitiators include Irgacure and camphorquinone.

- **Ionic Cross-Linking**

Description: Ionic cross-linking relies on the interaction between charged groups in the polymer chains and oppositely charged ions in the solution.

Techniques: For example, alginate hydrogels can be cross-linked using calcium ions ($Ca^{2+}$), which bind to the carboxyl groups on alginate chains, forming a stable gel structure.

- **Enzymatic Cross-Linking**

Description: Enzymatic cross-linking involves the use of enzymes to catalyze cross-linking reactions.

Techniques: Transglutaminase is an example of an enzyme used for this purpose. It can catalyze the formation of covalent bonds between specific amino acid residues in proteins, leading to hydrogel formation.

Each cross-linking technique has its advantages and limitations, making them suitable for different applications (Fig. 8.4).

#### 8.1.5.3 Encapsulation Methods

Encapsulation methods play a vital role in the synthesis and fabrication of hydrogels, enabling the incorporation of various substances within the gel matrix. Here are some common encapsulation methods:

- **Physical Entrapment**

Description: In physical entrapment, substances to be encapsulated are simply mixed into the hydrogel precursor solution before gel formation. The substances become trapped within the gel structure as it forms [28].

Applications: This method is suitable for encapsulating drugs, nanoparticles, or cells within the hydrogel matrix.

- **Coacervation**

Description: Coacervation involves the phase separation of two polymer solutions, one of which forms droplets that encapsulate the desired substances. These droplets can be further cross-linked to form a solid hydrogel.

Applications: Coacervation is used for encapsulating sensitive compounds or bioactive agents.

- **Emulsification**

Description: Emulsification techniques, like double emulsion or solvent evaporation, involve mixing two immiscible phases, typically an aqueous phase and an organic phase, to form droplets containing the substance to be encapsulated [29].

Applications: Emulsification is often used for encapsulating hydrophobic drugs or proteins within hydrophilic hydrogel matrices.

- **Electrostatic Gelation**

Description: Electrostatic gelation utilizes the electrostatic interactions between oppositely charged substances. For example, positively charged substances can be encapsulated within negatively charged hydrogel nanoparticles.

Applications: This method is used for controlled release applications and gene therapy.

- **Layer-by-Layer Assembly**

Description: Layer-by-layer assembly involves the sequential deposition of polyelectrolyte layers onto a substrate, with each layer encapsulating substances of interest [30]. The hydrogel can be formed by incorporating these layers.

Applications: This technique is versatile and can be used for encapsulating a wide range of materials, including drugs, proteins, and DNA.

- **Microfluidics**

Description: Microfluidic devices enable precise control over the encapsulation process by creating droplets of desired size and composition within a continuous flow.

Applications: Microfluidics is used for encapsulating cells, nanoparticles, and drugs with high precision.

- **Spray Drying**

Description: Spray drying involves atomizing a solution containing the substance to be encapsulated into fine droplets, which are then dried to form solid particles.

Applications: It is commonly used for encapsulating pharmaceuticals and food ingredients [31].

### 8.1.6 *Drug Loading and Release*

Drug loading and release are fundamental aspects of pharmaceutical research and development, playing a pivotal role in the design of drug delivery systems that enhance therapeutic efficacy while minimizing adverse effects. These processes represent a delicate balance between encapsulating therapeutic agents within carriers, such as hydrogels, nanoparticles, or liposomes, and subsequently controlling their release into the body [32]. Understanding and manipulating drug loading and release mechanisms are crucial for tailoring drug delivery systems to meet the specific needs of various medical treatments. Drug loading refers to the incorporation of pharmaceutical compounds into a carrier system, which can be a solid matrix, liquid, or gel-like structure. This step is essential for ensuring the stability and targeted delivery of drugs, particularly those with limited solubility, rapid degradation, or toxic side effects. By effectively loading drugs into carriers, researchers can improve drug bioavailability, extend circulation time, and protect sensitive compounds from degradation, thereby optimizing therapeutic outcomes.

On the other hand, drug release focuses on the controlled and sustained delivery of encapsulated drugs from the carrier to the target site within the body. Achieving precise control over the rate and timing of drug release is paramount in maximizing therapeutic benefits and minimizing potential side effects [33]. Different drug delivery systems employ a variety of mechanisms, such as diffusion, degradation of carriers, or external triggers like pH or temperature changes, to regulate drug release kinetics. This delicate interplay between drug loading and release is of paramount importance in fields like oncology, where targeted therapies seek to deliver potent drugs selectively to cancerous cells while sparing healthy tissue [34, 35]. It is also critical in chronic disease management, where maintaining consistent drug levels in the bloodstream is essential for achieving therapeutic goals and patient compliance.

The exploration of innovative drug delivery systems and the understanding of drug loading and release mechanisms continue to be at the forefront of pharmaceutical research. This introduction sets the stage for a comprehensive exploration of these vital processes, their underlying principles, and their applications in improving the effectiveness and safety of drug therapies.

### 8.1.7 *Loading Strategies in Drug Delivery Using Hydrogels*

Drug delivery systems utilizing hydrogels as carriers have gained significant attention due to their potential for precise, controlled, and targeted drug release. To optimize the therapeutic benefits of hydrogel-based drug delivery, various loading strategies have been developed. These strategies determine how drugs are incorporated into the hydrogel matrix, influencing drug release kinetics, bioavailability, and overall efficacy [36]. One prominent loading strategy is "Physical Entrapment." In this method, drugs are enclosed within the hydrogel matrix during gel formation. The mesh size of the hydrogel plays a critical role in determining drug entrapment efficiency. This approach is particularly effective for hydrophilic drugs but may have limitations with hydrophobic compounds.

"Chemical Conjugation" is another strategy wherein hydrogels are chemically modified to introduce functional groups that enable covalent attachment of drug molecules. This ensures controlled drug loading and release, making it especially valuable for precision medicine applications [37]. However, careful consideration of linker chemistry and drug-conjugate stability is necessary. For hydrophobic drugs, "Hydrophobic Interactions" can be harnessed. Hydrogels with hydrophobic domains or modified with hydrophobic moieties can facilitate drug loading through hydrophobic interactions. This approach enhances loading efficiency for lipophilic compounds while preserving the hydrogel's overall hydrophilic nature. "Electrostatic Interactions" leverage the electrostatic forces between charged drugs and oppositely charged hydrogel components. Polyelectrolyte hydrogels, for instance, can encapsulate charged drugs through electrostatic attraction. This approach allows efficient loading and controlled release of ionic compounds [38]. The incorporation of "Microparticles and Nanoparticles" within hydrogel matrices is a widely employed strategy. These drug-loaded particles can be suspended or dispersed in the hydrogel precursor solution, offering precise control over drug content and release kinetics.

"In Situ Gelation" strategies involve the simultaneous formation of the hydrogel network and drug loading. This approach is particularly valuable for injectable hydrogels used in minimally invasive drug delivery procedures [39]. It ensures drug incorporation during gelation, simplifying the process. "Responsive Hydrogels" are designed to respond to specific environmental cues, such as changes in pH or temperature. These hydrogels can trigger drug release in response to physiological conditions, enabling targeted therapies. For example, pH-responsive hydrogels can release drugs selectively in response to variations in acidity, a valuable feature for certain medical applications. Advanced techniques like "Magnetic and Ultrasound-Driven Loading" employ external forces such as magnetic fields or ultrasound to drive drug-loaded nanoparticles into hydrogels. This provides precise spatial and temporal control over drug distribution within the hydrogel matrix, making it valuable for site-specific drug delivery. Selecting the appropriate loading strategy is pivotal in the development of hydrogel-based drug delivery systems. The choice depends on factors like the drug's properties, desired release profile, and the specific requirements of the application. These loading strategies are instrumental in harnessing the full potential of hydrogels, ultimately improving the precision and effectiveness of drug delivery in various therapeutic contexts.

### 8.1.8 Release Kinetics in Drug Delivery Using Hydrogels

Release kinetics in drug delivery using hydrogels is a crucial aspect that governs how drugs are released from the hydrogel matrix over time. Understanding and manipulating these kinetics are vital for tailoring drug delivery systems to meet specific therapeutic needs. Different mechanisms and mathematical models are employed to describe drug release profiles from hydrogels, ensuring the controlled and sustained delivery of drugs to achieve optimal therapeutic outcomes.

- One of the fundamental mechanisms of drug release from hydrogels is "Diffusion-Controlled Release." In this process, drug molecules diffuse through the hydrogel matrix, driven by concentration gradients. The rate of diffusion depends on various factors, including the drug's molecular weight, size, and the mesh size of the hydrogel [40]. This mechanism is suitable for hydrophilic drugs that can readily diffuse through the aqueous hydrogel network.
- "Swelling-Controlled Release" is another prominent release mechanism. As the hydrogel absorbs water from its surroundings, it swells, creating pores and channels within the matrix. Drugs are released as the hydrogel swells and the drug molecules move through these channels [41]. The rate of swelling and release is influenced by factors like hydrogel composition and environmental conditions.
- "Matrix Erosion-Controlled Release" occurs when the hydrogel undergoes degradation or erosion over time. The degradation of the hydrogel matrix leads to the release of encapsulated drugs [42]. This mechanism is particularly relevant for biodegradable hydrogels, which gradually break down in physiological conditions, releasing drugs in a controlled manner.
- "Chemical Reaction-Controlled Release" involves drug release triggered by specific chemical reactions within the hydrogel matrix. For example, pH-sensitive hydrogels can release drugs in response to changes in pH, such as those occurring in different body compartments [32]. Chemical reactions can alter the hydrogel's structure, leading to controlled drug release.

Hydrogels can also exhibit "Swell-Deswell" behavior, where they undergo cyclic swelling and deswelling in response to environmental stimuli. This behavior can be harnessed for on-demand drug release. For instance, temperature-responsive hydrogels can release drugs upon temperature changes, making them suitable for applications requiring precise control. The "Higuchi Model" is a commonly used mathematical model to describe drug release kinetics from hydrogels. It is based on Fick's first law of diffusion and is suitable for systems where drug release is primarily diffusion controlled. This model helps predict drug release rates over time based on factors like the initial drug concentration and diffusion coefficient [43]. Another mathematical model is the "Zero-Order Kinetics" model, which describes drug release at a constant rate independent of the drug's concentration. This model is applicable when drug release is primarily controlled by factors other than concentration gradients, such as erosion or chemical reactions. Release kinetics in drug delivery using hydrogels

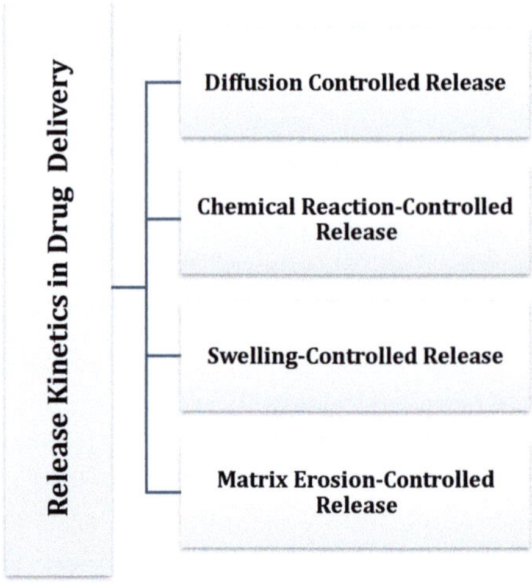


**Fig. 8.5** Different types of drug release mechanism

are multifaceted and influenced by various factors, including hydrogel composition, drug properties, and environmental conditions [32, 43]. Understanding these mechanisms and employing appropriate mathematical models allows researchers to design hydrogel-based drug delivery systems with precise control over drug release, enabling targeted and sustained therapeutic outcomes in a wide range of medical applications (Fig. 8.5).

**Exercise 2.**

**Fill Ups**

**(Polymer concentration, Degradation, Hydrogels, Drug encapsulation, Biomacromolecules)**

1. ...................... are formed through the crosslinking of polymer chains, resulting in a three-dimensional network structure capable of retaining a large amount of water.
2. The structure of hydrogels can be tuned by adjusting parameters such as ........................, crosslinking density, and environmental conditions, allowing for control over properties such as swelling behavior and mechanical strength.
3. ...............................in hydrogels involves the entrapment or loading of therapeutic agents within the hydrogel matrix, enabling controlled delivery and release of the drugs.
4. Hydrogels offer a versatile platform for drug encapsulation due to their ability to accommodate a wide range of drug molecules, including small molecules and large ......................., within their porous structure.
5. .................. of the drug during the encapsulation process can reduce encapsulation efficiency.

### 8.1.9 *Clinical Applications of Hydrogels in Drug Delivery*

#### 8.1.9.1 Wound Healing

Drug delivery systems play a pivotal role in advancing wound care and enhancing the healing process. These systems enable the controlled release of therapeutic agents directly to the wound site, promoting tissue regeneration, reducing inflammation, and preventing infection [44]. The clinical applications of drug delivery in wound healing are as follows:

- **Antimicrobial Agents**: Drug delivery systems can release antimicrobial agents, such as antibiotics or silver nanoparticles, to combat or prevent infections in chronic wounds, surgical incisions, or burns.
- **Growth Factors**: Controlled delivery of growth factors like platelet-derived growth factor (PDGF) and transforming growth factor-beta (TGF-β) can stimulate tissue repair, angiogenesis, and collagen synthesis, expediting wound closure.
- **Pain Management**: Localized drug delivery can administer analgesics or anesthetics directly to the wound area, effectively managing pain and discomfort associated with wounds or surgical procedures.
- **Anti-Inflammatory Agents**: Controlled release of anti-inflammatory drugs, such as corticosteroids or non-steroidal anti-inflammatory drugs (NSAIDs), reduces inflammation at the wound site, promoting a more favorable healing environment.
- **Hemostatic Agents**: Drug delivery systems can deliver hemostatic agents to control bleeding in traumatic wounds, surgeries, or vascular procedures.
- **Exudate Management**: Hydrogels and other wound dressings with drug delivery capabilities can absorb excess wound exudate while releasing therapeutic agents, maintaining a moist wound environment conducive to healing.
- **Tissue Engineering**: Advanced drug delivery systems support tissue engineering approaches by delivering stem cells, growth factors, and scaffolding materials directly to the wound, facilitating the regeneration of functional tissue.
- **Bioactive Dressings**: Modern wound dressings incorporate drug delivery components to provide a combination of moisture management, infection control, and controlled release of therapeutic agents in a single application.
- **Chronic Wound Care**: In the case of chronic wounds, such as diabetic ulcers or pressure sores, drug delivery systems enable long-term, targeted treatment to address underlying issues impeding healing.
- **Scar Management**: Drug delivery systems can release scar-reducing agents like silicone gels or corticosteroids to minimize scarring and enhance cosmetic outcomes after surgical procedures or injury.
- **Biofilm Disruption**: For wounds with biofilm-associated infections, drug delivery can target biofilm disruption agents to eradicate resistant microorganisms and promote wound healing.
- **Controlled Release Bandages**: Bandages equipped with drug delivery capabilities are used to provide continuous and controlled release of medications, ensuring sustained therapeutic effects.

Drug delivery in wound healing has revolutionized clinical approaches to managing a wide range of wounds and injuries. These systems offer targeted and personalized treatment, optimizing the healing process while minimizing complications and discomfort for patients.

#### 8.1.9.2 Ophthalmic Drug Delivery

Drug delivery systems in ophthalmology have transformed the treatment of various eye conditions by providing controlled and sustained release of therapeutic agents directly to the ocular tissues. These advanced delivery methods enhance patient compliance, reduce side effects, and improve the effectiveness of treatments. Here are some clinical applications of drug delivery in ophthalmic care:

- **Glaucoma Management**: Drug delivery systems, such as punctal plugs and drug-eluting contact lenses, are used to provide continuous and controlled release of intraocular pressure-lowering medications for the treatment of glaucoma.
- **Dry Eye Syndrome**: Ocular inserts and artificial tear formulations with drug delivery capabilities help manage the symptoms of dry eye syndrome by providing sustained lubrication and delivering anti-inflammatory agents to the ocular surface.
- **Retinal Diseases**: Intravitreal injections and implants enable targeted drug delivery to the retina for conditions like age-related macular degeneration (AMD) and diabetic retinopathy, reducing the need for frequent injections and improving vision outcomes.
- **Antibacterial and Antifungal Treatment**: Ophthalmic drug delivery systems deliver antibiotics or antifungal agents to treat ocular infections, such as conjunctivitis and keratitis, ensuring effective local therapy.
- **Post-Operative Care**: Drug-eluting implants and formulations are used to provide anti-inflammatory and anti-infective medications following ocular surgeries, such as cataract surgery and corneal transplantation.
- **Allergic Conjunctivitis**: Topical ophthalmic solutions with controlled drug release are employed to manage allergic conjunctivitis by delivering antihistamines and mast cell stabilizers directly to the eye.
- **Inflammatory Eye Disorders**: Drug delivery systems deliver corticosteroids or non-steroidal anti-inflammatory drugs (NSAIDs) for the treatment of inflammatory conditions like uveitis and scleritis, reducing inflammation and pain.
- **Pain Management**: Ocular inserts and drug-eluting contact lenses can provide sustained release of analgesics to alleviate pain and discomfort after eye surgeries or injuries.
- **Treatment of Ocular Neoplasms**: Intraocular implants loaded with chemotherapeutic agents are used to treat ocular tumors and retinoblastoma while minimizing systemic side effects.

- **Retinal Vein Occlusion**: Drug delivery devices are employed to administer anti-VEGF agents directly into the vitreous cavity to treat conditions like retinal vein occlusion and associated macular edema.
- **Corneal Disorders**: Ophthalmic drug delivery systems are used to address corneal disorders, including corneal ulcers and dystrophies, by delivering medications that promote healing and reduce pain.
- **Pediatric Eye Care**: Ocular inserts and specialized formulations are designed for pediatric patients to enhance treatment adherence and reduce discomfort during ophthalmic therapies.

In conclusion, drug delivery in ophthalmic care offers precise and targeted treatment options for a wide range of eye conditions. These systems not only improve therapeutic outcomes but also enhance patient comfort and compliance, making them integral to modern ophthalmology practice.

#### 8.1.9.3 Vaccine Delivery Using Hydrogels

Hydrogels have emerged as promising vehicles for vaccine delivery due to their unique properties. This application is critical in ensuring the effective administration of vaccines. Key points include:

- **Stabilization of Vaccines**: Hydrogels can encapsulate and protect fragile vaccine components, such as proteins or antigens, from degradation, ensuring vaccine stability during storage and transport.
- **Sustained Release**: Hydrogels can provide a controlled and sustained release of antigens, which enhances the immune response by prolonging exposure to the immune system [45].
- **Needle-Free Delivery**: Some hydrogel-based vaccine formulations can be administered orally or through transdermal patches, eliminating the need for traditional needle injections and reducing pain and anxiety associated with vaccinations.
- **Adjuvant Properties**: Hydrogels can be engineered to include adjuvants, substances that enhance the immune response to vaccines, further boosting vaccine efficacy.
- **Improved Patient Compliance**: Needle-free or painless administration options offered by hydrogel-based vaccines may improve patient compliance, particularly in pediatric and needle-phobic populations.
- **Tailored Release Profiles**: Hydrogel formulations can be tailored to release antigens in response to specific physiological conditions, ensuring optimal vaccine delivery to target cells or tissues.
- **Multivalent Vaccines**: Hydrogels allow for the incorporation of multiple vaccine components, enabling the development of multivalent vaccines that protect against several diseases in a single administration.
- **Enhanced Vaccine Efficacy**: The controlled release and presentation of antigens by hydrogels can enhance the activation of immune cells, leading to a more robust and long-lasting immune response.

- **Customizable Delivery Routes**: Hydrogels offer versatility in delivery routes, allowing for oral, nasal, subcutaneous, or intradermal administration based on the specific vaccine and patient requirements.
- **Potential for Thermostability**: Certain hydrogel formulations can help maintain vaccine stability at higher temperatures, critical for vaccine distribution in resource-limited regions with inadequate cold chain infrastructure.

Hydrogels play a pivotal role in vaccine delivery by stabilizing vaccines, providing controlled release, and offering alternative, patient-friendly administration methods. These advantages contribute to improved vaccine efficacy and accessibility.

#### 8.1.9.4 Cancer Therapy

Hydrogels have demonstrated significant clinical applications in drug delivery within the context of cancer therapy. One of the primary challenges in cancer treatment is the need for targeted drug delivery to minimize systemic toxicity while maximizing therapeutic efficacy. Hydrogels offer a promising solution to address this challenge.

In cancer therapy, hydrogels can be loaded with chemotherapeutic agents and designed to release these drugs at the tumor site. This localized drug delivery approach minimizes exposure to healthy tissues, reducing the severe side effects associated with conventional systemic chemotherapy. Additionally, hydrogels can be engineered to respond to specific physiological cues, such as pH or enzyme activity in the tumor microenvironment, ensuring controlled and precise drug release at the cancer site. Furthermore, hydrogels can be administered through various routes, including injection or implantation near or within the tumor. This flexibility allows for tailored treatment strategies based on the tumor's location and characteristics. Moreover, hydrogel-based drug delivery systems can provide sustained drug release, maintaining effective drug concentrations within the tumor over an extended period. This sustained exposure enhances the cytotoxic effects on cancer cells, potentially improving treatment outcomes. Hydrogels also enable the encapsulation of a wide range of therapeutic agents, including small molecules, biologics, and nanoparticles, making them versatile carriers for diverse cancer therapies [46]. These hydrogel-based drug delivery systems are under active research and development, with the aim of improving the efficacy of cancer treatments, reducing side effects, and enhancing the overall quality of life for cancer patients.

The clinical application of hydrogels in drug delivery for cancer therapy offers a promising approach to enhance the specificity, efficacy, and tolerability of cancer treatments. The ability to precisely target tumors and provide controlled drug release represents a significant advancement in the field of oncology, offering new hope in the fight against cancer.

#### 8.1.9.5 Diabetes Management

Hydrogels have found valuable clinical applications in drug delivery, particularly in the context of diabetes management. Diabetes is a chronic condition that often requires frequent and precise administration of insulin to regulate blood glucose levels. Hydrogels offer innovative solutions to improve the delivery of insulin and enhance the quality of life for individuals with diabetes. One notable application is the development of insulin-loaded hydrogels. These hydrogels can encapsulate insulin and release it gradually in response to changing glucose levels. Such controlled release systems mimic the physiological function of the pancreas, which secretes insulin in response to glucose fluctuations. By delivering insulin in a more natural and responsive manner, hydrogels help maintain stable blood sugar levels, reducing the risk of hypoglycemia and hyperglycemia [47].

Moreover, hydrogel-based insulin delivery can offer convenience and improve patient compliance. Some formulations allow for less frequent injections, reducing the burden of daily insulin administration. This can lead to enhanced treatment adherence and better glycemic control. Hydrogels are also advantageous for protecting insulin from enzymatic degradation in the gastrointestinal tract, making oral insulin delivery a possibility. This route, if successfully developed, could eliminate the need for injections altogether, further improving the patient experience. In addition to insulin, hydrogels can accommodate other diabetes medications, expanding their potential applications. For example, glucagon-like peptide-1 receptor agonists (GLP-1 RAs), which stimulate insulin secretion and reduce appetite, can be delivered using hydrogel-based systems to manage blood glucose and body weight effectively [47]. The clinical use of hydrogels in diabetes management represents a significant step forward in personalized and patient-centered care. These innovative drug delivery systems have the potential to enhance the precision, convenience, and effectiveness of diabetes treatments, ultimately improving the overall well-being of individuals living with diabetes.

#### 8.1.9.6 Postoperative Pain Management

Hydrogels have emerged as valuable tools in the clinical realm of postoperative pain management. After surgical procedures, patients often experience discomfort and require pain relief. Hydrogels offer several advantages in this context, contributing to improved patient comfort and recovery. One key application of hydrogels in post-operative pain management is the development of localized drug delivery systems. These hydrogels can be loaded with analgesic medications, such as opioids or non-steroidal anti-inflammatory drugs (NSAIDs), and applied directly to the surgical site [48]. By doing so, they provide targeted and sustained pain relief precisely where it is needed, minimizing the need for systemic pain medications that can have side effects.

The controlled release properties of hydrogels allow for a gradual and prolonged release of pain-relieving drugs, ensuring consistent pain management over an extended period. This is particularly beneficial during the initial stages of post-operative recovery when patients may experience acute pain. Hydrogel-based drug delivery systems can help maintain adequate pain relief while reducing the risk of opioid-related side effects and dependency. Furthermore, hydrogels can be designed to adhere to the surgical site, forming a protective barrier against infection and promoting wound healing. This dual function enhances the overall postoperative recovery process by not only managing pain but also reducing the risk of complications [48]. Hydrogel-based dressings are also employed in postoperative wound care. These dressings can maintain a moist environment, which is conducive to optimal wound healing, while also providing pain relief through drug-loaded hydrogels. This combination of features can expedite the healing process and improve patient comfort during the critical postoperative phase.

Hydrogels have made significant contributions to postoperative pain management by providing localized, controlled, and sustained drug delivery at surgical sites. This approach enhances patient comfort, reduces the reliance on systemic pain medications, and supports a smoother recovery process after surgical procedures.

#### 8.1.9.7 Intra-Articular Injections

These injections target joints and are essential for treating various joint-related conditions, including osteoarthritis and rheumatoid arthritis. The use of hydrogels in this area offers several advantages. One key benefit of using hydrogels in intra-articular injections is their ability to provide sustained and controlled drug release within the joint space [49]. Joint tissues are enclosed by a synovial membrane, which makes drug penetration challenging. Hydrogels can encapsulate therapeutic agents and slowly release them over time, ensuring that the medication stays within the joint, where it's needed the most. This controlled release minimizes the need for frequent injections and helps maintain therapeutic drug levels, improving patient compliance. Moreover, hydrogels can be tailored to respond to the specific physiological conditions of the joint. For example, they can be designed to release drugs in response to changes in pH or enzyme activity within the synovial fluid. This responsiveness ensures that the drug is delivered precisely when and where it's most effective, optimizing therapeutic outcomes while reducing potential side effects.

Hydrogel-based intra-articular injections can also improve the overall patient experience. These injections are often used to provide pain relief and reduce inflammation associated with joint disorders. By delivering drugs directly to the affected joint, hydrogels can enhance the efficacy of pain management while minimizing systemic exposure to medications [49]. This can lead to reduced side effects and improved quality of life for patients with joint conditions.

Hydrogels are making a significant impact on drug delivery in intra-articular injections by providing controlled, sustained, and targeted drug release within the joint space. This approach offers benefits such as improved patient compliance, enhanced therapeutic outcomes, and reduced systemic side effects, making it a valuable tool in the management of joint-related disorders.

#### 8.1.9.8 Oral Drug Delivery

This mode of drug administration is widely used for its convenience and patient compliance, and hydrogels play a crucial role in enhancing its effectiveness. One key advantage of using hydrogels in oral drug delivery is their ability to protect drugs from the harsh conditions of the gastrointestinal tract. Hydrogels can encapsulate drugs, shielding them from acidic environments in the stomach and enzymatic degradation in the intestines. This protective barrier ensures the drug's stability and bioavailability, allowing it to reach its intended target in a more effective and intact form. Moreover, hydrogel-based drug delivery systems can provide controlled release of medications. By adjusting the composition of the hydrogel, drug release can be tailored to specific requirements. This means that drugs can be released gradually over an extended period, reducing the need for frequent dosing and providing a consistent therapeutic effect [50]. Patients benefit from improved compliance, as they don't have to remember to take medications multiple times a day.

Hydrogels also offer versatility in drug delivery for different patient populations. They can be formulated to release drugs in response to specific physiological conditions, such as changes in pH or enzyme activity in the gastrointestinal tract. This targeted release ensures that the drug is delivered where it's most needed, enhancing its efficacy while minimizing side effects. In addition to their drug protection and release capabilities, hydrogels can improve the overall patient experience. They are often used to create easy-to-swallow formulations for individuals who have difficulty swallowing traditional solid tablets or capsules. This makes medication administration more accessible and comfortable, especially for pediatric and geriatric patients.

Hydrogels are valuable tools in oral drug delivery, offering drug protection, controlled release, and enhanced patient compliance. Their ability to navigate the challenges of the gastrointestinal tract while providing precise drug delivery makes them an essential component in improving the effectiveness of oral medications (Fig. 8.6).

#### 8.1.9.9 Sustained Drug Release

The controlled and prolonged release of medications is crucial for various therapeutic purposes, and hydrogels excel in providing this benefit.

One of the key advantages of hydrogels in sustained drug release is their ability to encapsulate drugs and release them gradually over an extended period. This controlled release mechanism is especially valuable in chronic conditions where maintaining consistent drug levels is essential for therapeutic efficacy. By releasing drugs slowly, hydrogels help reduce the need for frequent dosing, improve patient compliance, and minimize potential side effects associated with fluctuating drug concentrations [51]. Hydrogels can be tailored to release drugs in response to specific environmental cues. For instance, pH-responsive hydrogels can release drugs in response to changes in acidity, such as those found in specific body tissues or

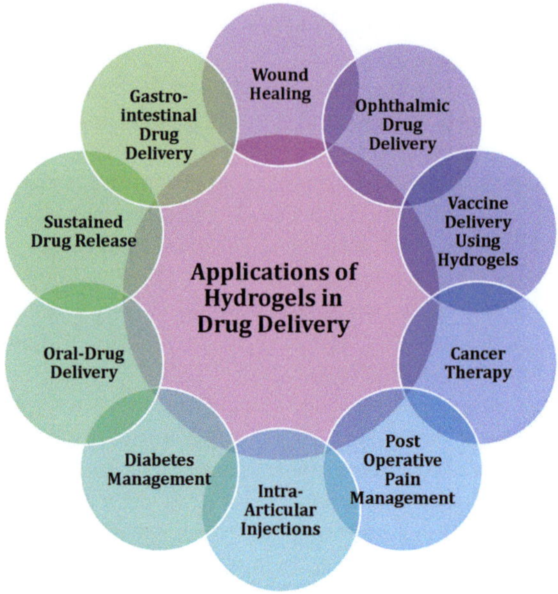


**Fig. 8.6** Applications of hydrogels in drug delivery

compartments. This feature ensures that medications are released precisely where they are needed, enhancing therapeutic outcomes while minimizing off-target effects. Furthermore, the biocompatibility of hydrogels is a critical factor in their clinical success. They are generally well-tolerated by the body, reducing the risk of adverse reactions. This makes hydrogel-based drug delivery systems suitable for a wide range of therapeutic applications, including pain management, hormone replacement therapy, and the treatment of various chronic diseases. Hydrogels are also versatile in terms of drug types that can be delivered. They can encapsulate small molecules, proteins, peptides, and nanoparticles, allowing for the sustained release of a wide variety of therapeutic agents. This adaptability broadens their utility across different medical disciplines.

The clinical application of hydrogels in sustained drug release is pivotal for managing chronic conditions and improving patient outcomes. Their controlled drug release, responsiveness to physiological cues, biocompatibility, and versatility make hydrogels valuable tools in delivering therapeutic agents over extended periods, ultimately enhancing the effectiveness and convenience of medical treatments.

#### 8.1.9.10 Gastrointestinal Drug Delivery

The gastrointestinal tract poses unique obstacles to effective drug absorption, but hydrogels offer innovative solutions. One key advantage of hydrogels in gastrointestinal drug delivery is their ability to protect drugs from harsh acidic environments in the stomach. Hydrogels can encapsulate drugs, shielding them from degradation and ensuring their stability as they pass through the stomach. This protection is crucial for drugs that are sensitive to gastric acidity, allowing them to reach the intestines in an intact and active form. Hydrogels can be designed to respond to specific pH levels within the gastrointestinal tract. pH-responsive hydrogels can release drugs in response to changes in acidity, ensuring targeted drug delivery to specific regions of the gut. This feature is particularly beneficial when drugs need to be released in the small intestine or colon for optimal absorption or local treatment. Hydrogel-based drug delivery systems also offer the advantage of controlled drug release. By adjusting the composition of the hydrogel, drug release rates can be tailored to match the drug's pharmacokinetics and therapeutic requirements. This controlled release minimizes rapid drug fluctuations in the bloodstream, improving drug efficacy and reducing potential side effects. Additionally, hydrogels are versatile carriers for a wide range of drugs, including small molecules, biologics, and nanoparticles. This versatility makes them suitable for delivering a broad spectrum of medications used in gastrointestinal conditions, from oral antibiotics to targeted therapies for gastrointestinal diseases like Crohn's disease.

Hydrogels play a crucial role in gastrointestinal drug delivery by protecting drugs from stomach acidity, providing controlled release, and enabling targeted drug delivery to specific regions of the gastrointestinal tract. Their adaptability and biocompatibility make them valuable tools for enhancing drug absorption and improving therapeutic outcomes in the complex environment of the digestive system.

#### 8.1.9.11 Recent Advances and Future Trends

Recent advances in hydrogels for drug delivery have paved the way for exciting future trends in this field, offering innovative solutions to enhance therapeutic outcomes and patient experiences.

- **Smart Hydrogels**: Recent research has focused on the development of smart hydrogels that respond to specific physiological cues, such as pH, temperature, or enzyme activity. These hydrogels enable precise control over drug release, ensuring medications are delivered where and when they are needed most [12, 52].
- **Nanotechnology Integration**: Integration with nanotechnology has led to the development of hydrogel nanoparticles and nanocomposites. These nanoscale drug carriers offer improved drug loading capacities, sustained release profiles, and the potential for targeted delivery to specific cells or tissues.
- **Personalized Medicine**: Advances in 3D printing and tissue engineering have allowed for the creation of patient-specific hydrogel-based drug delivery systems. Tailored to individual needs, these systems optimize treatment regimens and improve therapeutic efficacy.
- **Biodegradable Hydrogels**: Biodegradable hydrogels have gained attention for their ability to gradually degrade in the body, eliminating the need for removal or retrieval after drug release. This minimizes the invasiveness of drug delivery systems [14].

- **Combination Therapy**: Hydrogels are increasingly being used for combination therapy, where multiple drugs or therapeutic agents are delivered simultaneously. This approach is particularly promising in cancer treatment, infectious diseases, and chronic conditions.
- **Immunotherapy Delivery**: Hydrogels are being explored as carriers for immunotherapeutic agents, such as checkpoint inhibitors and immune-stimulating molecules. These systems aim to enhance the immune response against diseases like cancer.
- **Targeted Delivery Systems**: Advances in hydrogel design have enabled the development of highly targeted delivery systems. These systems can guide drugs to specific cells, tissues, or organs, reducing off-target effects and improving therapeutic outcomes.
- **Remote-Controlled Drug Release**: Research into remotely controlled drug release systems, where external stimuli like ultrasound, magnetic fields, or light trigger drug release from hydrogels, is an emerging trend. This approach allows for precise and on-demand drug delivery.
- **Long-Acting Implants**: Hydrogel-based implants are being explored for long-acting drug delivery. These implants can provide sustained release of medications over weeks, months, or even years, offering convenience and improved patient compliance.
- **Environmental Sensing Hydrogels**: Hydrogels that can sense changes in the environment, such as glucose levels in diabetes management or infection markers in wound care, are under development [53]. These hydrogels can trigger drug release in response to specific conditions.

Recent advances in hydrogel-based drug delivery systems are driving innovative trends in personalized medicine, nanotechnology integration, and smart drug release. These developments hold great promise for improving drug efficacy, reducing side effects, and enhancing patient outcomes in various medical applications. Future research in this field will likely continue to revolutionize the way medications are delivered and administered.

**Exercise 3.**

**Choose The Correct Option**

1. What are hydrogels primarily composed of?
   a) Metals
   b) Polymers
   c) Ceramics
   d) Liquids

2. What is a characteristic property of hydrogels?
   a) Low swelling capacity
   b) Brittle nature
   c) High water content
   d) Low porosity

3. Which of the following is NOT a potential application of hydrogels in drug delivery?
   a) Sustained release
   b) Targeted delivery
   c) Rapid drug degradation
   d) Controlled release

4. How do hydrogels release drugs?
   a) Through a single burst release
   b) Through slow diffusion
   c) Through chemical reaction
   d) Through physical erosion

5. Which of the following factors influences the drug release rate from hydrogels?
   a) Temperature
   b) pH
   c) Presence of enzymes
   d) All of the above

6. Which type of drug molecules can be incorporated into hydrogels for controlled release?
   a) Only hydrophobic drugs
   b) Only hydrophilic drugs
   c) Both hydrophobic and hydrophilic drugs
   d) Only small molecules

7. How does the structure of hydrogels contribute to their ability to hold and release drugs?
   a) Hydrogels have a rigid structure
   b) Hydrogels have a porous structure
   c) Hydrogels have a smooth surface
   d) Hydrogels have a dense structure

8. Which of the following methods can be used to prepare hydrogels for drug delivery?
   a) Solvent casting
   b) Freeze-drying
   c) Electrospinning
   d) All of the above

9. What is the advantage of using hydrogels for drug delivery?
   a) Limited biocompatibility
   b) Rapid drug release
   c) Reduced side effects
   d) Low stability

10. What is the future outlook for hydrogels in drug delivery?
   a) Limited applications
   b) Declining interest
   c) Continued research and development for enhanced drug delivery systems
   d) Replacement by traditional drug delivery methods

## 8.2 Summary

In the chapter on "Hydrogels in Drug Delivery," the versatile and evolving role of hydrogels in the pharmaceutical field is explored in depth. Hydrogels, 3D networks of hydrophilic polymers, have garnered significant attention due to their unique properties that make them ideal candidates for drug delivery systems. One of the standout features of hydrogels is their high water content and biocompatibility, which closely resemble the physiological environment of living tissues. This characteristic allows hydrogels to interact favorably with biological systems, reducing the risk of adverse reactions and making them suitable for various drug delivery applications. Hydrogels excel in the controlled release of drugs, a critical aspect of effective therapy. By encapsulating drugs within the hydrogel matrix, they can be released gradually over time, maintaining therapeutic drug levels in the body. This controlled release not only enhances drug efficacy, but also reduces the frequency of dosing, leading to improved patient compliance. Moreover, the responsive nature of hydrogels has opened up new avenues in drug delivery innovation. Smart hydrogels can be engineered to respond to specific environmental cues, such as changes in pH, temperature, or enzyme activity. This responsiveness allows for precise and targeted drug release, minimizing off-target effects and improving the overall therapeutic impact. In recent years, the integration of nanotechnology with hydrogels has further expanded their capabilities. Nanoscale hydrogel carriers offer enhanced drug loading capacities, sustained release profiles, and the potential for targeted delivery to specific cells or tissues. This has significantly broadened the spectrum of drugs that can be effectively delivered using hydrogel-based systems.

Overall, the chapter on "Hydrogels in Drug Delivery" underscores the pivotal role of hydrogels in advancing drug delivery technologies. Their adaptability, biocompatibility, and ability to provide controlled and targeted release make them invaluable tools in improving the efficacy and safety of drug therapies across various medical applications. As research and innovation continue, hydrogels are likely to remain at the forefront of drug delivery advancements, offering hope for more effective and personalized treatments.

## Points to Remember

- **Hydrogels possess remarkable self-healing properties, meaning they can autonomously repair damage to their structure. This feature allows for sustained drug release over extended periods, enhancing the efficacy and longevity of drug delivery systems.**
- **Certain hydrogels exhibit temperature-responsive behavior, undergoing reversible changes in swelling and drug release in response to variations in temperature. This property enables precise control over drug release kinetics, optimizing therapeutic outcomes.**
- **Hydrogels can be engineered to deliver multiple drugs simultaneously, either sequentially or in combination. This capability opens up new possibilities for the treatment of complex medical conditions and diseases that require synergistic drug combinations.**
- **Injectable hydrogels offer a minimally invasive approach to drug delivery, allowing for targeted delivery to specific sites within the body. These hydrogels can be injected as a liquid and undergo gelation in situ, conforming to the shape of the target tissue and releasing drugs locally.**
- **Hydrogels can be designed to respond to various stimuli, such as pH, enzymes, light, or magnetic fields, triggering drug release in a controlled manner. This spatiotemporal control over drug release enhances therapeutic efficacy while minimizing side effects.**
- **Some hydrogels exhibit bio adhesive properties, allowing them to adhere to biological surfaces such as mucosal membranes or injured tissues. This adhesive behavior enables prolonged contact and sustained drug release at the target site, improving treatment outcomes.**
- **Hydrogels can encapsulate living cells for therapeutic purposes, such as cell therapy and regenerative medicine. These cell-laden hydrogels provide a supportive environment for cell growth and function, facilitating tissue repair and regeneration.**
- **Hydrogels are used in biosensing devices for real-time monitoring of biomarkers and analytes in the body. By incorporating sensing elements into the hydrogel matrix, these devices can detect specific molecules and trigger drug release in response to physiological changes.**

## References


1. Vigata M, Meinert C, Hutmacher DW, Bock N (2020) Hydrogels as drug delivery systems: a review of current characterization and evaluation techniques. Pharmaceutics 12(12):1188. https://doi.org/10.3390/pharmaceutics12121188
2. Li J, Mooney DJ (2016) Designing hydrogels for controlled drug delivery. Nat Rev Mater 1(12):16081. https://doi.org/10.1038/natrevmats.2016.81
3. Del Valle L, Díaz A, Puiggalí J (2018) Hydrogels for biomedical applications: cellulose, chitosan, and protein/peptide derivatives. Gels 3(3):28. MDPI AG. https://doi.org/10.3390/gels3030028
4. Narayanaswamy R, Torchilin VP (2019) Hydrogels and their applications in targeted drug delivery. Molecules (Basel, Switzerland) 24(3):603. https://doi.org/10.3390/molecules24030603

5. Hoare TR, Kohane DS (2008) Hydrogels in drug delivery: Progress and challenges. Polymer 49(8):1993–2008
6. Ho TC, Chang CC, Chan HP, Chung TW, Shu CW, Chuang KP, Duh TH, Yang MH, Tyan YC (2022) Hydrogels: properties and applications in biomedicine. Molecules (Basel, Switzerland) 28(9):2902. https://doi.org/10.3390/molecules28092902
7. Wang Y (2018) Programmable hydrogels. Biomaterials 188:663–680. https://doi.org/10.1016/j.biomaterials.2018.03.008
8. Song F, Li X, Wang Q, Liao L, Zhang C (2015) Nanocomposite hydrogels and their applications in drug delivery and tissue engineering. J Biomed Nanotechnol 11(1):40–52. https://doi.org/10.1166/jbn.2015.1962
9. Kashyap N, Kumar N, Kumar MN (2005) Hydrogels for pharmaceutical and biomedical applications. Crit Rev Ther Drug Carrier Syst 22(2):108–149. https://doi.org/10.1615/critrevtherdrugcarriersyst.v22.i2.10
10. Ahmed EM (2015) Hydrogel: preparation, characterization, and applications: a review. J Adv Res 6(2):105–121. https://doi.org/10.1016/j.jare.2013.08.006
11. Guo B, Liang Y, Dong R (2023) Physical dynamic double-network hydrogels as dressings to facilitate tissue repair. Nat Protoc https://doi.org/10.1038/s41596-023-00888-9
12. Gu D, O'Connor AJ, GH Qiao G, Ladewig K (2018) Hydrogels with smart systems for delivery of hydrophobic drugs. Expert Opin Drug Deliv 14(8):889–895. https://doi.org/10.1080/18425248.2018.1245290
13. Vasile C, Pamfil D, Stoleru E, Baican M (2020) New developments in medical applications of hybrid hydrogels containing natural polymers. Molecules (Basel, Switzerland) 25(8):1539. https://doi.org/10.3390/molecules25081539
14. Ozcelik B (2016) Degradable hydrogel systems for biomedical applications
15. Lee JH (2018) Injectable hydrogels delivering therapeutic agents for disease treatment and tissue engineering. Biomater Res 22:28. https://doi.org/10.1186/s40824-018-0138-6
16. Pishavar E, Khosravi F, Naserifar M, Rezvani Ghomi E, Luo H, Zavan B, Seifalian A et al (2021) Multifunctional and self-healable intelligent hydrogels for cancer drug delivery and promoting tissue regeneration in vivo. Polymers 13(16):2680. MDPI AG. https://doi.org/10.3390/polym13162680
17. Guiseppi-Elie A (2010) Electroconductive hydrogels: synthesis, characterization and biomedical applications. Biomaterials 31(10):2801–2816. https://doi.org/10.1016/j.biomaterials.2009.12.052
18. Corkhill PH, Hamilton CJ, Tighe BJ (1989) Synthetic hydrogels. VI. Hydrogel composites as wound dressings and implant materials. Biomaterials 10(1):3–10. https://doi.org/10.1016/0142-9612(89)90002-1
19. Kopecek J (2008) Hydrogel biomaterials: a smart future? Biomaterials 28(34):5185–5192. https://doi.org/10.1016/j.biomaterials.2008.08.044
20. Bustamante-Torres M, Romero-Fierro D, Arcentales-Vera B, Palomino K, Magaña H, Bucio E (2021) Hydrogels classification according to the physical or chemical interactions and as stimuli-sensitive materials. Gels 8(4):182. MDPI AG. https://doi.org/10.3390/gels8040182
21. Onaciu A, Munteanu RA, Moldovan AI, Moldovan CS, Berindan-Neagoe I (2019) Hydrogels based drug delivery synthesis, characterization and administration. Pharmaceutics 11(9):432. https://doi.org/10.3390/pharmaceutics11090432
22. Jacob S, Nair AB, Shah J, Sreeharsha N, Gupta S, Shinu P (2021) Emerging role of hydrogels in drug delivery systems, tissue engineering and wound management. Pharmaceutics 13(3):358. https://doi.org/10.3390/pharmaceutics13030358
23. Khan F, Atif M, Haseen M, Kamal S, Khan MS, Shahid S, Nami SAA (2022) Synthesis, classification and properties of hydrogels: their applications in drug delivery and agriculture. J Mater Chem B 10(2):180–203. https://doi.org/10.1039/d1tb01345a
24. Yang JM, Olanrele OS, Zhang X, Hsu CC (2018) Fabrication of hydrogel materials for biomedical applications. Adv Exp Med Biol 1088:198–224. https://doi.org/10.1008/988-981-13-0948-2_12

25. Yigit S, Sanyal R, Sanyal A (2011) Fabrication and functionalization of hydrogels through "click" chemistry. Chem Asian J 6(10):2648–2659. https://doi.org/10.1002/asia.201100440
26. Gul K, Gan RY, Sun CX, Jiao G, Wu DT, Li HB, Kenaan A, Corke H, Fang YP (2022) Recent advances in the structure, synthesis, and applications of natural polymeric hydrogels. Crit Rev Food Sci Nutr 62(14):3818–3832. https://doi.org/10.1080/10408398.2020.1880034
27. Ungerleider JL, Johnson TD, Rao N, Christman KL (2015) Fabrication and characterization of injectable hydrogels derived from decellularized skeletal and cardiac muscle. Methods (San Diego, Calif.) 84:53–59. https://doi.org/10.1016/j.ymeth.2015.03.024
28. Pérez-Luna V, González-Reynoso O (2018) Encapsulation of biological agents in hydrogels for therapeutic applications. Gels 4(3):61. MDPI AG. https://doi.org/10.3390/gels4030061
29. Jamshaid U, Anton N, Elhassan M, Conzatti G, Vandamme TF (2024) Novel hydrogels based on the Nano-Emulsion formulation process: development, rheological characterization, and study as a drug delivery system. Pharmaceutics 16(6):812. https://doi.org/10.3390/pharmaceutics16060812
30. Seidi F, Zhao W, Xiao H, Jin Y, Zhao C (2020) Layer-by-Layer assembly for surface tethering of Thin-Hydrogel films: design strategies and applications. Chem Rec 20(8): 857–881. https://doi.org/10.1002/tcr.202000007
31. Anandharamakrishnan C, Ishwarya SP (2015) Spray drying techniques for food ingredient encapsulation. https://doi.org/10.1002/9781118863985
32. Kim SW, Bae YH, Okano T (1992) Hydrogels: swelling, drug loading, and release. Pharm Res 9(3):283–290. https://doi.org/10.1023/a:1015888213431
33. Dong K, Dong Y, You C, Xu W, Huang X, Yan Y, Zhang L, Wang K, Xing J (2016) Assessment of the drug loading, in vitro and in vivo release behavior of novel pH-sensitive hydrogel. Drug Deliv 23(1):184–184. https://doi.org/10.3109/10818544.2014.908329
34. Li J, Weber E, Guth-Gundel S, Schuleit M, Kuttler A, Halleux C, Accart N, Doelemeyer A, Basler A, Tigani B, Wuersch K, Fornaro M, Kneissel M, Stafford A, Freedman BR, Mooney DJ (2018) Tough composite hydrogels with high loading and local release of biological drugs. Adv Healthc Mater 8(9):e1801393. https://doi.org/10.1002/adhm.201801393
35. Caccavo D, Cascone S, Lamberti G, Barba AA (2015) Controlled drug release from hydrogel-based matrices: experiments and modeling. Int J Pharm 486(1–2):144–152. https://doi.org/10.1016/j.ijpharm.2015.03.054
36. Raina N, Pahwa R, Bhattacharya J, Paul AK, Nissapatorn V, de Lourdes Pereira M, Oliveira SMR et al (2022) Drug delivery strategies and biomedical significance of hydrogels: translational considerations. Pharmaceutics 14(3):584. MDPI AG. https://doi.org/10.3390/pharmaceutics14030584
37. Gavan A, Colobatiu L, Hanganu D, Bogdan C, Olah N, Achim M, Mirel S (2022) Development and evaluation of hydrogel wound dressings loaded with herbal extracts. Processes 10(2):242. MDPI AG. https://doi.org/10.3390/pr10020242
38. Vesvoranan O, Liu BS, Zheng Y, Wagner WL, Sutlive J, Chen Z, Khalil HA et al (2022) Active loading of pectin hydrogels for targeted drug delivery. Polymers 15(1):92. MDPI AG. https://doi.org/10.3390/polym15010092
39. Alonso JM, Andrade del Olmo J, Perez Gonzalez R, Saez-Martinez V (2021) Injectable hydrogels: from laboratory to industrialization. Polymers 13(4):650. MDPI AG. https://doi.org/10.3390/polym13040650
40. Mercer IG, Italiano AN, Gazaryan IG, Steiner AB, Kazakov SV (2023) Degradation kinetics of disulfide cross-linked microgels: real-time monitoring by confocal microscopy. Gels 9(10):882. MDPI AG. https://doi.org/10.3390/gels9100882
41. Mortensen N, Toews P, Bates J (2022) Crosslinking-dependent drug kinetics in hydrogels for ophthalmic delivery. Polymers 14:248. https://doi.org/10.3390/polym14020248
42. Wang Y, Tu S, Pinchuk AN, Xiong MP (2013) Active drug encapsulation and release kinetics from hydrogel-in-liposome nanoparticles. J Colloid Interface Sci 406:248–255. https://doi.org/10.1016/j.jcis.2013.05.081
43. Kim S, Lee K, Cha C (2016) Refined control of thermoresponsive swelling/deswelling and drug release properties of poly(N-isopropylacrylamide) hydrogels using hydrophilic polymer

crosslinkers. J Biomater Sci Polym Ed 28(18):1698–1811. https://doi.org/10.1080/09205063.2016.1230933
44. Li J, Chen Q, Wang J, Pan X, Zhang J (2021) Insight into bioactive hydrogels for wound healing and drug delivery systems. Curr Med Chem 28(42):8692–8810. https://doi.org/10.2184/0929868328666210405125348
45. Basu P, Saha N, Saha T, Saha P (2021) Polymeric hydrogel based systems for vaccine delivery: a review. Polymer 230:124088. https://doi.org/10.1016/j.polymer.2021.124088
46. Sun Z, Song C, Wang C, Hu Y, Wu J (2020) Hydrogel-based controlled drug delivery for cancer treatment: a review. Mol Pharm 18(2):383–391. https://doi.org/10.1021/acs.molpharmaceut.9b01020
47. Fuchs S, Ernst AU, Wang LH, Shariati K, Wang X, Liu Q, Ma M (2021) Hydrogels in emerging technologies for type 1 diabetes. Chem Rev 121(18):11458–11526. https://doi.org/10.1021/acs.chemrev.0c01062
48. Cai J, Guo J, Wang S (2023) Application of polymer hydrogels in the prevention of postoperative adhesion: a review. Gels (Basel, Switzerland) 9(2):98. https://doi.org/10.3390/gels9020098
49. Song P, Cui Z, Hu L (2022) Applications and prospects of intra-articular drug delivery system in arthritis therapeutics. J Controll Release: Off J Controll Release Soc 352:946–960. https://doi.org/10.1016/j.jconrel.2022.11.018
50. Sharpe LA, Daily AM, Horava SD, Peppas NA (2014) Therapeutic applications of hydrogels in oral drug delivery. Expert Opin Drug Deliv 11(6):901–915. https://doi.org/10.1518/18425248.2014.902048
51. Peers S, Montembault A, Ladavière C (2020) Chitosan hydrogels for sustained drug delivery. J Controll Release: Off J Controll Release Soc 326:150–163. https://doi.org/10.1016/j.jconrel.2020.06.012
52. Dos Santos AM, Carvalho SG, Araujo VHS, Carvalho GC, Gremião MPD, Chorilli M (2020) Recent advances in hydrogels as strategy for drug delivery intended to vaginal infections. Int J Pharm 590:119868. https://doi.org/10.1016/j.ijpharm.2020.119868
53. Yu Y, Cheng Y, Tong J, Zhang L, Wei Y, Tian M (2021) Recent advances in thermo-sensitive hydrogels for drug delivery. J Mater Chem B 9(13):2989–2992. https://doi.org/10.1039/d0tb02888k

# Chapter 9
# Biomimetic Materials

## 9.1 Introduction

Biomimetic materials, at the nexus of biology and materials science, represent a burgeoning field driven by the inspiration drawn from natural systems. At its core, biomimicry seeks to emulate the efficiency, adaptability, and sustainability inherent in biological structures and processes. By mimicking nature's design principles, researchers aim to develop innovative materials with tailored properties and functions for various applications. This interdisciplinary approach harnesses insights from biology, chemistry, physics, and engineering to create biomaterials that exhibit remarkable performance and versatility. Central to biomimetic materials is the concept of nature-inspired design, wherein researchers study biological systems to understand their underlying mechanisms and translate them into synthetic materials [1]. This approach not only leverages billions of years of evolution, but also fosters a deeper appreciation for the intricacies of natural structures and processes. By emulating nature's strategies, biomimetic materials can achieve remarkable feats, such as self-healing, self-cleaning, and adaptive behavior, which are often elusive in conventional materials. One of the fascinating phenomena observed in nature is biomineralization, where organisms produce mineralized structures with remarkable precision and efficiency. Biomineralization serves as a blueprint for biomimetic mineralization techniques, enabling the synthesis of complex inorganic materials under mild conditions [2]. By understanding the biological templates and processes involved in biomineralization, researchers can design novel materials with tailored structures and properties, ranging from lightweight composites to high-strength ceramics.

In the realm of bio ceramics, biomimicry plays a pivotal role in developing materials for orthopedic applications. Bio ceramics, such as hydroxyapatite and bioactive glasses, mimic the composition and structure of natural bone, promoting osseointegration and bone regeneration. These biomimetic materials find widespread use



D. Suhag et al., *Handbook of Biomaterials for Medical Applications, Volume 1*,
Biomedical Materials for Multi-functional Applications,
https://doi.org/10.1007/978-981-97-4818-1_9

in orthopedic implants, bone grafts, and tissue engineering scaffolds, offering superior biocompatibility and mechanical properties compared to traditional materials. The application of biomimetic materials in orthopedics represents a paradigm shift in the field, offering innovative solutions to address challenges such as implant failure and tissue rejection. By mimicking the hierarchical structure and chemical composition of bone, biomimetic materials facilitate seamless integration with the host tissue, reducing the risk of adverse reactions and enhancing long-term implant success [1, 2]. Moreover, biomimetic approaches enable the development of personalized implants tailored to the unique anatomical and physiological characteristics of individual patients. As the field of biomimetic materials continues to evolve, researchers are exploring new avenues to expand the scope of biomimicry and unlock novel functionalities. From bioinspired adhesives to smart materials capable of sensing and responding to environmental cues, biomimetic materials hold tremendous promise for diverse applications, including medicine, energy, and environmental remediation. However, realizing this potential requires addressing challenges such as scalability, reproducibility, and ethical considerations, underscoring the need for interdisciplinary collaboration and ethical stewardship in biomaterials research.

Biomimetic materials represent a transformative approach to materials design and synthesis, inspired by nature's elegance and efficiency. By harnessing the principles of biomimicry, researchers are unlocking new possibilities for creating advanced materials with unprecedented properties and functionalities.

## *9.1.1 Biomineralization*

Biomineralization, a captivating process ubiquitous in nature, entails the controlled formation of inorganic minerals within organic matrices by living organisms. This intricate phenomenon, honed over millions of years of evolution, underpins the creation of diverse mineralized structures found in shells, bones, teeth, and other biological tissues [1, 2]. At its essence, biomineralization embodies the remarkable synergy between organic molecules and mineral precursors, orchestrated by intricate biological machinery. By unraveling the underlying mechanisms of biomineralization, scientists seek to glean insights into the design and synthesis of biomimetic materials with tailored properties and functionalities. This interdisciplinary endeavor draws upon principles from biology, chemistry, materials science, and engineering, offering profound implications for fields ranging from biomedicine to materials engineering.

### 9.1.1.1 Natural Mineralization Processes

Natural mineralization processes encompass a diverse array of mechanisms employed by living organisms to synthesize inorganic minerals with remarkable precision and efficiency. One such process is biologically controlled mineralization, where organisms exert precise control over the nucleation, growth, and organization of minerals within specialized structures [3]. This intricate regulation, orchestrated by organic molecules such as proteins and polysaccharides, enables the formation of complex mineral architectures tailored to specific biological functions. Examples of biologically controlled mineralization abound in nature, from the intricate shells of mollusks to the intricate skeletons of corals, showcasing the exquisite interplay between organic and inorganic components. Another prevalent mineralization process is templated mineralization, where pre-existing organic templates serve as scaffolds for mineral deposition. Organisms utilize these templates, often composed of proteins or polysaccharides, to guide the nucleation and growth of minerals, resulting in hierarchical structures with tailored properties. Templated mineralization is exemplified in the formation of bone, where collagen fibers provide a template for the deposition of hydroxyapatite crystals, imparting strength and resilience to the skeletal framework [4]. This templated approach allows organisms to achieve precise control over mineral morphology and distribution, facilitating the creation of functional materials optimized for diverse biological roles. In addition to biologically controlled and templated mineralization, organisms employ strategies such as self-assembly and biomineral precipitation to synthesize mineralized structures. Self-assembly processes involve the spontaneous organization of mineral precursors into ordered structures under the influence of biological molecules or environmental cues. This self-organization enables the formation of intricate mineral architectures with tailored properties, such as the iridescent hues of butterfly wings produced through the self-assembly of photonic crystals. Biomineral precipitation, on the other hand, involves the direct precipitation of minerals from solution under the influence of biological factors such as pH, temperature, and ion concentration [1, 2]. This process is exemplified in the formation of calcium carbonate deposits by marine organisms such as corals and mollusks, which play crucial roles in reef-building and shell formation.

Organisms often employ strategies such as mineral phase transformation and ion transport to modulate mineral composition and properties. Mineral phase transformation involves the conversion of one mineral phase into another through chemical reactions or physical processes, resulting in alterations in material properties and functionalities. This dynamic process is essential for remodeling bone tissue, where calcium phosphate minerals undergo phase transformations to adapt to changing mechanical and physiological demands [2]. Ion transport mechanisms facilitate the uptake, transport, and deposition of mineral ions within biological tissues, enabling organisms to regulate mineralization processes with precision. These intricate mechanisms underscore the remarkable adaptability and resilience of natural mineralization processes, providing inspiration for the design of biomimetic materials with enhanced properties and functionalities.

#### 9.1.1.2 Biomimetic Mineralization Techniques

Biomimetic mineralization techniques strive to emulate the precision and efficiency of natural mineralization processes by leveraging principles inspired by biology. One such approach involves the use of biomolecules, such as proteins and peptides, as templates or nucleators for mineral deposition [5]. By mimicking the role of organic molecules in biomineralization, researchers can guide the formation of mineralized structures with controlled morphology and composition. These biomolecules provide precise spatial organization and chemical cues, facilitating the nucleation and growth of minerals in a manner akin to natural mineralization processes. In addition to biomolecular templates, biomimetic mineralization techniques often employ synthetic polymers as scaffolds for mineral deposition. These polymers can be designed to mimic the structure and functionality of natural extracellular matrices, providing a biomimetic environment conducive to mineral formation [2, 3]. By tuning the physical and chemical properties of the polymer scaffolds, researchers can control parameters such as mineral nucleation kinetics, crystal orientation, and hierarchical organization. This approach enables the fabrication of mineralized materials with tailored properties for applications ranging from tissue engineering to catalysis.

Biomimetic mineralization techniques frequently utilize biomimetic additives or modifiers to regulate mineralization processes. These additives, which may include organic molecules, ions, or small molecules, mimic the function of biological factors involved in natural mineralization. By modulating factors such as solution chemistry, surface interactions, and crystal growth kinetics, biomimetic additives can influence the morphology, phase composition, and mechanical properties of mineralized materials. This approach allows researchers to fine-tune mineralization processes to achieve desired outcomes, such as the synthesis of bone-like apatite coatings for orthopedic implants or the fabrication of bioinspired ceramics with enhanced mechanical properties [5]. Biomimetic mineralization techniques often leverage bottom-up assembly strategies to fabricate hierarchical mineralized structures. By combining molecular self-assembly with mineral deposition, researchers can create materials with complex architectures reminiscent of those found in natural tissues. This bottom-up approach enables precise control over the spatial organization and interfacial interactions within mineralized materials, leading to enhanced mechanical, optical, and biological properties. Examples include the fabrication of bone-mimetic composites with hierarchical structures and the synthesis of photonic materials inspired by iridescent biological tissues.

Biomimetic mineralization techniques offer a versatile toolkit for the design and fabrication of mineralized materials inspired by nature [6]. By drawing inspiration from natural mineralization processes and leveraging principles from biology, chemistry, and materials science, researchers can create materials with tailored properties and functionalities for a wide range of applications.

#### 9.1.1.3 Examples of Biomineralized Structures

- **Mollusk Shells**: Mollusks, such as oysters, clams, and snails, produce intricate shells composed primarily of calcium carbonate.
  - These shells exhibit complex structures, including layered architectures and iridescent nacre, or mother-of-pearl, in the case of some species.
- **Bone Tissue**: Bone tissue in vertebrates is a biomineralized structure composed of a composite material made of collagen fibers and hydroxyapatite crystals.
  - This hierarchical structure provides both strength and flexibility, essential for supporting the body and facilitating movement.
- **Corals**: Coral reefs are formed by colonies of coral polyps that secrete calcium carbonate skeletons, known as coral skeletons or corallite.
  - These skeletons provide structural support and protection for the coral polyps, forming the foundation of coral reef ecosystems.
- **Teeth**: Teeth in vertebrates, including humans, are biomineralized structures composed primarily of hydroxyapatite crystals embedded in a matrix of collagen and other proteins.
  - Tooth enamel, the outermost layer of teeth, is one of the hardest substances in the body, thanks to its high mineral content.
- **Eggshells**: Bird eggs are enclosed in mineralized shells composed primarily of calcium carbonate.
  - These shells provide protection for developing embryos while allowing for gas exchange and moisture regulation.
- **Sea Urchin Spines**: Sea urchins possess spiny skeletons composed of calcite crystals arranged in intricate patterns.
  - These mineralized spines serve as defense mechanisms against predators and provide structural support for the organism.
- **Silica Diatom Frustules**: Diatoms are single-celled algae that produce intricate silica shells, called frustules, through biomineralization.
  - These frustules exhibit diverse shapes and patterns, contributing to the ecological success of diatoms and their role in global carbon cycling.
- **Sponge Spicules**: Sponges, or poriferans, produce mineralized structures called spicules, which provide support and rigidity to their soft bodies.
  - These spicules are composed of various minerals, including calcium carbonate and silica, depending on the species.

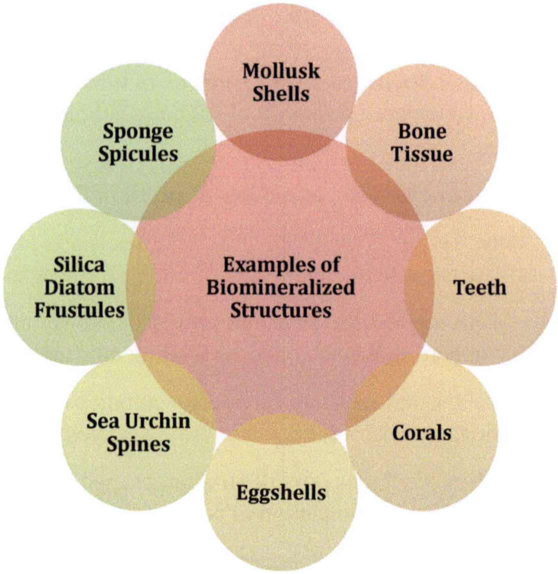


**Fig. 9.1** Examples of biomineralized structures

These examples highlight the diverse array of biomineralized structures found in nature, each finely tuned to serve specific biological functions and ecological roles (Fig. 9.1).

**Exercise 1.**

**Fill Ups**

**(Hydroxyapatite, Bone graft, Biomineralization, Silica shells, Calcium carbonate)**

**1.** ......................is the process by which living organisms produce and control the growth of minerals within their tissues.

**2.** In biomineralization, organisms such as mollusks and corals secrete organic molecules that serve as templates for the formation of ........................

**3.** The mineralization of bones and teeth in vertebrates involves the deposition of ...............within a collagen matrix, resulting in strong and durable skeletal structures.

**4.** Biomineralization plays a crucial role in the formation of ...................in organisms such as diatoms and algae, providing structural support and protection.

**5.** Understanding the mechanisms of biomineralization has inspired the development of biomimetic materials for various applications, including ..................substitutes and biomedical implants.

## 9.1.2 *Bio Ceramics*

Bio ceramics represent a class of ceramic materials specifically designed for biomedical applications, ranging from orthopedic implants to dental restorations. These materials exhibit unique properties such as biocompatibility, bioactivity, and mechanical strength, making them well-suited for use in biological environments [7]. Bio ceramics are often inspired by the composition and structure of natural bone tissue, with calcium phosphates being a common component due to their similarity to the mineral phase of bone. Their ability to integrate with surrounding tissues and promote bone regeneration has led to widespread use in orthopedics, where they serve as bone grafts, coatings for implants, and scaffolds for tissue engineering.

### 9.1.2.1 Properties and Characteristics

Bio ceramics possess a unique set of properties and characteristics that make them highly suitable for biomedical applications (Fig. 9.2). One key property is biocompatibility, which refers to the ability of a material to interact favorably with biological systems without eliciting adverse reactions. Bio ceramics exhibit excellent biocompatibility, as they are often derived from naturally occurring minerals that closely resemble components found in the human body, such as calcium phosphates [8]. This property is crucial for ensuring the compatibility of bio ceramic implants with surrounding tissues, minimizing the risk of inflammation, rejection, or allergic reactions. Another important characteristic of bio ceramics is bioactivity, which refers to their ability to form a bond with living tissues and promote biological responses at the material-tissue interface. Bio ceramics, particularly calcium phosphate-based materials like hydroxyapatite, possess inherent bioactivity due to their similarity to the mineral phase of bone tissue. When implanted in the body, bio ceramics can stimulate the formation of new bone tissue through processes such as osteoconduction and osteointegration, facilitating the integration of implants with native bone and promoting long-term stability and functionality. In addition to biocompatibility and bioactivity, bio ceramics exhibit excellent mechanical properties, including high strength, hardness, and fracture resistance. These materials can withstand physiological loads and stresses encountered in the body, making them suitable for load-bearing applications such as orthopedic implants and dental restorations [9]. Moreover, bio ceramics often possess tailored porosity and surface characteristics that can be engineered to modulate their mechanical properties and enhance tissue ingrowth and integration. This versatility allows for the design of bio ceramic implants with optimized mechanical performance for specific clinical applications.

Bo ceramics demonstrate excellent chemical stability and resistance to degradation in biological environments, ensuring their long-term functionality and biocompatibility. Unlike organic materials, which may degrade over time or undergo enzymatic breakdown, bio ceramics maintain their structural integrity and properties,

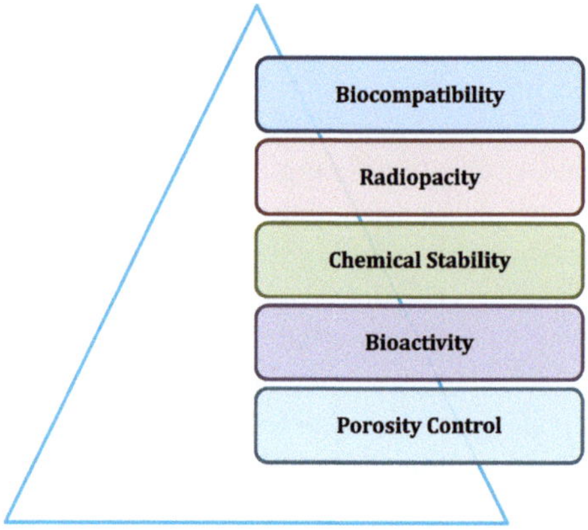


**Fig. 9.2** Properties of bio ceramics

providing durable solutions for biomedical implants and devices [10]. This chemical stability also allows bio ceramics to serve as effective carriers for drug delivery systems, enabling controlled release of therapeutic agents to target sites within the body while minimizing systemic side effects. Overall, the properties and characteristics of bio ceramics make them indispensable materials for a wide range of biomedical applications, including orthopedics, dentistry, and tissue engineering. With their combination of biocompatibility, bioactivity, mechanical strength, and chemical stability, bio ceramics offer versatile solutions for restoring function and improving the quality of life for patients suffering from musculoskeletal disorders, dental conditions, and other medical ailments.

#### 9.1.2.2 Types of Bio Ceramics

Bio ceramics encompasses a diverse range of materials tailored for biomedical applications, each offering unique properties and functionalities suited to specific clinical needs (Fig. 9.3). One common type of bio ceramic is calcium phosphate-based ceramics, which mimic the mineral composition of natural bone tissue. Hydroxyapatite (HA) and tricalcium phosphate (TCP) are two widely used calcium phosphate ceramics known for their excellent biocompatibility and bioactivity. HA, in particular, closely resembles the mineral phase of bone and exhibits superior osteoconductivity, making it ideal for bone grafts, coatings for implants, and scaffolds for tissue engineering applications. Another type of bio ceramic is bioactive glasses, which are silica-based materials doped with various ions to enhance their bioactivity and degradation properties. Bioactive glasses stimulate the formation of hydroxyapatite on their surface when in contact with biological fluids, promoting bone bonding and integration [11]. These materials are used in bone regeneration procedures, such as filling bone defects and repairing fractures, due to their ability to support new bone growth and remodeling. In addition to calcium phosphate ceramics and bioactive glasses, alumina ($Al_2O_3$) and zirconia ($ZrO_2$) are commonly employed as bio

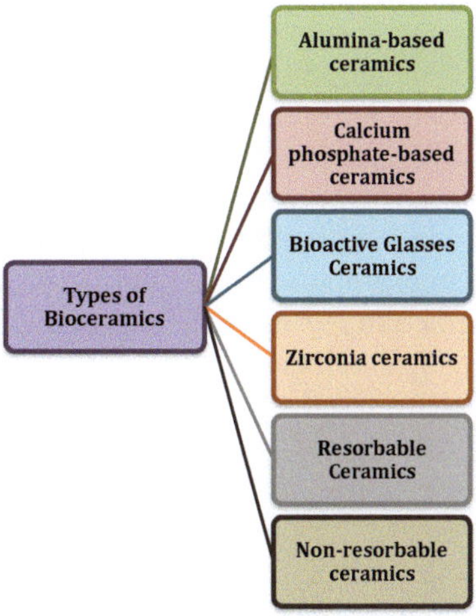


**Fig. 9.3** Types of bio ceramics

ceramics, particularly in orthopedic and dental applications. Alumina-based ceramics exhibit high mechanical strength [12], wear resistance, and biocompatibility, making them suitable for load-bearing implants such as hip prostheses and dental crowns. Zirconia ceramics, on the other hand, offer excellent aesthetic properties, biocompatibility, and low thermal conductivity, making them ideal for dental restorations, dental implants, and orthopedic components.

Bio ceramics can be classified based on their resorbability into resorbable or bioresorbable ceramics and non-resorbable ceramics. Bioresorbable ceramics, such as calcium phosphates and bioactive glasses, gradually degrade in the body over time and are replaced by natural bone tissue during the healing process. These materials eliminate the need for implant removal surgeries and offer advantages in tissue regeneration and remodeling. Non-resorbable ceramics, such as alumina and zirconia, maintain their structural integrity over the long term and are used in permanent implants where mechanical stability and durability are paramount.

Overall, the diversity of bio ceramics allows for tailored solutions to meet the specific requirements of different biomedical applications, ranging from bone repair and regeneration to dental restoration and implantology. By leveraging the unique properties of various bio ceramic materials, clinicians and researchers can develop innovative treatments and devices to address the growing demands of modern healthcare and improve patient outcomes [11, 12]. Continued advancements in bio ceramic design, fabrication techniques, and clinical applications hold promise for further enhancing the efficacy and versatility of these materials in biomedicine.

#### 9.1.2.3 Applications in Orthopedics

Bio ceramics have revolutionized the field of orthopedics, offering innovative solutions for the treatment of musculoskeletal disorders and injuries. These materials, including calcium phosphate ceramics, bioactive glasses, alumina, and zirconia, exhibit properties that make them highly suitable for orthopedic applications. Bio ceramics are utilized in orthopedics for a wide range of purposes, including bone grafting, joint replacement, and tissue engineering. Their excellent biocompatibility, bioactivity, mechanical strength, and chemical stability make them ideal for restoring function and stability to damaged or diseased bones and joints [13, 14]. As such, bio ceramics play a crucial role in improving patient outcomes and quality of life in orthopedic surgery, offering durable and effective solutions for bone repair, regeneration, and replacement.

### Bone Grafts and Scaffolds

Bone grafts and scaffolds represent critical applications of bio ceramics in orthopedics, offering solutions for bone repair, regeneration, and augmentation. Bone grafts are used to replace missing bone tissue resulting from trauma, disease, or surgical procedures, facilitating the healing process and restoring skeletal integrity [10, 11]. Bio ceramics, such as calcium phosphate ceramics and bioactive glasses, serve as ideal materials for bone grafts due to their biocompatibility, bioactivity, and similarity to the mineral phase of natural bone. These materials provide a scaffold for new bone formation, promoting osteoconduction and osteointegration while gradually resorbing and being replaced by native bone tissue over time. Moreover, bio ceramic scaffolds play a crucial role in tissue engineering approaches for bone regeneration, providing a 3D framework for cell attachment, proliferation, and differentiation. These scaffolds mimic the extracellular matrix of bone tissue, offering mechanical support and biochemical cues to guide the growth and development of new bone. By incorporating bioactive factors such as growth factors, peptides, or drugs into the scaffold matrix, researchers can enhance the regenerative capacity and efficacy of bio ceramic scaffolds, promoting accelerated bone healing and integration [15].

The design and fabrication of bio ceramic scaffolds can be tailored to match the anatomical and mechanical requirements of specific bone defects or injuries. 3D printing and additive manufacturing techniques enable the customization of scaffold architecture, pore size, and distribution, allowing for precise control over scaffold properties and performance. This personalized approach ensures optimal fit and functionality, minimizing complications and improving patient outcomes in bone reconstruction surgeries [12, 13]. Bio ceramic scaffolds can be functionalized or surface-modified to enhance their biological and mechanical properties. Surface treatments such as coating deposition, plasma spraying, or chemical modification can improve scaffold bioactivity, biodegradability, and osteoconductivity, facilitating cell attachment, proliferation, and differentiation. These modifications also enable the incorporation of antibacterial agents or drug delivery systems into the scaffold matrix, reducing the risk of infection and inflammation while promoting tissue regeneration. Bone grafts and scaffolds represent versatile applications of bio ceramics in orthopedics, offering solutions for bone repair, regeneration, and augmentation. By harnessing the unique properties of bio ceramic materials and leveraging advanced fabrication techniques, researchers and clinicians can develop customized grafts and scaffolds tailored to the specific needs of patients. These innovative approaches hold promise for improving the efficacy, safety, and long-term outcomes of orthopedic surgeries, ultimately enhancing the quality of life for individuals suffering from bone injuries or diseases.

### Joint Replacement Materials

Bio ceramics play a vital role in the development of joint replacement materials, offering solutions for restoring mobility and relieving pain in individuals with degenerative joint diseases such as osteoarthritis. Total joint replacement surgeries involve the removal of damaged or diseased joint components and their replacement with prosthetic implants made from biocompatible materials [15]. Bio ceramics, including alumina, zirconia, and certain calcium phosphate ceramics, are commonly used in joint replacement materials due to their excellent mechanical properties, wear resistance, and biocompatibility. One of the key advantages of bio ceramics in joint replacement materials is their similarity to natural bone tissue, which minimizes the risk of adverse reactions and promotes long-term implant stability. Alumina and zirconia ceramics, in particular, exhibit high strength, hardness, and biocompatibility, making them suitable for load-bearing applications such as hip and knee replacements. These materials offer excellent wear resistance and low friction properties, reducing the risk of implant wear and osteolysis, which are common complications associated with metal-on-metal or metal-on-polyethylene implants. Bio ceramics can be surface-modified or coated to further enhance their performance and biocompatibility in joint replacement materials [16]. Surface treatments such as plasma spraying, chemical vapor deposition, or ion implantation can improve ceramic implant osseointegration, reducing the risk of implant loosening and improving long-term implant survival rates. The incorporation of bioactive coatings or growth factors into ceramic implant surfaces can promote bone ingrowth and accelerate the healing process, leading to faster recovery and improved implant outcomes. The design and fabrication of bio ceramic joint replacement materials can be tailored to match the anatomical and mechanical requirements of specific joints and patients. Advanced manufacturing techniques such as computer-aided design (CAD) and additive manufacturing enable the customization of implant geometry, size, and surface topography, ensuring optimal fit and functionality [15, 16]. This personalized approach minimizes complications and improves patient satisfaction by optimizing implant performance and longevity.

Bio ceramics offer versatile solutions for joint replacement materials, providing durable and biocompatible implants for restoring function and mobility in individuals with degenerative joint diseases. With their excellent mechanical properties, wear resistance, and biocompatibility, bio ceramics play a crucial role in improving the outcomes of total joint replacement surgeries, reducing complications, and enhancing the quality of life for patients.

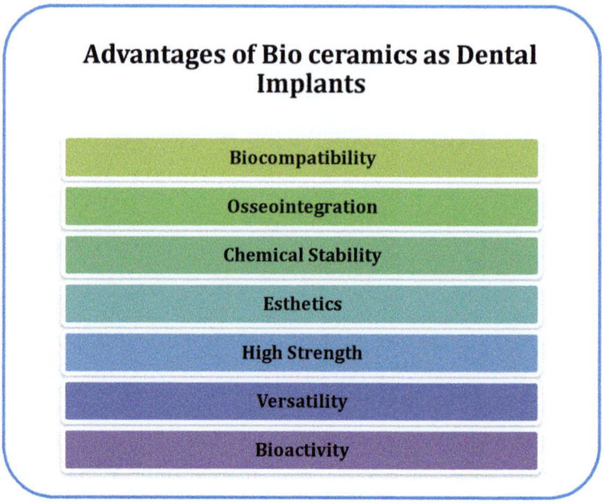


**Fig. 9.4** Various advantages of bio ceramics as dental implants

Dental Implants

Bio ceramics have revolutionized the field of dental implants, offering durable and biocompatible solutions for replacing missing teeth and restoring oral function (Fig. 9.4). Dental implants are prosthetic devices used to replace natural teeth roots and support artificial teeth, such as crowns, bridges, or dentures [17]. Bio ceramics, including titanium-based ceramics, zirconia, and hydroxyapatite, are commonly used in dental implants due to their excellent biocompatibility, mechanical properties, and osseointegration capabilities. One of the key advantages of bio ceramics in dental implants is their ability to promote osseointegration, the process by which the implant fuses with surrounding bone tissue. Titanium-based ceramics, such as titanium alloy and titanium oxide, exhibit superior osseointegration properties, allowing for stable and long-lasting implant fixation. This ensures the durability and longevity of dental implants, reducing the risk of implant failure and complications such as peri-implantitis, which can lead to bone loss and implant loosening. Moreover, bio ceramic dental implants offer aesthetic advantages over traditional metal implants, particularly in the case of anterior or visible tooth replacements. Zirconia ceramics, in particular, closely resemble the natural color and translucency of tooth enamel, providing a more natural and lifelike appearance. This aesthetic compatibility enhances patient satisfaction and confidence, making zirconia implants an increasingly popular choice for cosmetic dental procedures [17, 18].

**Exercise 2**

**Fill Ups**

**(Zirconia, Biocompatibility, Osseointegration, Bio ceramics, Cell adhesion)**

1. ...................... are a class of materials specifically designed for use in medical and dental applications.
2. One of the key characteristics of bio ceramics is their excellent .................., which ensures compatibility with living tissue.
3. Hydroxyapatite, a mineral similar to the composition of natural bone, is commonly used in bio ceramics for promoting ....................between the implant and surrounding tissue.
4. Bio ceramics such as ............... are often used for dental implants due to their high strength, biocompatibility, and aesthetic properties.
5. Bio ceramics can be engineered to have tailored surface properties, such as enhanced ........................... or reduced bacterial adhesion, to improve their performance in medical applications.

Bio ceramic dental implants can be customized to match the unique anatomical and mechanical requirements of individual patients. Advanced computer-aided design (CAD) and computer-aided manufacturing (CAM) techniques enable the precise fabrication of implant components, ensuring optimal fit and function. This personalized approach minimizes surgical complications and maximizes implant success rates, leading to improved patient outcomes and satisfaction. In addition to their osseointegration and aesthetic properties, bio ceramic dental implants offer advantages in terms of biocompatibility and bioactivity. Hydroxyapatite-coated implants, for example, promote bone growth and remodeling by releasing calcium and phosphate ions, mimicking the mineral composition of natural bone tissue. This bioactive behavior enhances implant stability and integration while reducing the risk of post-operative complications, making hydroxyapatite-coated implants a preferred choice for certain dental implant procedures.

### *9.1.3 Biomimetic Materials in Orthopedics*

Biomimetic materials have emerged as innovative solutions in the field of orthopedics, offering tailored approaches inspired by nature to address the complexities of musculoskeletal disorders and injuries. These materials draw inspiration from biological systems, mimicking the structure, composition, and functionality of natural tissues to enhance compatibility and effectiveness in orthopedic applications [19]. By harnessing principles from biology, chemistry, and materials science, biomimetic materials aim to improve outcomes in bone repair, joint replacement, and tissue engineering. Through advanced fabrication techniques and bioactive properties, biomimetic materials hold promise for promoting regeneration, integration, and long-term functionality in orthopedic implants and devices, ultimately improving patient care and quality of life.

#### 9.1.3.1 Role of Biomimetic Materials in Bone Regeneration

The role of biomimetic materials in bone regeneration is paramount, offering innovative approaches to promote healing and restore function in damaged or diseased bone tissue. These materials leverage insights from natural bone formation processes to create scaffolds, implants, and biomaterials that mimic the composition, structure, and bioactivity of native bone [19]. By replicating the microenvironment of bone tissue, biomimetic materials provide a supportive framework for cell attachment, proliferation, and differentiation, facilitating the regeneration of new bone tissue. One key aspect of biomimetic materials in bone regeneration is their ability to promote osteoconduction, the process by which bone-forming cells migrate into and proliferate within the scaffold matrix. Biomimetic scaffolds often incorporate bioactive molecules or surface modifications that enhance cell adhesion and signaling, stimulating the recruitment and differentiation of osteogenic cells [20]. This promotes the formation of new bone tissue, allowing the scaffold to gradually degrade and be replaced by native bone over time. Biomimetic materials play a crucial role in modulating the local microenvironment to promote bone healing and remodeling. By incorporating growth factors, cytokines, or extracellular matrix proteins into the scaffold matrix, researchers can create bioactive materials that mimic the biochemical cues present during natural bone regeneration processes. These biomimetic signals help regulate cell behavior, matrix deposition, and tissue organization, guiding the formation of functional bone tissue with proper architecture and mechanical properties. Biomimetic materials offer advantages in terms of biocompatibility, mechanical properties, and degradation kinetics, ensuring optimal performance and integration in the body. Bio ceramics, such as calcium phosphate-based materials and bioactive glasses, closely resemble the mineral composition of natural bone tissue, promoting favorable interactions with host cells and tissues. The design and fabrication of biomimetic scaffolds can be tailored to match the anatomical and mechanical requirements of specific bone defects or injuries, maximizing the efficacy and success of bone regeneration therapies.

Biomimetic materials hold tremendous promise for advancing bone regeneration strategies, offering tailored solutions to address the complex challenges of bone repair and healing [21]. By harnessing the principles of biomimicry and leveraging advanced materials science techniques, researchers can develop innovative scaffolds and implants that promote the formation of functional bone tissue and improve patient outcomes in orthopedic surgery and regenerative medicine.

#### 9.1.3.2 Enhanced Biocompatibility and Tissue Integration

Enhanced biocompatibility and tissue integration are critical aspects of biomimetic materials, contributing to their effectiveness in biomedical applications such as tissue engineering and regenerative medicine. Biocompatibility refers to the ability of a material to interact favorably with biological systems without eliciting adverse reactions or toxicity. Biomimetic materials are designed to closely mimic the properties of natural tissues, reducing the risk of immune responses, inflammation, and rejection when implanted in the body. This enhanced biocompatibility promotes the integration of biomimetic materials with surrounding tissues, facilitating the regeneration of functional tissue and promoting long-term implant success. One key strategy for enhancing biocompatibility and tissue integration is the incorporation of bioactive molecules or surface modifications into biomimetic materials [19, 20]. These biomimetic signals, which may include growth factors, extracellular matrix proteins, or cell-adhesive peptides, mimic the biochemical cues present in natural tissue microenvironments. By providing cues for cell attachment, proliferation, and differentiation, biomimetic materials promote the recruitment of host cells and facilitate tissue ingrowth and remodeling. This biomimetic approach fosters seamless integration between the implant and surrounding tissues, improving biocompatibility and promoting tissue regeneration.

Biomimetic materials often exhibit tailored surface properties and topographies designed to enhance cell-material interactions and tissue integration. Surface modifications such as nanostructuring, micro texturing, or functionalization with bioactive coatings can influence cellular responses and promote tissue integration. These surface modifications mimic the nanoscale features and biochemical cues present in natural extracellular matrices, facilitating cell adhesion, spreading, and migration on the material surface. This enhanced interaction between cells and biomimetic materials promotes tissue ingrowth and vascularization, leading to improved tissue integration and long-term implant stability [22]. Moreover, the degradation kinetics of biomimetic materials can be tailored to match the rate of tissue regeneration and remodeling, further enhancing tissue integration and biocompatibility. Biodegradable biomimetic scaffolds gradually degrade over time as new tissue forms, allowing for seamless integration and remodeling of the implant site. By controlling the degradation rate and degradation of products of biomimetic materials, researchers can minimize inflammatory responses and optimize tissue regeneration outcomes. This biomimetic approach ensures that the implant provides mechanical support and structural integrity during the healing process while gradually being replaced by native tissue over time.

The design and fabrication of biomimetic materials can be optimized to match the anatomical and mechanical requirements of specific tissue types and implant sites. Advanced manufacturing techniques such as 3D printing and electrospinning [23] enable the precise fabrication of biomimetic scaffolds with customizable architectures, pore sizes, and mechanical properties. This personalized approach ensures optimal fit and functionality, promoting tissue integration and improving long-term implant success rates.

#### 9.1.3.3 Advantages Over Traditional Orthopedic Materials

Biomimetic materials offer several advantages over traditional orthopedic materials, making them increasingly preferred choices for orthopedic implants and devices. One significant advantage is enhanced biocompatibility, as biomimetic materials closely mimic the properties of natural tissues, reducing the risk of adverse reactions and inflammatory responses when implanted in the body. This improved biocompatibility promotes better tissue integration and reduces the likelihood of implant rejection or complications, ultimately leading to improved patient outcomes and long-term implant success rates. Biomimetic materials often exhibit superior mechanical properties compared to traditional orthopedic materials, such as metals or polymers [24]. By leveraging advanced fabrication techniques and biomimetic design principles, researchers can create biomaterials with optimized mechanical strength, flexibility, and durability. These materials provide better support and stability for orthopedic implants, reducing the risk of implant failure, loosening, or fracture, particularly in load-bearing applications such as joint replacements or spinal implants.

Biomimetic materials offer advantages in terms of bioactivity and tissue regeneration, facilitating the healing process and promoting the formation of functional tissue at the implant site. Unlike inert materials that remain unchanged after implantation, biomimetic materials actively interact with the surrounding biological environment, releasing bioactive molecules, ions, or growth factors that stimulate cell proliferation and differentiation [25]. This bioactive behavior promotes tissue ingrowth, vascularization, and remodeling, leading to improved tissue integration and long-term implant success. Biomimetic materials often exhibit tailored surface properties and topographies that enhance cell-material interactions and tissue integration. Surface modifications such as nano structuring, micro texturing, or bioactive coatings promote cell adhesion, spreading, and migration on the material surface, facilitating tissue ingrowth and vascularization. This enhanced tissue integration reduces the risk of implant loosening, migration, or fibrous encapsulation, improving the stability and longevity of orthopedic implants and devices. Biomimetic materials offer opportunities for personalized medicine and patient-specific treatments, allowing for customized implants and devices tailored to match the anatomical and mechanical requirements of individual patients [26]. Advanced manufacturing techniques such as 3D printing and computer-aided design (CAD) enable the precise fabrication of biomimetic implants with customizable architectures, pore sizes, and mechanical properties. This personalized approach ensures optimal fit, functionality, and compatibility, minimizing complications and improving patient satisfaction in orthopedic surgery and rehabilitation.

Biomimetic materials offer numerous advantages over traditional orthopedic materials, including enhanced biocompatibility, superior mechanical properties, improved tissue regeneration, tailored surface properties, and opportunities for personalized medicine [27]. These advantages make biomimetic materials increasingly preferred choices for orthopedic implants and devices, offering improved outcomes, reduced complications, and enhanced patient satisfaction in orthopedic surgery and rehabilitation.

### 9.1.4 Emerging Trends in Biomimetic Materials Research

Emerging trends in biomimetic materials research are driving innovation and shaping the future of various fields, from biomedicine to materials science. One prominent trend is the integration of multi-scale and hierarchical structures inspired by nature into biomimetic materials design. By mimicking the complex organization found in natural tissues and organisms, researchers aim to create materials with enhanced functionalities and properties for diverse applications. This approach involves combining different length scales, from nanostructures to macroscale architectures, to achieve superior mechanical, optical, and biological performance in biomimetic materials. Another emerging trend is the development of smart and responsive biomimetic materials capable of dynamic self-regulation and adaptation to changing environmental conditions. Inspired by biological systems such as living organisms and tissues, researchers are designing materials that can sense and respond to external stimuli such as light, temperature, pH, or mechanical forces. These stimuli-responsive materials hold promise for applications in drug delivery, tissue engineering, and biosensing, where precise control over material properties and functions is essential for achieving desired therapeutic outcomes [28]. Biomimetic materials research is increasingly focusing on bioinspired fabrication techniques that enable the scalable production of complex structures with tailored properties. Techniques such as additive manufacturing, self-assembly, and microfluidics allow researchers to create biomimetic materials with precise control over composition, structure, and morphology. By leveraging these advanced fabrication methods, researchers can overcome traditional manufacturing limitations and unlock new possibilities for biomimetic materials in areas such as regenerative medicine, bioelectronics, and energy storage.

Biomimetic materials research is exploring novel bioinspired functionalities inspired by natural systems, such as self-healing, self-cleaning, and anti-fouling properties. By mimicking biological processes such as wound healing, lotus leaf-inspired surface structures, and marine organism-inspired anti-fouling strategies, researchers are developing materials with enhanced durability, longevity, and performance in harsh or dynamic environments. These bioinspired functionalities have applications in various fields, including biomedical implants, coatings, membranes, and environmental remediation technologies [29]. Biomimetic materials research is increasingly interdisciplinary, with collaborations between scientists, engineers, biologists, and clinicians driving innovation and cross-pollination of ideas. By bringing together expertise from diverse disciplines, researchers can gain new insights into biological systems and translate these insights into practical solutions for real-world challenges.

#### 9.1.4.1 Overcoming Hurdles in Biomimetic Material Development

Overcoming hurdles in biomimetic material development requires addressing various challenges related to material design, fabrication, and translation into practical applications. One significant hurdle is achieving a comprehensive understanding of the complex structure-function relationships present in natural biological systems. Biomimetic materials aim to replicate the intricate organization and functionalities found in living organisms, which requires a deep understanding of biological principles at multiple length scales. Overcoming this hurdle involves interdisciplinary collaboration between biologists, materials scientists, engineers, and clinicians to decipher biological mechanisms and translate them into biomimetic design principles.

Another challenge in biomimetic material development is achieving precise control over material properties and functionalities at multiple scales. Biomimetic materials often involve complex hierarchical structures and nanoscale features that are challenging to replicate using traditional fabrication techniques [29, 30]. Overcoming this hurdle requires the development of advanced manufacturing methods, such as additive manufacturing, self-assembly, and bio fabrication, capable of producing biomimetic materials with tailored properties and architectures. These advanced fabrication techniques enable researchers to mimic the complexity and functionality of natural tissues and organisms more effectively. Translating biomimetic materials from the laboratory to practical applications presents challenges related to scalability, reproducibility, and regulatory approval. Many biomimetic materials are still in the early stages of development, and scaling up production to meet commercial demand while maintaining consistency and quality is a significant hurdle. Navigating regulatory pathways for biomimetic materials in healthcare applications requires rigorous testing, validation, and approval processes to ensure safety and efficacy. Overcoming these hurdles requires collaboration between academia, industry, and regulatory agencies to establish standards and guidelines for biomimetic material development and translation [31]. Moreover, biomimetic material development faces challenges related to cost-effectiveness and sustainability. Many advanced fabrication techniques used to create biomimetic materials are costly and resource-intensive, limiting their widespread adoption and commercial viability [32, 33]. Sourcing biomimetic materials from sustainable and renewable sources is essential to minimize environmental impact and ensure long-term sustainability. Overcoming these hurdles involves optimizing fabrication processes, exploring alternative materials and production methods, and considering the lifecycle impacts of biomimetic materials from production to disposal.

Overcoming hurdles in biomimetic material development requires interdisciplinary collaboration, advanced fabrication techniques, regulatory compliance, cost-effectiveness, and sustainability considerations [34, 35]. By addressing these challenges, researchers can unlock the full potential of biomimetic materials to revolutionize healthcare, technology, and sustainability in the years to come. Continued advancements in biomimetic material development hold promise for addressing pressing societal challenges and unlocking new opportunities for innovation and discovery.

**Exercise 3**
**Multiple Choice Questions**

1. Which of the following best describes biomimetic materials?
   a) Materials derived from natural sources
   b) Materials that mimic the structure and function of natural tissues
   c) Materials used exclusively in medical applications
   d) Materials with high mechanical strength

2. What is the primary goal of biomimetic materials in medicine?
   a) To replace natural tissues entirely
   b) To repair damaged tissues by promoting regeneration
   c) To prevent biological responses to foreign materials
   d) To improve the aesthetics of medical devices

3. Which component is often incorporated into biomimetic bone graft substitutes?
   a) Silicon
   b) Hydroxyapatite
   c) Polyethylene
   d) Titanium

4. What is the purpose of biomimetic coatings on orthopedic implants?
   a) To reduce the cost of implants
   b) To increase the weight of implants
   c) To improve the integration of implants with surrounding tissues
   d) To change the color of implants for aesthetic purposes

5. Which type of biomimetic material mimics the extracellular matrix of cartilage?
   a) Hydrogels
   b) Metals
   c) Polymers
   d) Ceramics

6. How do biomimetic implants improve patient outcomes in orthopedic surgery?
   a) By causing adverse reactions in the body
   b) By increasing the risk of implant failure
   c) By mimicking the mechanical properties of natural tissues
   d) By reducing the need for surgery

7. What is the significance of osseointegration in dental implants?
   a) It prevents bone growth around the implant
   b) It causes inflammation in the surrounding tissues
   c) It promotes direct bonding of the implant with the bone
   d) It decreases the stability of the implant

8. Which property of bio ceramics makes them suitable for dental implants?
   a) High electrical conductivity
   b) Low biocompatibility
   c) Resistance to corrosion and degradation
   d) Inability to bond with bone tissue

9. What role do biomimetic materials play in drug delivery systems?
   a) They accelerate the degradation of drugs
   b) They prevent drugs from reaching target tissues
   c) They promote tissue regeneration and reduce inflammation
   d) They increase the risk of adverse drug reactions

10. Which of the following is not an advantage of biomimetic materials in medicine?
   a) Biocompatibility
   b) Osseointegration
   c) Corrosion resistance
   d) High toxicity

## 9.2 Summary

This chapter explores the fascinating world of biomimetic materials, highlighting their significance in various fields such as orthopedics, tissue engineering, and regenerative medicine. We began by discussing the principles of biomimicry, which involve drawing inspiration from nature to design materials with enhanced functionalities and properties. Through examples such as bone grafts, joint replacements, and dental implants, we saw how biomimetic materials offer superior biocompatibility, mechanical properties, and tissue integration compared to traditional materials [1, 2, 7]. Furthermore, the chapter delved into the mechanisms and fabrication techniques used to create biomimetic materials, including additive manufacturing, surface modification, and biomolecular engineering. By replicating the hierarchical structures and biochemical cues found in natural tissues, researchers can create biomimetic materials capable of promoting tissue regeneration, modulating cellular behavior, and responding to environmental stimuli. These advanced fabrication techniques enable precise control over material properties and functionalities, driving innovation in biomaterials science and engineering [9].

The implications of biomimetic materials for the future of materials science are profound. As researchers continue to unravel the mysteries of biological systems and develop new fabrication methods, biomimetic materials hold promise for revolutionizing various industries, from healthcare and biotechnology to energy and environmental remediation. The development of smart and responsive biomimetic materials capable of self-regulation and adaptation to changing conditions opens up new possibilities for applications such as drug delivery, biosensing, and tissue regeneration [12, 13]. The interdisciplinary nature of biomimetic materials research fosters collaboration between scientists, engineers, biologists, and clinicians, driving innovation and cross-pollination of ideas. By integrating expertise from diverse disciplines, researchers can gain new insights into biological systems and translate these insights into practical solutions for real-world challenges. This collaborative approach not only advances biomimetic materials research but also accelerates progress in other fields such as materials science, biology, and medicine. The development of biomimetic materials has implications for addressing global challenges such as healthcare disparities, environmental sustainability, and resource depletion. By creating materials that mimic the efficiency and resilience of natural systems, researchers can develop sustainable solutions for healthcare, technology, and the environment. Biomimetic materials offer opportunities for creating biodegradable, recyclable, and eco-friendly alternatives to traditional materials, reducing environmental impact and promoting a more sustainable future.

## Points to Remember

- Biomimetic materials draw inspiration from biological structures and processes, such as bone, skin, and collagen, to design materials with similar functionalities.
- These materials are often biocompatible, meaning they are well-tolerated by the body, reducing the risk of rejection or adverse reactions.
- Biomimetic materials play a crucial role in tissue engineering by providing scaffolds that support cell growth and tissue regeneration, ultimately helping in the repair or replacement of damaged tissues and organs.
- They are used to develop advanced drug delivery systems that can mimic biological processes for targeted and controlled release of therapeutic agents, improving efficacy and reducing side effects.
- Biomimetic materials can replicate the mechanical properties of natural tissues, providing strength, flexibility, and elasticity similar to those found in the body.
- Some biomimetic materials incorporate natural antimicrobial properties found in certain organisms, helping to prevent infections in biomedical devices or implants.
- Inspired by self-repair mechanisms in living organisms, biomimetic materials with self-healing capabilities are being developed, which can repair damage autonomously, extending the lifespan of biomedical devices.
- Advances in nanotechnology enable the development of biomimetic materials at the nanoscale, allowing precise control over properties such as surface texture, porosity, and chemical composition.

## References

1. Xu A, Ma Y, Cölfen H (2007) Biomimetic mineralization. J Mater Chem 17(5)415–449. https://doi.org/10.1039/b611918m
2. Boskey AL (2003) Biomineralization: an overview. Connect Tissue Res 44(Suppl 1):5–9

3. Nie Z, Zhang Y, Tang R, Wang X (2022) Biomimetic mineralization: an emerging organism engineering strategy for biomedical applications. J Inorg Biochem 232:111815. https://doi.org/10.1016/j.jinorgbio.2022.111815
4. Zhou Y, Liu K, Zhang H (2023) Biomimetic mineralization: from microscopic to macroscopic materials and their biomedical applications. ACS Appl Bio Mater 6(9):3516–3531. https://doi.org/10.1021/acsabm.3c00109
5. Wang Q, Hu L, Wang X, Tang R (2024) Expanding from materials to biology inspired by biomineralization. Interdisc Mater. https://doi.org/10.1002/idm2.12144
6. Kovacs CS, Chaussain C, Osdoby P, Brandi ML, Clarke B, Thakker RV (2021) The role of biomineralization in disorders of skeletal development and tooth formation. Nat Rev Endocrinol 17(6):336–349. https://doi.org/10.1039/s41574-021-00499-z
7. Cui W, Cao L, Zhang X (2022) Bioceramics: materials, properties, and applications. In: Elsevier eBooks (pp. 65–110). https://doi.org/10.1016/b978-0-323-89956-7.00017-6
8. Yamamuro T (2004) Bioceramics. In: Poitout DG (eds) Biomechanics and biomaterials in orthopedics. Springer, London. https://doi.org/10.1007/978-1-4471-3774-0_3
9. Fopase R, Pandey LM (2022) Bioceramics for biomedical applications. In: Nanoscale engineering of biomaterials: properties and applications (pp. 543–565). https://doi.org/10.1007/978-981-16-3667-7_19
10. Park J (2008) Bioceramics. In: Springer eBooks. https://doi.org/10.1007/978-0-387-09545-5. PMID: 35297994
11. Gul H, Khan M, Khan AS (2020) Bioceramics. In: Elsevier eBooks (pp. 53–83).https://doi.org/10.1016/b978-0-08-102834-6.00003-3
12. Vaiani L, Boccaccio A, Uva AE, Palumbo G, Piccininni A, Guglielmi P, Cantore S, Santacroce L, Charitos IA, Ballini A (2023) Ceramic materials for biomedical applications: an overview on properties and fabrication processes. J Funct Biomater 14(3):146. https://doi.org/10.3390/jfb14030146
13. Radunović A, Radunović O, Vulović M, Aksić M (2023) Bioceramics in orthopedic surgery. In: Engineering materials (pp. 53–65). https://doi.org/10.1007/978-3-031-17269-4_3
14. Shekhawat D, Singh A, Banerjee M, Singh T, Patnaik A (2021) Bioceramic composites for orthopaedic applications: a comprehensive review of mechanical, biological, and microstructural properties. Ceram Int 47(3):3013–3030. https://doi.org/10.1016/j.ceramint.2020.09.214
15. Piconi C, Maccauro G (2016) Perspective and trends on bioceramics in joint Replacement. In: Springer eBooks (pp. 821–857). https://doi.org/10.1007/978-3-319-12460-5_41
16. Gao C, Yao M, Shuai C, Feng P, Peng S (2020) Advances in biocermets for bone implant applications. Bio-Des Manuf 3(4):307–330. https://doi.org/10.1007/s42242-020-00087-3
17. Jayaswal GP, Dange SP, Khalikar AN (2010) Bioceramic in dental implants: a review. J Indian Prosthodont Soc 10(1):8–12. https://doi.org/10.1007/s13191-010-0002-4
18. Mussano F, Genova T, Munaron L, Faga MG, Carossa S (2016) Ceramic biomaterials for dental implants: current use and future perspectives. InTech. https://doi.org/10.5772/62701
19. Rathore R, Suhag D, Wan F, Verma R, Thakur A, Thakur P (2023) Functionalization of biomaterials. In: Suhag D, Thakur A, Thakur P (eds) Integrated nanomaterials and their applications. Springer, Singapore. https://doi.org/10.1007/979-991-99-6105-4_11
20. Gilbert PUPA, Bergmann KD, Boekelheide N, Tambutté S, Mass T, Marin F, Adkins JF, Erez J, Gilbert B, Knutson V, Cantine M, Hernández JO, Knoll AH (2022) Biomineralization: integrating mechanism and evolutionary history. Sci Adv 9(10):eabl9653. https://doi.org/10.1126/sciadv.abl9653
21. Kim HD, Amirthalingam S, Kim SL, Lee SS, Rangasamy J, Hwang NS (2017) Biomimetic materials and fabrication approaches for bone tissue engineering. Adv Healthc Mater/Adv Healthc Mater 6(23). https://doi.org/10.1002/adhm.201700612
22. Jiang S, Wang M, He J (2020) A review of biomimetic scaffolds for bone regeneration: toward a cell-free strategy. Bioeng Transl Med 6(2):e10206. https://doi.org/10.1002/btm2.10206
23. Zhu Y, Joralmon D, Shan W, Chen Y, Rong J, Zhao H, Xiao S, Li X (2021) 3D printing biomimetic materials and structures for biomedical applications. Bio-Des Manuf 4(2):405–428. https://doi.org/10.1007/s42242-020-00117-0

24. Fernandez-Yague MA, Abbah SA, McNamara L, Zeugolis DI, Pandit A, Biggs MJ (2015) Biomimetic approaches in bone tissue engineering: integrating biological and physicomechanical strategies. Adv Drug Deliv Rev 84:1–29. https://doi.org/10.1016/j.addr.2014.09.005
25. Nitti P, Narayanan A, Pellegrino R, Villani S, Madaghiele M, Demitri C (2023) Cell-tissue interaction: the biomimetic approach to design tissue engineered biomaterials. Bioengineering 10(10):1122. https://doi.org/10.3390/bioengineering10101122
26. Aguado BA, Grim JC, Rosales AM, Watson-Capps JJ, Anseth KS (2018) Engineering precision biomaterials for personalized medicine. Sci Transl Med 10(424). https://doi.org/10.1126/scitranslmed.aam8645
27. Gareev KG, Grouzdev DS, Koziaeva VV, Sitkov NO, Gao H, Zimina TM, Shevtsov M (2022) Biomimetic nanomaterials: diversity, technology, and biomedical applications. Nanomaterials (Basel, Switzerland) 12(14):2485. https://doi.org/10.3390/nano12142485
28. Suhag D, Bhatia R, Das S, Shakeel A, Ghosh A, Singh A, Sinha OP, Chakrabarti S, Mukherjee M (2015) Physically cross-linked pH-responsive hydrogels with tunable formulations for controlled drug delivery. RSC Adv 5(66):53963–53972. https://doi.org/10.1039/c5ra07424j
29. Fujita S, Sakairi N (2022) Bio-inspired materials for environmental remediation. In: The handbook of environmental chemistry (pp. 507–537). https://doi.org/10.1007/698_2021_820
30. Rathore R, Suhag D, Wan F, Thakur A, Thakur P (2023) Toxicology and toxicity studies of nano-biomaterials. In: Suhag D, Thakur A, Thakur P (eds) Integrated nanomaterials and their applications. Springer, Singapore. https://doi.org/10.1007/979-991-99-6105-4_20
31. Ebrahimi M, Botelho M, Lu W, Monmaturapoj N (2019) Synthesis and characterization of biomimetic bioceramic nanoparticles with optimized physicochemical properties for bone tissue engineering. J Biomed Mater Res Part A 107(9):1654–1666. https://doi.org/10.1002/jbm.a.36691
32. Ciulla MG, Massironi A, Sugni M, Ensign MA, Marzorati S, Forouharshad M (2023) Recent advances in the development of biomimetic materials. Gels (Basel, Switzerland) 9(10):933. https://doi.org/10.3390/gels9100933
33. Hwang J, Jeong Y, Park JM, Lee KH, Hong JW, Choi J (2015) Biomimetics: forecasting the future of science, engineering, and medicine. Int J Nanomed 10:5701–5713. https://doi.org/10.2147/IJN.S93642
34. Vijayan PP, Puglia D (2019) Biomimetic multifunctional materials: a review. Emergent Mater 2(4):391–415. https://doi.org/10.1007/s42247-019-00051-7
35. Bhushan B (2009) Biomimetics: lessons from nature–an overview. Philoso Trans R Soc A Math Phys Eng Sci 367(1893):1445–1486

# Chapter 10
# Theranostics: Combining Diagnosis and Therapy

## 10.1 Introduction

Theranostics represents a revolutionary approach in modern medicine, bridging the gap between diagnosis and therapy to offer a comprehensive and personalized treatment paradigm. By seamlessly integrating diagnostic assessments with therapeutic interventions, theranostics aims to tailor treatment strategies to the individual molecular characteristics of each patient, thereby maximizing efficacy while minimizing adverse effects. This innovative approach holds immense promise for revolutionizing patient care across a spectrum of medical disciplines, offering new avenues for precision medicine and improved clinical outcomes. In the era of precision medicine, characterized by tailored therapeutic approaches based on an individual's unique genetic makeup, environment, and lifestyle factors, theranostics assumes paramount importance [1]. By leveraging advanced imaging technologies and molecularly targeted therapies, theranostics enables clinicians to identify disease biomarkers, assess treatment response, and guide therapeutic decision-making with unprecedented precision. This personalized approach holds the potential to revolutionize patient care by optimizing treatment efficacy, minimizing adverse effects, and improving overall clinical outcomes [2]. Moreover, the integration of diagnostic and therapeutic modalities within a single platform streamlines the clinical workflow, facilitating rapid translation of research findings into clinical practice and enhancing the delivery of precision medicine to patients worldwide.

### *10.1.1 Historical Overview*

The roots of theranostics can be traced back to the early developments in nuclear medicine and molecular imaging. In the mid-twentieth century, the advent of

 
D. Suhag et al., *Handbook of Biomaterials for Medical Applications, Volume 1*,
Biomedical Materials for Multi-functional Applications,
https://doi.org/10.1007/978-981-97-4818-1_10

radionuclide-based imaging techniques laid the foundation for non-invasive visualization of physiological processes within the body. Concurrently, the emergence of targeted radiotherapy enabled precise delivery of therapeutic radiation to malignant cells, marking the inception of theranostic principles [1, 2]. Over the decades, advancements in imaging technology, molecular biology, and nanotechnology have propelled theranostics into a dynamic and rapidly evolving field, with multifunctional nanoparticles and molecular probes serving as the cornerstone of modern theranostic approaches.

### *10.1.2 Significance in Precision Medicine*

Precision medicine, characterized by the customization of healthcare based on individual variability in genes, environment, and lifestyle, represents a paradigmatic shift from the traditional one-size-fits-all approach. In this context, theranostics emerges as a quintessential tool in the arsenal of precision medicine, offering tailored diagnostic and therapeutic interventions that account for the unique molecular signatures of each patient. By harnessing the power of molecular imaging and targeted therapy, theranostics enables clinicians to make informed decisions regarding treatment selection, monitor therapeutic response in real-time, and optimize patient outcomes with unparalleled precision and efficacy [3]. Moreover, the integration of diagnostic and therapeutic functionalities within a single platform not only streamlines patient care but also minimizes treatment-associated toxicity and enhances therapeutic efficacy, underscoring the transformative potential of theranostics in shaping the future of medicine.

### *10.1.3 Theranostic Nanoparticles*

Theranostic nanoparticles represent a cutting-edge innovation in biomedical engineering, offering multifunctional platforms that combine diagnostic imaging capabilities with targeted therapeutic interventions (Fig. 10.1). These nanoparticles, typically ranging in size from 1 to 100 nm, are designed with precise control over their composition, size, and surface properties to enable specific targeting of diseased tissues and cells [4]. By incorporating imaging agents and therapeutic payloads into a single nanoscale carrier, theranostic nanoparticles facilitate real-time monitoring of disease progression and treatment response, while also delivering therapeutic agents with enhanced precision and efficacy. This convergence of diagnostic and therapeutic functionalities holds tremendous potential for advancing personalized medicine and improving patient outcomes across a wide range of diseases, from cancer to cardiovascular and neurological disorders.

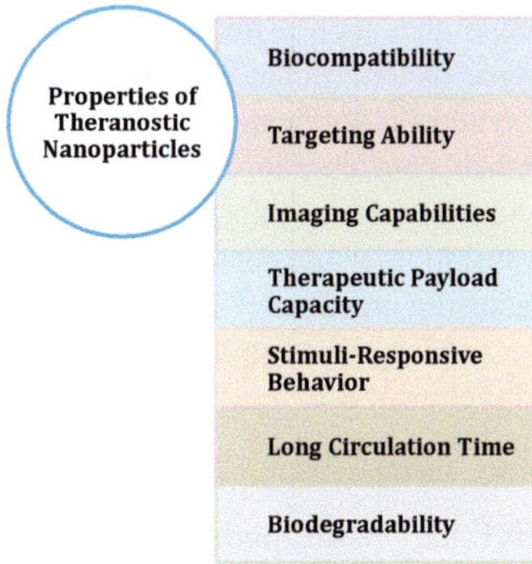


**Fig. 10.1** Properties of theranostics nanoparticles

#### 10.1.3.1 Nanoparticle Design and Fabrication

Designing and fabricating nanoparticles with precision is a cornerstone of theranostics, enabling the development of versatile platforms for targeted diagnosis and therapy. At the core of nanoparticle design is the selection of materials tailored to the intended application, such as biocompatible polymers, metals, or lipids. These materials are chosen based on factors such as stability, biodegradability, and compatibility with imaging agents and therapeutic payloads. The size and shape of nanoparticles play a crucial role in their behavior within biological systems, influencing parameters such as circulation time, cellular uptake, and biodistribution [2, 3]. Various fabrication techniques are employed to produce nanoparticles with controlled size, shape, and surface properties. One common approach is nanoprecipitation, where polymer or lipid components are dissolved in an organic solvent and rapidly mixed with an aqueous phase containing the desired payload. This process leads to the formation of nanoparticles through the spontaneous assembly of amphiphilic molecules at the liquid–liquid interface. Alternatively, techniques such as emulsion-based methods, solvent evaporation, and self-assembly can be utilized to generate nanoparticles with tailored characteristics. Surface functionalization of nanoparticles is another critical aspect of their design, allowing for the attachment of targeting ligands, imaging probes, and therapeutic agents. Functional groups such as carboxyl, amino, or thiol moieties can be incorporated into nanoparticle formulations to facilitate conjugation with biomolecules or chemical modifications. This enables precise control over the interactions between nanoparticles and biological targets, enhancing their specificity and therapeutic efficacy.

In addition to traditional fabrication methods, advancements in nanotechnology have enabled the development of novel approaches for nanoparticle synthesis. For instance, bottom-up techniques like molecular self-assembly and DNA origami allow for the precise assembly of nanoparticles from molecular building blocks, offering unprecedented control over their structure and properties [4]. Similarly, techniques such as microfluidics and electrospray enable the production of monodisperse nanoparticles with tunable size and composition, opening new avenues for theranostic applications. The design and fabrication of nanoparticles also encompasses considerations related to scalability, reproducibility, and regulatory compliance. Manufacturing processes must be scalable to meet the demands of clinical translation, while ensuring batch-to-batch consistency and adherence to good manufacturing practices (GMP). Moreover, efforts to optimize nanoparticle formulations for stability, biocompatibility, and safety are paramount to their successful translation into clinical use.

#### 10.1.3.2 Multimodal Imaging Capabilities

Multimodal imaging capabilities represent a transformative advancement in medical imaging, offering the ability to acquire complementary anatomical and functional information using a single imaging modality. By combining multiple imaging techniques such as magnetic resonance imaging (MRI), positron emission tomography (PET), single-photon emission computed tomography (SPECT), fluorescence imaging, and ultrasound, multimodal imaging enables comprehensive assessment of biological processes at various spatial and temporal scales [5]. This synergistic approach not only enhances diagnostic accuracy and sensitivity but also provides invaluable insights into disease pathophysiology and therapeutic response. With its versatility and potential for real-time monitoring, multimodal imaging holds great promise for advancing precision medicine and improving patient outcomes across a wide range of clinical applications.

##### MRI Contrast Agents

Magnetic resonance imaging (MRI) contrast agents play a pivotal role in enhancing the sensitivity and specificity of MRI by altering the magnetic properties of tissues and highlighting specific anatomical structures or pathological features. These contrast agents contain paramagnetic or superparamagnetic elements that interact with the surrounding magnetic field, leading to changes in relaxation times and signal intensity. Gadolinium-based contrast agents are the most widely used for clinical MRI due to their high relaxivity and biocompatibility. These agents shorten the T1 relaxation time of tissues, resulting in increased signal intensity on T1-weighted images and improved visualization of blood vessels, tumors, and inflammatory lesions. The design and development of MRI contrast agents involve careful consideration of factors such as biocompatibility, stability, and relaxivity [6]. Gadolinium chelates, such as gadoterate and gadobenate, are commonly used to encapsulate gadolinium ions and minimize their toxicity while enhancing their stability and tissue-specific targeting. Surface modifications, such as the attachment of targeting ligands or polymers, can further improve the pharmacokinetics and biodistribution of contrast agents, enabling selective accumulation at disease sites while minimizing off-target effects.

Recent advancements in MRI contrast agent technology have focused on improving their relaxivity, specificity, and multiplexing capabilities. Novel contrast agents incorporating paramagnetic metals, such as manganese and iron oxide nanoparticles, offer unique advantages such as prolonged retention in tissues and enhanced contrast enhancement [7]. The development of responsive contrast agents that undergo changes in relaxivity or signal intensity in response to specific biological stimuli holds promise for applications such as molecular imaging and image-guided therapy.

### PET/SPECT Imaging Probes

Positron emission tomography (PET) and single-photon emission computed tomography (SPECT) imaging probes are indispensable tools in molecular imaging, providing non-invasive visualization and quantification of biological processes at the molecular and cellular levels. PET probes typically consist of radiolabeled molecules, known as radiotracers, that emit positrons upon decay, while SPECT probes utilize radiopharmaceuticals emitting gamma rays. These probes are designed to selectively bind to specific molecular targets, such as receptors, enzymes, or transporters, allowing for the detection and characterization of disease-related biomarkers in vivo [8]. The development of PET and SPECT imaging probes involves the synthesis and radiolabeling of small molecules, peptides, antibodies, or nanoparticles with radionuclides suitable for imaging purposes. Commonly used radionuclides for PET include fluorine-18, carbon-11, and oxygen-15, whereas technetium-1010m and iodine-123 are frequently employed in SPECT imaging. Radiolabeling techniques must ensure high radiochemical purity and stability of the probes to enable accurate and reproducible imaging results. PET and SPECT imaging probes have diverse applications in clinical and preclinical research, ranging from oncology and neurology to cardiology and infectious diseases. In oncology, PET tracers targeting glucose metabolism (e.g., 18F-fluorodeoxyglucose) or specific cancer biomarkers (e.g., 68Ga-PSMA for prostate cancer) are used for tumor detection, staging, and monitoring of treatment response. Similarly, SPECT tracers targeting myocardial perfusion (e.g., 1010mTc-sestamibi) aid in the diagnosis of coronary artery disease and evaluation of cardiac function [9].

Advancements in radiopharmaceutical design and imaging technology continue to expand the repertoire of PET and SPECT probes, enabling more precise and comprehensive characterization of disease processes. The integration of PET/SPECT imaging with other modalities such as MRI and CT further enhance diagnostic accuracy and provides complementary anatomical information [7, 8]. Moreover, the emergence of theranostic approaches combining imaging and targeted therapy holds promise for personalized medicine, where imaging probes can guide the selection and optimization of therapeutic interventions based on individual patient characteristics.

### Fluorescence and Bioluminescence Tags

Fluorescence and bioluminescence tags are essential tools in molecular imaging, offering non-invasive visualization and tracking of biological processes at the cellular and molecular levels. Fluorescent tags consist of fluorophores that emit light upon excitation by a specific wavelength of light, whereas bioluminescent tags utilize enzymes such as luciferase to generate light through chemical reactions. These tags can be conjugated to molecules of interest, such as antibodies, peptides, or small molecules, enabling targeted imaging of specific biomolecules or cellular structures in living organisms. The design and selection of fluorescence and bioluminescence tags depend on factors such as brightness, photostability, spectral properties, and compatibility with imaging instruments [10]. Fluorescent tags encompass a wide range of fluorophores, including organic dyes, quantum dots, and fluorescent proteins such as green fluorescent protein (GFP) and its variants. Each fluorophore has unique characteristics that determine its suitability for specific imaging applications, such as single-molecule tracking, super-resolution microscopy, or in vivo imaging. Bioluminescent tags, on the other hand, rely on the enzymatic reaction between luciferase and its substrate, resulting in the emission of light. Commonly used luciferases include firefly luciferase, Renilla luciferase, and NanoLuc, each with distinct emission spectra and kinetics [11]. Bioluminescent tags offer advantages such as low background signal, high signal-to-noise ratio, and minimal phototoxicity, making them ideal for longitudinal imaging studies in live animals and deep-tissue imaging applications.

Fluorescence and bioluminescence tags have diverse applications in biomedical research, including cell tracking, gene expression analysis, protein–protein interactions, and drug discovery. In cell biology, fluorescent tags enable visualization of subcellular structures, dynamic processes, and signaling pathways in real-time. Bioluminescent tags, on the other hand, are widely used for monitoring gene expression, protein–protein interactions, and intracellular signaling events in live cells and animals [10, 11]. The combination of fluorescence and bioluminescence imaging modalities offers complementary strengths for multiplexed imaging and quantitative analysis of biological phenomena. Moreover, advancements in imaging instrumentation and molecular biology techniques continue to expand the versatility and utility of fluorescence and bioluminescence tags in biomedical research and drug development.

### Photoacoustic Imaging Nanoparticles

Photoacoustic imaging nanoparticles are innovative agents that enable high-resolution imaging of biological tissues by harnessing the photoacoustic effect. These nanoparticles, typically composed of optically absorbing materials such as gold nanorods or carbon nanotubes, exhibit strong light absorption properties in the near-infrared (NIR) region. When irradiated with pulsed laser light, the absorbed energy is converted into heat, leading to rapid thermoelastic expansion and emission of acoustic waves that can be detected by ultrasound transducers [8, 10]. This process allows

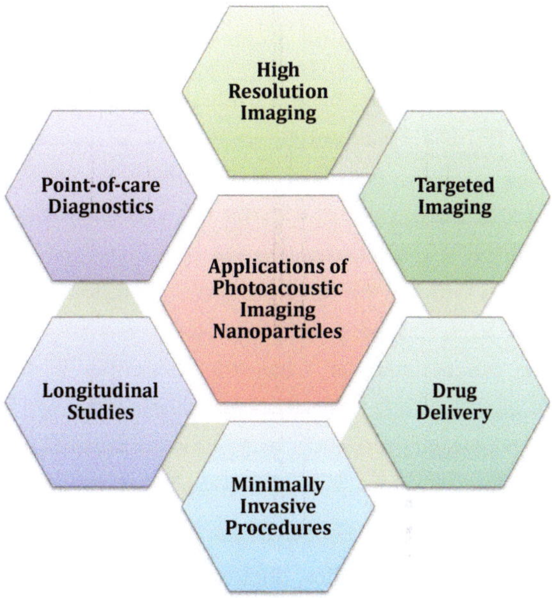


**Fig. 10.2** Applications of photoacoustic imaging nanoparticles

for deep tissue penetration and high spatial resolution imaging with minimal tissue scattering, making photoacoustic imaging nanoparticles promising tools for a wide range of biomedical applications (Fig. 10.2). The design and synthesis of photoacoustic imaging nanoparticles involves careful consideration of factors such as optical absorption properties, biocompatibility, and surface chemistry. Gold nanoparticles, in particular, are favored for their strong plasmonic absorption in the NIR region and tunable optical properties. Surface modifications with biocompatible coatings, such as polyethylene glycol (PEG) or silica, enhance stability, reduce nonspecific interactions, and enable functionalization with targeting ligands or imaging agents for specific biomedical applications. Photoacoustic imaging nanoparticles have shown great potential for molecular imaging, tumor targeting, and theranostic applications [12]. By conjugating targeting ligands or molecular probes onto the nanoparticle surface, researchers can achieve selective accumulation and imaging of specific molecular targets in vivo. The integration of therapeutic agents, such as chemotherapeutic drugs or photothermal agents, within the nanoparticles enables simultaneous imaging and therapy, paving the way for personalized medicine and image-guided interventions.

**Exercise 1.**

**Fill Ups**

**(Gold, Therapeutic, Tailored treatments, Ligands, Diagnostic, Monitoring)**

**1.** Theranostic nanoparticles are designed to simultaneously perform ............... and .....................functions within the body.

**2.** The nanoparticles often consist of a core material, such as.............., that can carry therapeutic agents or imaging agents.

**3.** The surface of theranostic nanoparticles is typically functionalized with ................that enable targeting specific cells or tissues.

**4.** One advantage of theranostic nanoparticles is their ability to provide real-time .........................of treatment response during therapy.

**5.** Theranostic nanoparticles hold promise for personalized medicine approaches by allowing for .....................based on individual patient characteristics and treatment responses.

In preclinical studies, photoacoustic imaging nanoparticles have been used for non-invasive monitoring of disease progression, evaluation of treatment response, and assessment of drug delivery kinetics. Their ability to provide functional and molecular information, coupled with high spatial resolution and deep tissue penetration, makes them valuable tools for studying complex biological processes in vivo.

## *10.1.4 Therapeutic Payload Delivery*

Therapeutic payload delivery represents a crucial aspect of modern medicine, offering targeted and efficient delivery of therapeutic agents to diseased tissues while minimizing systemic toxicity. This approach involves the encapsulation or conjugation of drugs, nucleic acids, or other therapeutic molecules within carrier systems such as nanoparticles, liposomes, or polymer micelles, designed to enhance their pharmacokinetics, biodistribution, and cellular uptake [12]. By exploiting the unique properties of these carrier systems, therapeutic payload delivery enables precise control over drug release kinetics, tissue targeting, and intracellular delivery, ultimately improving the efficacy and safety of therapeutic interventions across a wide range of diseases.

### 10.1.4.1 Drug Loading Strategies

Drug loading strategies play a pivotal role in the development of effective therapeutic delivery systems, ensuring efficient encapsulation or conjugation of drugs within carrier vehicles while maintaining their stability and bioactivity [13]. One commonly employed strategy involves physical encapsulation, where drugs are physically entrapped within the core or matrix of carrier materials such as nanoparticles, liposomes, or polymer micelles. This approach relies on factors such as the size, shape, and surface properties of the carrier system to facilitate drug loading and retention, with parameters such as drug-carrier compatibility and solubility playing crucial roles in determining loading efficiency. Chemical conjugation represents another versatile approach for drug loading, involving covalent attachment of drugs to carrier materials through chemical linkers or functional groups. This strategy offers precise control over drug-carrier interactions and release kinetics, allowing for tailored modifications of drug pharmacokinetics and biodistribution [14]. Moreover, conjugation facilitates targeted delivery of drugs to specific tissues or cell types by exploiting ligand-receptor interactions or stimuli-responsive mechanisms, thereby enhancing therapeutic efficacy, and reducing off-target effects.

Hybrid drug loading strategies combine elements of both physical encapsulation and chemical conjugation to achieve synergistic advantages in drug delivery. For instance, drug-loaded nanoparticles may be functionalized with targeting ligands or stimuli-responsive moieties to enhance their specificity and controllability. Similarly, prodrug approaches involve chemically modifying drugs to enable their activation or release in response to specific biological stimuli or environmental cues, thereby enhancing drug stability and bioavailability while minimizing systemic toxicity [12, 13]. Advancements in drug loading strategies continue to drive innovation in therapeutic delivery systems, offering tailored solutions for addressing challenges such as poor drug solubility, low bioavailability, and off-target effects.

#### 10.1.4.2 Controlled Release Mechanisms

Controlled release mechanisms play a crucial role in therapeutic delivery systems, allowing for precise modulation of drug release kinetics and spatiotemporal distribution to optimize therapeutic outcomes while minimizing side effects. One common approach to controlled release involves the use of stimuli-responsive materials that undergo structural changes or degradation in response to specific environmental cues, such as pH, temperature, or enzyme activity. These materials can be incorporated into carrier systems such as nanoparticles [14], hydrogels, or microparticles, enabling triggered release of encapsulated drugs at target sites within the body. pH-responsive systems are particularly versatile for controlled drug release, exploiting variations in pH levels between different physiological compartments or diseased tissues. For example, pH-sensitive polymers such as poly (lactic-co-glycolic acid) (PLGA) or polyethylene glycol (PEG) can be engineered to undergo conformational changes or dissolution in acidic environments, facilitating the release of encapsulated drugs in acidic tumor microenvironments or intracellular compartments such as endosomes or lysosomes. Temperature-responsive materials offer another approach to controlled drug release, leveraging temperature-dependent phase transitions or solubility changes to modulate drug release kinetics. Thermosensitive polymers such as poly(N-isopropylacrylamide) (PNIPAAm) can undergo reversible coil-to-globule transitions in response to changes in temperature [15], allowing for on-demand drug release triggered by local hyperthermia or external stimuli such as infrared light.

Enzyme-responsive systems exploit the presence of specific enzymes in diseased tissues or physiological fluids to trigger drug release through enzymatic degradation or cleavage of polymer chains. For instance, peptide-based hydrogels containing enzyme-cleavable peptide sequences can be designed to undergo proteolytic degradation by tumor-associated proteases, enabling site-specific drug release within the tumor microenvironment while sparing healthy tissues [15]. Overall, controlled release mechanisms offer versatile strategies for optimizing the pharmacokinetics, biodistribution, and therapeutic efficacy of drugs in therapeutic delivery systems.

#### 10.1.4.3 Targeting Ligands for Specific Cell Recognition

Targeting ligands play a crucial role in therapeutic delivery systems by facilitating specific recognition and binding to target cells or tissues, thereby enhancing the selectivity and efficacy of drug delivery. These ligands, typically peptides, antibodies, aptamers, or small molecules, are designed to interact with cell surface receptors or biomarkers that are overexpressed or uniquely expressed on target cells, while minimizing nonspecific interactions with healthy tissues. By harnessing the molecular specificity of targeting ligands, therapeutic agents can be delivered directly to diseased cells [16], enabling localized treatment and minimizing systemic toxicity. Peptide ligands are commonly used for cell recognition due to their versatility, stability, and ease of synthesis. Peptides can be designed to mimic natural ligands or receptor-binding motifs, allowing for selective targeting of specific cell types or signaling pathways. Moreover, peptide sequences can be optimized through combinatorial screening or computational design to enhance binding affinity and specificity, enabling precise control over drug delivery and therapeutic outcomes. Antibodies represent another powerful class of targeting ligands, offering high specificity and affinity for cell surface receptors or antigens. Monoclonal antibodies can be generated against a wide range of targets, including cancer-specific antigens, inflammatory markers, or pathogen-associated molecules, providing a versatile platform for targeted drug delivery in cancer therapy [17], immunotherapy, and infectious disease treatment. Antibody-drug conjugates (ADCs) combine the targeting specificity of antibodies with the cytotoxic potency of chemotherapeutic drugs, enabling selective killing of tumor cells while sparing healthy tissues. Aptamers, synthetic nucleic acid molecules selected through systematic evolution of ligands by exponential enrichment (SELEX), offer unique advantages as targeting ligands [18] due to their high specificity, affinity, and stability. Aptamers can be engineered to bind to a wide range of targets, including proteins, nucleic acids, or small molecules, with exquisite selectivity and sensitivity. Aptamers can be easily conjugated to nanoparticles, liposomes, or other carrier systems, enabling targeted drug delivery to specific cell types or tissues.

Overall, targeting ligands for specific cell recognition represents essential components of therapeutic delivery systems, enabling precise and efficient delivery of therapeutic agents to diseased cells while minimizing off-target effects.

**Exercise 2**

**Fill Ups**

**(Off-target effects, Small size, Drug release, Therapeutic molecules, Therapeutic agents)**

1. Therapeutic payload delivery involves the transport of ..................to specific target sites in the body for treatment.
2. Nanoparticles are commonly used as carriers for therapeutic payloads due to their ......................, and ability to evade immune detection.
3. The payload delivered by nanoparticles can include various types of......................, such as drugs, nucleic acids, or proteins.
4. One advantage of targeted therapeutic payload delivery is the ability to minimize ...............................by concentrating treatment at the site of disease.
5. Therapeutic payload delivery systems can be engineered to respond to specific stimuli, such as changes in pH or temperature, to trigger ..................at the target site.

## *10.1.5 Integration of Diagnosis and Treatment*

The integration of diagnosis and treatment represents a pivotal advancement in modern medicine, where diagnostic assessments inform and guide therapeutic interventions to achieve personalized and precise patient care. This approach seeks to seamlessly combine the capabilities of diagnostic imaging modalities with therapeutic strategies, enabling clinicians to not only identify disease pathology but also to deliver targeted treatments directly to the affected tissues or cells [17, 18]. By integrating diagnosis and treatment, healthcare providers can tailor therapeutic interventions to individual patient characteristics, optimizing efficacy while minimizing adverse effects.

### 10.1.5.1 Real-Time Imaging-Guided Therapy

Real-time imaging-guided therapy represents a cutting-edge approach in modern medicine, where diagnostic imaging modalities are utilized to guide therapeutic interventions in real-time, ensuring precision and accuracy during treatment delivery. This approach enables clinicians to visualize anatomical structures, monitor disease progression, and dynamically adjust treatment parameters to optimize therapeutic outcomes [19, 20]. By integrating imaging technology with therapeutic interventions, real-time imaging-guided therapy enhances procedural safety, efficacy, and patient outcomes across a wide range of medical specialties. One key application of real-time imaging-guided therapy is in interventional radiology, where imaging modalities such as fluoroscopy, ultrasound, and MRI are used to guide minimally invasive procedures such as biopsies, ablations, and catheter-based interventions. These imaging techniques provide live visualization of internal organs and tissues, allowing clinicians to precisely target pathological lesions while avoiding damage to surrounding healthy tissue. Real-time feedback from imaging also enables immediate assessment of treatment response, facilitating on-the-fly adjustments to optimize procedural outcomes. In the field of oncology, real-time imaging-guided therapy has emerged as a valuable tool for delivering precise and targeted cancer treatments, such as radiotherapy and brachytherapy [21]. Imaging modalities such as MRI, CT, and PET are used to accurately delineate tumor boundaries, track tumor motion, and monitor treatment response in real-time, allowing for adaptive radiation therapy techniques that adjust treatment plans based on changes in tumor size, shape, and position. This approach maximizes therapeutic efficacy while minimizing radiation exposure to surrounding healthy tissues, reducing the risk of treatment-related complications.

Moreover, real-time imaging-guided therapy holds promise for advancing the field of image-guided surgery, where intraoperative imaging techniques such as intraoperative CT, MRI, and fluorescence imaging are employed to provide surgeons with real-time visualization and guidance during surgical procedures. These imaging modalities enable precise localization of tumors, identification of critical structures, and navigation of complex anatomical regions, enhancing surgical accuracy and reducing the risk of complications.

#### 10.1.5.2 Responsive Therapeutic Nanoparticles

Responsive therapeutic nanoparticles represent a dynamic approach to drug delivery, where the release of therapeutic agents is triggered by specific biological stimuli or environmental cues (Fig. 10.3). These nanoparticles are designed to respond to changes in physiological parameters such as pH, temperature, enzyme activity, or redox potential, enabling controlled release of drugs at target sites within the body [22]. By harnessing the body's natural cues, responsive nanoparticles can deliver therapeutic agents with spatial and temporal precision, enhancing therapeutic efficacy while minimizing off-target effects and systemic toxicity. One common strategy for designing responsive therapeutic nanoparticles involves the use of pH-responsive polymers or materials that undergo conformational changes or degradation in response to variations in pH levels. In acidic tumor microenvironments or intracellular compartments such as endosomes or lysosomes, pH-sensitive nanoparticles can disassemble or undergo protonation, triggering the release of encapsulated drugs. This approach enables targeted drug delivery to tumor cells while sparing healthy tissues, enhancing the therapeutic index of anticancer agents and reducing side effects. Temperature-responsive nanoparticles offer another versatile platform

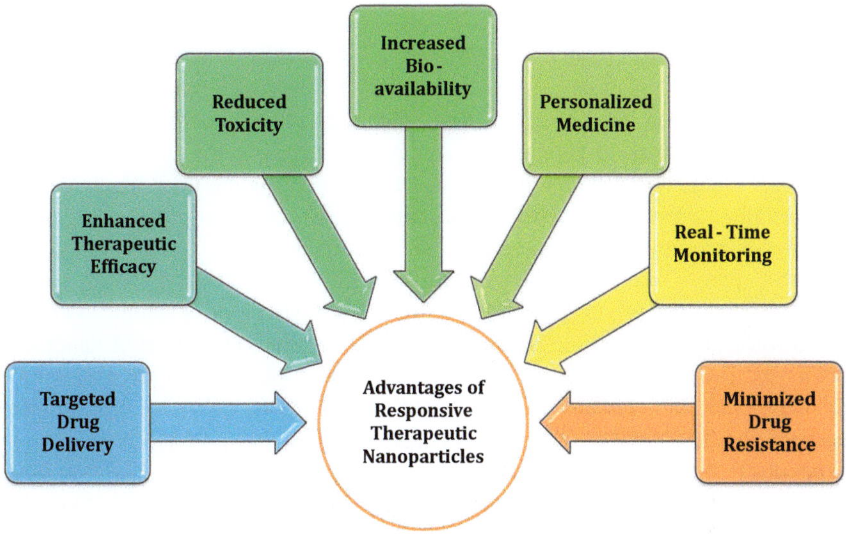


**Fig. 10.3** Advantages of responsive therapeutic nanoparticles

for controlled drug release, leveraging temperature-dependent phase transitions or solubility changes to modulate drug release kinetics. Thermosensitive polymers such as poly(N-isopropylacrylamide) (PNIPAAm) can undergo reversible coil-to-globule transitions in response to changes in temperature, enabling on-demand drug release triggered by local hyperthermia or external stimuli such as infrared light [23]. This temperature-responsive behavior can be exploited for targeted drug delivery in hyperthermia-based cancer therapies or for thermally triggered release of therapeutic agents in response to localized heating.

Enzyme-responsive nanoparticles are designed to exploit the presence of specific enzymes in diseased tissues or physiological fluids to trigger drug release through enzymatic degradation or cleavage of polymer chains. Peptide-based hydrogels containing enzyme-cleavable peptide sequences, for example, can be engineered to undergo proteolytic degradation by tumor-associated proteases, enabling site-specific drug release within the tumor microenvironment while sparing healthy tissues.

#### 10.1.5.3 Targeted Drug Delivery Systems

Targeted drug delivery systems represent a significant advancement in pharmaceutical research, offering the potential to improve therapeutic efficacy while minimizing systemic side effects. These systems are designed to deliver therapeutic agents specifically to target cells, tissues, or organs, thereby enhancing drug concentration at the site of action while reducing exposure to healthy tissues. Targeted drug delivery systems utilize various strategies, including the conjugation of drugs to targeting ligands, encapsulation within nanoparticles, or activation in response to specific physiological cues, to achieve precise and controlled drug release [24]. One common approach to targeted drug delivery involves the use of targeting ligands, such as antibodies, peptides, or aptamers, which selectively bind to receptors or biomarkers overexpressed on the surface of target cells. By conjugating drugs to these targeting ligands, researchers can achieve selective accumulation and internalization of therapeutic agents within diseased cells, enhancing therapeutic efficacy while minimizing off-target effects. Targeting ligand-based drug delivery systems have shown promise in cancer therapy, where they can target tumor-specific antigens or receptors to deliver cytotoxic drugs directly to cancer cells while sparing healthy tissues. Nanoparticle-based drug delivery systems offer another versatile platform for targeted drug delivery, leveraging the unique properties of nanoparticles to encapsulate, protect, and deliver therapeutic agents to specific sites within the body. Nanoparticles can be engineered to exploit the enhanced permeability and retention (EPR) effect, which allows them to accumulate preferentially in tumor tissues due to their leaky vasculature and impaired lymphatic drainage [25]. Nanoparticles can be functionalized by targeting ligands or responsive materials to enable active targeting or triggered drug release in response to specific biological cues, further enhancing their precision and efficacy. Moreover, stimuli-responsive drug delivery systems have emerged as a promising approach for targeted drug delivery, where drug release is triggered by changes in physiological parameters such as pH, temperature, enzyme activity, or redox potential. These systems utilize materials that undergo conformational changes, degradation, or disassembly in response to specific stimuli, enabling controlled release of therapeutic agents at target sites within the body. Stimuli-responsive drug delivery systems offer the potential for on-demand drug release, enhanced therapeutic efficacy, and reduced systemic toxicity, making them valuable tools for precision medicine and personalized therapeutics.

### *10.1.6 Clinical Applications of Theranostics*

Clinical applications of theranostics encompass a diverse array of medical interventions aimed at seamlessly integrating diagnosis and therapy to tailor treatments to individual patient needs. This innovative approach leverages advanced imaging modalities and therapeutic agents to not only detect and visualize disease but also deliver targeted treatments with precision and efficacy. Theranostics holds immense promise across various medical disciplines, ranging from oncology and cardiology to neurology and infectious diseases, by offering personalized [26] and comprehensive solutions to complex medical challenges. By bridging the gap between diagnosis and therapy, theranostics revolutionizes patient care, optimizing treatment outcomes while minimizing adverse effects, and paving the way for a new era of precision medicine.

#### 10.1.6.1 Cancer Theranostics

Cancer theranostics represents a paradigm shift in oncology, offering personalized treatment approaches that integrate diagnosis, therapy, and monitoring of treatment response and disease progression (Fig. 10.4). This innovative approach aims to tailor cancer treatments to individual patient characteristics, optimizing therapeutic outcomes while minimizing side effects. By combining diagnostic imaging modalities such as positron emission tomography (PET), magnetic resonance imaging (MRI), and computed tomography (CT) with targeted therapies and molecularly guided interventions, cancer theranostics enables clinicians to identify specific molecular targets, assess treatment response, and adapt therapeutic strategies in real-time [27]. One key aspect of cancer theranostics is the use of molecular imaging techniques to non-invasively visualize tumor biology and heterogeneity, providing insights into tumor aggressiveness, metastatic potential, and treatment resistance. PET imaging with radiolabeled tracers targeting cancer-specific biomarkers, such as glucose metabolism (18F-fluorodeoxyglucose) or prostate-specific membrane antigen (68Ga-PSMA), allows for accurate staging and characterization of tumors, guiding treatment decisions and predicting patient prognosis. Similarly, MRI and CT imaging provide detailed anatomical and functional information about tumor morphology, vascularity, and tissue perfusion, aiding in treatment planning and monitoring of disease progression.

Cancer theranostics encompasses targeted therapies and molecularly guided interventions that exploit the molecular characteristics of tumors to deliver personalized treatments. Targeted therapies, such as tyrosine kinase inhibitors, monoclonal antibodies, and immune checkpoint inhibitors, selectively inhibit or modulate specific molecular pathways involved in tumor growth and progression, leading to improved clinical outcomes in subsets of patients with molecularly defined cancers. Molecularly guided interventions, such as image-guided biopsies, radiofrequency ablation, and brachytherapy [25, 26], enable precise localization and targeting of tumor lesions while sparing surrounding healthy tissues, reducing treatment-related morbidity and

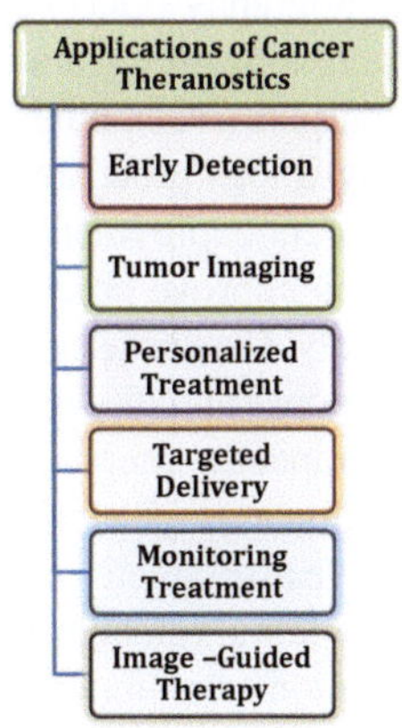


**Fig. 10.4** Various applications of cancer theranostics

improving quality of life. Cancer theranostics enables real-time monitoring of treatment response and disease progression, allowing clinicians to adapt therapeutic strategies based on individual patient responses. Imaging modalities such as PET, MRI, and CT provide quantitative measures of tumor size, metabolic activity, and functional changes over time, facilitating early detection of treatment response or disease recurrence. Emerging techniques such as radiomics and artificial intelligence (AI) analysis of imaging data offer opportunities to predict treatment outcomes, identify treatment-resistant phenotypes, and stratify patients for personalized therapeutic interventions.

Cancer theranostics represents a transformative approach in oncology, offering personalized treatment approaches and real-time monitoring of treatment response and disease progression [27]. By integrating diagnostic imaging modalities with targeted therapies and molecularly guided interventions, cancer theranostics enables clinicians to tailor treatments to individual patient characteristics, optimizing therapeutic outcomes and improving patient prognosis.

#### 10.1.6.2 Cardiovascular Theranostics

Cardiovascular theranostics represents a comprehensive approach to cardiovascular care, integrating diagnostic imaging biomarkers for cardiovascular risk assessment with interventional therapies and imaging guidance to optimize patient outcomes. Imaging biomarkers play a crucial role in assessing cardiovascular risk by providing insights into the structural and functional abnormalities of the heart and blood vessels. Techniques such as echocardiography, cardiac MRI, CT angiography, and nuclear imaging offer detailed assessments of cardiac morphology, function, perfusion, and viability, enabling clinicians to identify individuals at high risk for cardiovascular events and tailor preventive interventions accordingly [28]. In addition to risk assessment, cardiovascular theranostics leverages imaging biomarkers to guide interventional therapies for the treatment of cardiovascular diseases. Imaging modalities such as fluoroscopy, intravascular ultrasound (IVUS), and optical coherence tomography (OCT) provide real-time visualization of cardiovascular anatomy and pathology during minimally invasive procedures such as percutaneous coronary interventions (PCI), transcatheter aortic valve replacement (TAVR), and cardiac electrophysiology procedures [210]. By enabling precise localization of lesions, assessment of stent deployment, and identification of complications, imaging-guided interventions improve procedural success rates, reduce complications, and enhance patient safety. Cardiovascular theranostics facilitates image-based assessment of treatment response and disease progression, enabling clinicians to monitor the effectiveness of therapeutic interventions and adjust treatment strategies as needed. Serial imaging evaluations with techniques such as cardiac MRI or CT angiography allow for quantitative assessments of changes in cardiac structure, function, and perfusion over time, providing valuable insights into the efficacy of medical therapies, lifestyle interventions, or surgical procedures. Advanced imaging techniques such as myocardial perfusion imaging or late gadolinium enhancement MRI offer prognostic information by identifying myocardial ischemia, scar tissue, or microvascular dysfunction, aiding in risk stratification, and guiding clinical management decisions.

Cardiovascular theranostics harnesses the power of imaging biomarkers for cardiovascular risk assessment and interventional therapies, providing personalized and precision-driven care for patients with cardiovascular diseases [30]. By integrating diagnostic imaging with therapeutic interventions and imaging guidance, cardiovascular theranostics enhances patient outcomes, improves procedural success rates, and optimizes clinical decision-making in the management of cardiovascular conditions.

#### 10.1.6.3 Neurological Disorder Theranostics

Neurological disorder theranostics represents a multifaceted approach to managing conditions affecting the central nervous system (CNS), encompassing both diagnostic imaging markers for neurodegenerative diseases and innovative nanotherapeutics for brain tumor treatment. In the realm of neurodegenerative diseases such as Alzheimer's disease, Parkinson's disease, and multiple sclerosis, diagnostic imaging plays a crucial role in early detection, disease staging, and monitoring of disease progression [31]. Techniques such as structural and functional MRI, positron emission tomography (PET), and cerebrospinal fluid (CSF) biomarker analysis provide valuable insights into neuroanatomical changes, protein aggregation, and metabolic dysfunction associated with these conditions, facilitating accurate diagnosis and treatment planning. Neuroimaging biomarkers serve as invaluable tools for monitoring treatment response and disease progression in neurodegenerative diseases, enabling clinicians to evaluate the effectiveness of therapeutic interventions and adjust treatment strategies as needed. Advanced imaging techniques such as diffusion tensor imaging (DTI) and functional MRI (fMRI) offer insights into changes in brain connectivity [32], white matter integrity, and functional connectivity networks over time, providing valuable prognostic information and guiding personalized treatment approaches. Molecular imaging techniques using PET tracers targeting specific pathological hallmarks, such as amyloid-beta or tau protein deposition, offer opportunities for early disease detection, monitoring of disease progression, and assessment of treatment response in neurodegenerative disorders.

In the realm of brain tumor treatment, theranostic approaches leverage nanotherapeutics to deliver targeted treatments directly to tumor cells while sparing healthy brain tissue. Nanoparticle-based drug delivery systems, such as liposomes, polymeric nanoparticles, and lipid nanoparticles, offer unique advantages for brain tumor treatment, including enhanced blood–brain barrier (BBB) permeability [33], prolonged circulation time, and tumor-targeting capabilities. These nanoparticles can be functionalized with targeting ligands, such as antibodies or peptides, to selectively bind to receptors overexpressed on tumor cells, facilitating targeted drug delivery and minimizing systemic toxicity.

Nanotherapeutics for brain tumor treatment can be designed to respond to specific biological cues within the tumor microenvironment, enabling controlled drug release and enhanced therapeutic efficacy. Stimuli-responsive nanoparticles, such as pH-responsive polymers or temperature-sensitive liposomes, can be engineered to release therapeutic agents in response to changes in pH, temperature, or enzyme activity characteristic of the tumor microenvironment, maximizing drug delivery to tumor cells while minimizing off-target effects. Theranostic nanoparticles incorporating imaging agents, such as iron oxide nanoparticles or quantum dots, enable real-time imaging guidance during nanoparticle delivery and distribution within the brain, facilitating accurate tumor targeting and treatment monitoring.

Neurological disorder theranostics encompasses a comprehensive approach to managing neurodegenerative diseases and brain tumors, leveraging diagnostic imaging markers and innovative nanotherapeutics to optimize patient care [32, 33]. By integrating diagnostic imaging with targeted treatments and real-time imaging guidance, theranostic approaches offer personalized and precision-driven solutions for patients with neurological disorders, improving diagnostic accuracy, treatment efficacy, and patient outcomes.

### *10.1.7 Biocompatibility and Safety Concerns*

Biocompatibility and safety considerations are paramount in the development and clinical translation of theranostic technologies. Nanoparticles used in theranostics must be thoroughly evaluated for their interactions with biological systems to ensure minimal toxicity and immunogenicity. Studies investigating nanoparticle-cell interactions, biodistribution, and clearance mechanisms are essential for assessing their biocompatibility profiles. Understanding the long-term effects of nanoparticle exposure on organ function, immune responses, and systemic homeostasis is critical for evaluating their safety for clinical use. Addressing concerns regarding nanoparticle toxicity [34] and immunogenicity requires comprehensive preclinical evaluation, including in vitro cell culture studies, animal toxicity assessments, and biodistribution studies. These studies provide valuable insights into the biological fate of nanoparticles, potential adverse effects on cellular function and tissue integrity, and immune responses elicited by nanoparticle exposure. Advanced imaging techniques such as intravital microscopy and in vivo imaging enable real-time visualization of nanoparticle behavior in living organisms, aiding in the assessment of nanoparticle biocompatibility and safety. Regulatory guidelines and standards play a crucial role in ensuring the safety and efficacy of theranostic nanoparticles for clinical use. Regulatory agencies such as the U.S. Food and Drug Administration (FDA) and the European Medicines Agency (EMA) provide guidelines for the preclinical testing, manufacturing, and clinical evaluation of nanomedicines [34, 35]. Compliance with these regulations is essential for obtaining regulatory approval and ensuring the translation of theranostic technologies from bench to bedside. Moreover, adherence to Good Manufacturing Practices (GMP) and quality control standards is imperative for maintaining the safety and consistency of nanoparticle formulations throughout the manufacturing process. Translating theranostic nanoparticles from preclinical research to clinical practice poses significant challenges related to scale-up production, reproducibility, and cost-effectiveness. Establishing scalable manufacturing processes that ensure batch-to-batch consistency and product quality is essential for meeting regulatory requirements and ensuring clinical success. Economic considerations, including production costs, pricing strategies, and reimbursement policies, influence the commercial viability and accessibility of theranostic nanoparticles. Collaboration between academia, industry, and regulatory agencies is essential for overcoming these translational hurdles and facilitating the clinical adoption of theranostic technologies [35].

Looking ahead, advancements in theranostic technologies hold promise for addressing unmet medical needs and revolutionizing patient care. Emerging strategies for enhancing nanoparticle biocompatibility and safety, such as surface modification techniques, biodegradable materials, and immunomodulatory coatings, offer opportunities for improving the clinical utility and safety profiles of theranostic nanoparticles.

#### 10.1.7.1 Artificial Intelligence in Image Analysis and Treatment Planning

Artificial intelligence (AI) has emerged as a powerful tool in the field of theranostics, revolutionizing image analysis and treatment planning through its ability to process vast amounts of imaging data quickly and accurately (Fig. 10.5). In the context of medical imaging, AI algorithms can analyze images from various modalities, including MRI, CT, PET, and ultrasound, to extract quantitative metrics, detect abnormalities, and identify biomarkers indicative of disease presence or progression. By leveraging machine learning techniques, such as deep learning and convolutional neural networks (CNNs) [36], AI algorithms can learn from large datasets and improve their performance over time, enabling more accurate and efficient image analysis. One of the key applications of AI in theranostics is in disease diagnosis and risk stratification, where AI algorithms analyze imaging data to identify subtle patterns or features indicative of disease pathology. For example, in oncology, AI-based image analysis can help radiologists detect early-stage tumors, characterize tumor subtypes, and predict patient prognosis based on imaging biomarkers. Similarly, in cardiology, AI algorithms can analyze cardiac imaging data to assess myocardial function, detect coronary artery disease, and identify regions of myocardial ischemia or infarction, aiding in the diagnosis and management of cardiovascular conditions.

AI plays a crucial role in treatment planning and optimization by providing clinicians with quantitative data and predictive analytics to guide therapeutic interventions. In radiation oncology, for instance, AI algorithms can analyze medical imaging data and patient-specific parameters to optimize treatment plans, such as intensity-modulated radiation therapy (IMRT) or stereotactic radiosurgery (SRS), to maximize

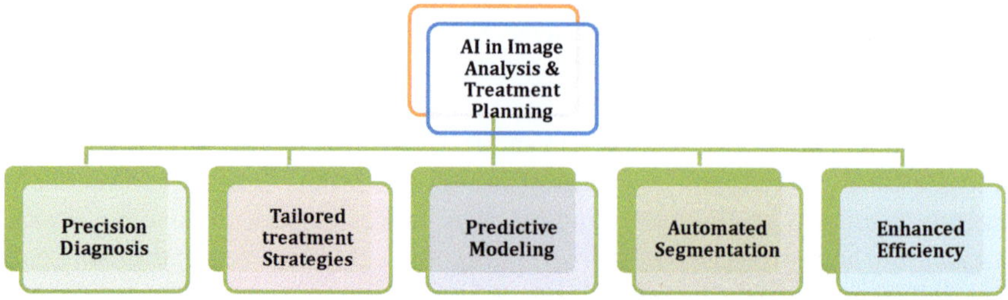


**Fig. 10.5** Advantages of AI in image analysis & treatment planning for theranostics

tumor coverage while minimizing radiation exposure to surrounding healthy tissues. Similarly, in neurosurgery, AI-based image analysis can assist surgeons in preoperative planning, tumor resection, and intraoperative navigation, improving surgical precision and patient outcomes.

AI facilitates personalized medicine by integrating imaging data with clinical, genetic, and demographic information to tailor treatment strategies to individual patient characteristics. By analyzing multimodal imaging datasets and patient-specific data, AI algorithms can generate predictive models and treatment recommendations customized to each patient's unique profile. This personalized approach to theranostics ensures that patients receive the most effective and appropriate treatments based on their specific disease characteristics, improving treatment outcomes and quality of life. Artificial intelligence is transforming image analysis and treatment planning in theranostics, enabling more accurate diagnosis, precise treatment delivery, and personalized medicine [2]. By harnessing the power of machine learning and deep learning algorithms, AI facilitates the analysis of complex imaging datasets, identifies disease biomarkers, and optimizes treatment strategies, ultimately improving patient care across a wide range of medical specialties.

**Exercise 3**
**Multiple Choice Questions**

1. What is the primary goal of theranostics?
   a) To diagnose diseases
   b) To treat diseases
   c) To combine diagnosis and therapy in a single approach
   d) To prevent diseases before symptoms appear

2. Which of the following imaging techniques is commonly used in theranostics?
   a) X-ray
   b) MRI (Magnetic Resonance Imaging)
   c) Electrocardiography (ECG)
   d) Blood tests

3. In theranostics, what role do nanoparticles often play?
   a) Targeting specific tissues
   b) Delivering drugs or therapeutic agents
   c) Providing diagnostic information
   d) All of the above

4. Which of the following statements about theranostics is true?
   a) It is only used for cancer treatment.
   b) It combines diagnostic and therapeutic agents in a sequential manner.
   c) It allows for real-time monitoring of treatment response.
   d) It does not involve personalized medicine approaches.

5. What is the advantage of using theranostics over traditional treatment approaches?
   a) It is less expensive.
   b) It is less invasive.
   c) It allows for personalized treatment strategies.
   d) It has fewer side effects.

6. Which type of imaging modality is commonly used for molecular imaging in theranostics?
   a) Computed Tomography (CT)
   b) Positron Emission Tomography (PET)
   c) Ultrasound
   d) Single Photon Emission Computed Tomography (SPECT)

7. What is the purpose of combining diagnosis and therapy in theranostics?
   a) To reduce the cost of healthcare
   b) To improve patient outcomes
   c) To simplify medical procedures
   d) To decrease the need for follow-up appointments

8. Which of the following diseases can be targeted using theranostic approaches?
   a) Cancer
   b) Cardiovascular diseases
   c) Neurological disorders
   d) All of the above

9. What is the term used to describe the process of selecting the most appropriate treatment based on diagnostic information in theranostics?
   a) Theragnostics
   b) Personalized medicine
   c) Combination therapy
   d) Biomarker analysis

10. Which of the following is an example of a theranostic agent?
    a) Chemotherapy drug
    b) Antibiotic
    c) Nanoparticle conjugated with a fluorescent dye and a drug
    d) Blood pressure medication

## 10.2 Summary

The chapter on "Theranostics: Combining Diagnosis and Therapy" explores the integration of diagnostic imaging and therapeutic interventions to advance personalized medicine and improve patient outcomes. Beginning with an introduction to theranostics, the chapter highlights its conceptual framework, historical context, and significance in precision medicine [1, 2]. It delves into the role of theranostic nanoparticles, detailing their design, fabrication, and multimodal imaging capabilities for targeted drug delivery and therapeutic payload delivery. Furthermore, the chapter discusses various imaging techniques in theranostics, including magnetic resonance imaging (MRI), positron emission tomography (PET), single-photon emission computed tomography (SPECT), optical imaging, and ultrasound imaging, emphasizing their applications in disease diagnosis, treatment planning, and monitoring. The chapter also addresses the integration of diagnosis and treatment in theranostics, focusing on real-time imaging-guided therapy, responsive therapeutic nanoparticles, and targeted drug delivery systems [3, 4]. It underscores the importance of image fusion and navigation systems, stimuli-responsive drug release mechanisms, and ligand-mediated targeting strategies in enhancing treatment precision and efficacy. Furthermore, the chapter explores the clinical applications of theranostics in cancer, cardiovascular, and neurological disorders, highlighting personalized treatment approaches, monitoring of treatment response, and disease progression.

Also, it discusses the challenges and future directions of theranostics, including biocompatibility and safety concerns, translational hurdles, and advancements in theranostic technologies. It emphasizes the importance of addressing nanoparticle toxicity, regulatory compliance, and economic considerations to facilitate the clinical adoption of theranostic nanoparticles. And highlights the potential of artificial intelligence in image analysis and treatment planning, underscoring its role in improving diagnostic accuracy, optimizing treatment strategies, and advancing personalized medicine.

## Points to Remember

- Theranostic nanoparticles can simultaneously deliver therapeutic agents and imaging contrast agents to target tissues, enabling real-time monitoring of treatment efficacy.
- Theranostic agents often utilize multiple imaging modalities such as MRI, PET, and fluorescence imaging, allowing for comprehensive visualization of disease progression and treatment response.
- Theranostic platforms can be engineered to selectively deliver therapeutic payloads to specific cells or tissues while sparing healthy ones, minimizing side effects and maximizing efficacy.
- Some theranostic systems are designed to respond to specific stimuli such as pH, temperature, or enzyme activity within the body, allowing for triggered release of therapeutic agents at the desired site.
- Peptide-based theranostic agents offer high specificity and affinity for target molecules, making them ideal for targeted imaging and therapy of cancer and other diseases.
- Theranostic vaccines combine diagnostic capabilities with immunotherapy, allowing for personalized cancer treatment strategies based on the patient's immune profile and tumor characteristics.
- Microfluidic devices integrated with diagnostic and therapeutic functionalities enable rapid and point-of-care analysis, offering potential applications in personalized medicine and remote healthcare settings.
- Liquid biopsy-based theranostics utilize circulating tumor cells, cell-free DNA, or exosomes for non-invasive diagnosis and monitoring of cancer, guiding treatment decisions and predicting treatment response.
- Theranostic approaches for neurological disorders combine imaging techniques with targeted drug delivery to the brain, enabling precise diagnosis and treatment of conditions such as Alzheimer's disease and brain tumors.
- Theranostic strategies for tissue regeneration involve combining imaging techniques with stem cell therapy or growth factor delivery, allowing for real-time monitoring of tissue repair and regeneration processes.

## References

1. Chen F, Ehlerding EB, Cai W (2014) Theranostic nanoparticles. J Nucl Med: Off Publ Soc Nucl Med 55(12):110110–11022. https://doi.org/10.21067/jnumed.114.1460110
2. Ramanathan S, Archunan G, Sivakumar M, Tamil Selvan S, Fred AL, Kumar S, Gulyás B, Padmanabhan P (2018) Theranostic applications of nanoparticles in neurodegenerative disorders. Int J Nanomed 13:5561–5576. https://doi.org/10.2147/IJN.S1410022
3. Suhag D, Bhatia R, Das S, Shakeel A, Ghosh AB, Singh A et al (2015) Physically cross-linked pH-responsive hydrogels with tunable formulations for controlled drug delivery. RSC Ad 5(66):531063–531072. https://doi.org/10.10310/c5ra07424j
4. Shakeel A, Singh A, Das S et al (2017) Synthesis and morphological insight of new biocompatible smart hydrogels. J Polym Res 24:113. https://doi.org/10.1007/s101065-017-1267-7
5. Wu W, Klockow JL, Mohanty S, Ku KS, Aghighi M, Melemenidis S, Chen Z, Li K, Morais GR, Zhao N, Schlegel J, Graves EE, Rao J, Loadman PM, Falconer RA, Mukherjee S, Chin FT,

Daldrup-Link HE (20110) Theranostic nanoparticles enhance the response of glioblastomas to radiation. Nanotheranostics 3(4):21010–31010. https://doi.org/10.7150/ntno.35342

6. Lin G, Zhang M (2023) Ligand chemistry in antitumor theranostic nanoparticles. Acc Chem Res 56(12):1578–15100. https://doi.org/10.1021/acs.accounts.3c00151
7. Ahmed N, Fessi H, Elaissari A (2012) Theranostic applications of nanoparticles in cancer. Drug Discov Today 17(17–18):1028–1034. https://doi.org/10.1016/j.drudis.2012.03.010
8. Kaur N, Sharma AK, Shakeel A, Kumar V, Singh A, Gupta A et al (2017) Therapeutic implications of superoxide dismutase and its importance in kinase drug discovery. Curr Top Med Chem 17(22). https://doi.org/10.2174/1568026617666170307112837
9. Shahdeo D, Kesarwani V, Suhag D et al (2022) Corrigendum to "Self-assembled chitosan polymer intercalating peptide functionalized gold nanoparticles as nanoprobe for efficient imaging of urokinase plasminogen activator receptor in cancer diagnostics" [Carbohydrate Polymers 266:118138 (2021)]. Carbohydr Polym 284:11102310. https://doi.org/10.1016/j.carbpol.2022.11102310. PMID: 35287884
10. Fernandes DA (2023) Review on metal-based theranostic nanoparticles for cancer therapy and imaging. Technol Cancer Res Treat 22:1533033823111014100. https://doi.org/10.1177/1533033823111014103
11. Pallares RM, Mottaghy FM, Schulz V, Kiessling F, Lammers T (2022) Nanoparticle diagnostics and theranostics in the clinic. J Nucl Med: Off Publ Soc Nucl Med 63(12):1802–1808. https://doi.org/10.21067/jnumed.122.2638105
12. Zavaleta C, Ho D, Chung EJ (2018) Theranostic nanoparticles for tracking and monitoring disease state. SLAS Technol 23(3):281–2103. https://doi.org/10.1177/247263031773861010
13. Suhag D, Chauhan M, Shakeel A, Das S (2020) Emerging trends in nanotheranostics. In: Saxena S, Khurana S (eds) NanoBioMedicine. Springer, Singapore. https://doi.org/10.1007/1078-1081-32-108108-10_14
14. Suhag D, Thakur P, Thakur A (2023) Introduction to nanotechnology. In: Suhag D, Thakur A, Thakur P (eds) Integrated nanomaterials and their applications. Springer, Singapore. https://doi.org/10.1007/1078-1081-1010-6105-4_1
15. Chen CH, Woo EM (2018) Polymer science and technology. Chem Phys 15(7):24105–32506
16. Alberti D, Protti N, Franck M, Stefania R, Bortolussi S, Altieri S, Deagostino A, Aime S, Geninatti Crich S (2017) Theranostic nanoparticles loaded with imaging probes and rubrocurcumin for combined cancer therapy by folate receptor targeting. ChemMedChem 12(7):502–5010. https://doi.org/10.1002/cmdc.2017000310
17. Janib SM, Moses AS, MacKay JA (2010) Imaging and drug delivery using theranostic nanoparticles. Adv Drug Deliv Rev 62(11):1052–1063. https://doi.org/10.1016/j.addr.2010.08.004
18. Shazly T, Smith A, Uline MJ, Spinale FG (2022) Therapeutic payload delivery to the myocardium: evolving strategies and obstacles. JTCVS Open 10:185–1104. https://doi.org/10.1016/j.xjon.2022.04.043
19. Moore C, Jokerst JV (2019) Strategies for image-guided therapy, surgery, and drug delivery using photoacoustic imaging. Theranostics 9(6):1550–1571. https://doi.org/10.7150/thno.32362
20. Shi Y, Zhang C, Liu C, Ma X, Liu Z (2023) Image-guided precision treatments. In: Advances in experimental medicine and biology (pp. 59–86). https://doi.org/10.1007/978-981-32-9902-3_4
21. Satyam A, Suhag D (2023) Nanomaterials for precision medicine. In: Suhag D, Thakur A, Thakur P (eds) Integrated nanomaterials and their applications. Springer, Singapore. https://doi.org/10.1007/1078-1081-1010-6105-4_13
22. Singh A, Suhag D (2023) Polymers and polymeric composites in nano/bio-medicine. In: Suhag D, Thakur A, Thakur P (eds) Integrated nanomaterials and their applications. Springer, Singapore. https://doi.org/10.1007/1078-1081-1010-6105-4_8
23. Rathore R, Suhag D, Wan F, Verma R, Thakur A, Thakur P (2023) Functionalization of biomaterials. In: Suhag D, Thakur A, Thakur P (eds) Integrated nanomaterials and their applications. Springer, Singapore. https://doi.org/10.1007/1078-1081-1010-6105-4_11

24. Tewabe A, Abate A, Tamrie M, Seyfu A, Abdela Siraj E (2021) Targeted drug delivery - from magic bullet to nanomedicine: principles, challenges, and future perspectives. J Multi Healthc 14:1711–1724. https://doi.org/10.2147/JMDH.S313968
25. Devarajan PV, Jain S (2015) Targeted drug delivery: concepts and design. In: Advances in delivery science and technology.https://doi.org/10.1007/978-3-319-11355-5
26. Xue Y, Gao Y, Meng F, Luo L (2021) Recent progress of nanotechnology-based theranostic systems in cancer treatments. Cancer Biol Med 18(2):336–351. Advance online publication. https://doi.org/10.20892/j.issn.2095-3941.2020.0510
27. Chen X, Wong ST (2014) Cancer theranostics. In: Elsevier eBooks (pp. 3–8). https://doi.org/10.1016/b978-0-12-407722-5.00001-3
28. Xia K, Zhong X, Zhang L, Wang J (20110) Optimization of diagnosis and treatment of chronic diseases based on association analysis under the background of regional integration. J Med Syst 43(3):46. https://doi.org/10.1007/s101016-0110-11610-10
29. Palekar-Shanbhag P, Jog SV, Chogale MM, Gaikwad SS (2013) Theranostics for cancer therapy. Curr Drug Deliv 10(3):357–362. https://doi.org/10.2174/1567201811310030013
30. Manners N, Priya V, Mehata AK, Rawat M, Mohan S, Makeen HA, Albratty M, Albarrati A, Meraya AM, Muthu MS (2022) Theranostic nanomedicines for the treatment of cardiovascular and related diseases: current strategies and future Perspectives. Pharmaceuticals 15(4):441. https://doi.org/10.3390/ph15040441
31. Kumar P, Zelena D, Gautam A (2023) Neurological disorders and challenges in their theranostics. In: Theranostic applications of nanotechnology in neurological disorders (pp. 1–29). https://doi.org/10.1007/978-981-99-9510-3_1
32. Sohail S, Fakhar-Ud-Din N (2021) Nanotheranostics: the future Remedy of Neurological Disorders. In: Nanotechnology in the life sciences (pp. 117–154). https://doi.org/10.1007/978-3-030-76263-6_5
33. Joshi R, Missong H, Mishra J, Kaur S, Saini S, Kandimalla R, Reddy PH, Babu A, Bhatti GK, Bhatti JS (2023) Nanotheranostics revolutionizing neurodegenerative diseases: From precision diagnosis to targeted therapies. J Drug Delivery Sci Technol 89:105067. https://doi.org/10.1016/j.jddst.2023.105067
34. Rathore R, Suhag D, Wan F, Thakur A, Thakur P (2023) Toxicology and toxicity studies of nano-biomaterials. In: Integrated Nanomaterials and their Applications (pp. 433–456). https://doi.org/10.1007/978-981-99-6105-4_20
35. Wang H D, Thakur P, Thakur, A (2023) Biocompatibility, bio-clearance, and toxicology. In: Integrated nanomaterials and their applications (pp. 201–221). https://doi.org/10.1007/978-981-99-6105-4_10
36. Belge Bilgin G, Bilgin C, Burkett BJ, Orme JJ, Childs DS, Thorpe MP, Halfdanarson TR, Johnson GB, Kendi AT, Sartor, O (2024) Theranostics and artificial intelligence: new frontiers in personalized medicine. Theranostics 14(6):2367–2378. https://doi.org/10.7150/thno.94788

# Chapter 11
# Drug Delivery via Lipid-Based Nanocarriers

## 11.1 Introduction

Drug delivery systems are instrumental in improving the efficacy and safety of pharmaceutical interventions by optimizing the delivery of therapeutic agents within the body. These systems encompass a diverse range of technologies and formulations designed to overcome biological barriers, control drug release kinetics, and target specific tissues or cells [1]. By encapsulating drugs within carrier systems such as nanoparticles, liposomes, or micelles, drug delivery systems can enhance drug stability, prolong circulation time, and facilitate site-specific accumulation, thereby maximizing therapeutic outcomes while minimizing systemic toxicity. The evolution of drug delivery systems has been driven by advancements in material science, nanotechnology, and pharmacology, enabling the development of increasingly sophisticated formulations with precise control over drug release profiles and tissue distribution [2]. Through the integration of innovative drug delivery strategies, such as stimuli-responsive or targeted delivery approaches, drug delivery systems can achieve spatiotemporal control over drug release and localization, thereby improving therapeutic efficacy and patient compliance. The interdisciplinary nature of drug delivery research, which draws upon expertise from chemistry, biology, engineering, and medicine, continues to foster innovation and drive the translation of novel drug delivery concepts from bench to bedside.

### *11.1.1 Advantages of Lipid-Based Nanocarriers*

Lipid-based nanocarriers offer numerous advantages as drug delivery vehicles, making them highly desirable for pharmaceutical applications (Fig. 11.1). One of their key attributes is their biocompatibility, which ensures minimal adverse reactions upon administration. Lipid-based nanocarriers are derived from natural or synthetic



D. Suhag et al., *Handbook of Biomaterials for Medical Applications, Volume 1*,
Biomedical Materials for Multi-functional Applications,
https://doi.org/10.1007/978-981-97-4818-1_11

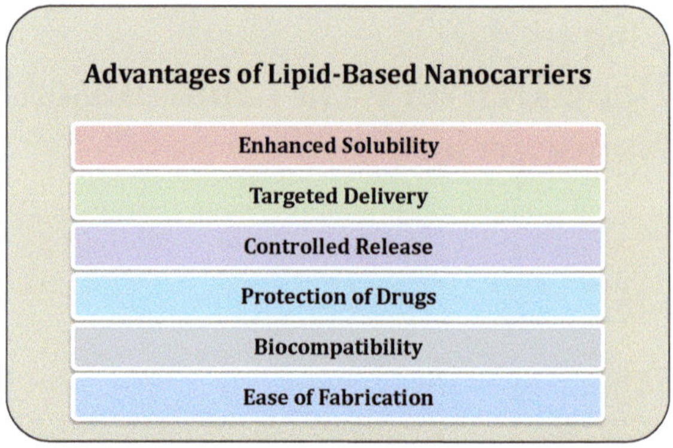


**Fig. 11.1** Advantages of lipid-based nanocarriers

lipids that closely resemble biological membranes, minimizing immunogenicity and toxicity. This biocompatibility not only enhances patient safety but also facilitates the biodegradation of lipid-based formulations within the body, reducing the risk of long-term accumulation or adverse effects. Lipid-based nanocarriers exhibit exceptional versatility in encapsulating a wide range of drugs, including hydrophilic and hydrophobic molecules, peptides, and nucleic acids [1, 2]. Their amphiphilic nature allows lipids to form stable colloidal dispersions, encapsulating drugs within their core or lipid bilayers. This flexibility enables the formulation of lipid-based nanocarriers tailored to specific drug properties and therapeutic requirements, enhancing drug solubility, stability, and bioavailability. Lipid-based formulations can be engineered to achieve controlled drug release kinetics, enabling sustained or triggered release profiles to optimize therapeutic efficacy while minimizing dosing frequency and systemic side effects.

#### 11.1.1.1 Historical Development

The development of lipid-based nanocarriers for drug delivery traces back to the early 1960s with the advent of liposomes by Alec D. Bangham and his colleagues. Liposomes, composed of phospholipid bilayers, were initially explored for their potential as model membrane systems but soon garnered attention for their drug delivery capabilities. Subsequent research efforts focused on refining liposome formulations and exploring their applications in drug delivery, leading to the commercialization of liposomal formulations such as Doxil® for cancer therapy. Concurrently, lipid nanoparticles emerged as another promising class of drug carriers, offering advantages such as enhanced stability and controlled release properties [3]. Over the decades, significant progress has been made in the design and optimization of lipid-based nanocarriers, driven by advances in materials science, nanotechnology, and pharmaceutical engineering. The introduction of novel lipid materials, including synthetic lipids and lipid derivatives, has expanded the versatility of lipid-based formulations and enabled the development of tailored delivery systems with enhanced performance characteristics. Moreover, advancements in manufacturing techniques, such as microfluidics and nanoprecipitation, have facilitated the scalable production of lipid-based nanocarriers with precise control over particle size and distribution [4]. Collectively, these developments have propelled lipid-based nanocarriers to the forefront of modern drug delivery research and positioned them as promising platforms for addressing unmet clinical needs in various therapeutic areas.

## *11.1.2 Lipid-Based Nanocarrier Formulation*

Formulating lipid-based nanocarriers involves a meticulous process aimed at harnessing the unique properties of lipids to encapsulate and deliver therapeutic agents effectively. Central to this process is the selection of appropriate lipid materials, considering factors such as biocompatibility, stability, and drug loading capacity. Lipids chosen for formulation must possess suitable physicochemical properties to facilitate the formation of stable nanocarriers while ensuring compatibility with the encapsulated drug [3, 4]. Lipid-based nanocarriers require precise control over particle size, morphology, and surface properties to optimize drug delivery performance. Various techniques, including lipid film hydration, solvent evaporation, and microfluidics, are employed to fabricate lipid-based nanocarriers with tailored characteristics. Through systematic optimization of formulation parameters, lipid-based nanocarriers can be engineered to achieve desired drug release kinetics, targeting specificity, and therapeutic outcomes, positioning them as promising platforms for advanced drug delivery applications.

### **11.1.2.1 Lipid Selection and Properties**

Selecting appropriate lipids for the formulation of lipid-based nanocarriers is a critical step that significantly influences the performance and functionality of the delivery system. The choice of lipids is guided by several key properties that impact the formulation process and the behavior of the resulting nanocarriers. Firstly, lipids must possess biocompatibility to ensure compatibility with biological systems and minimize potential adverse reactions upon administration. Natural lipids, such as phospholipids and cholesterol, are commonly employed due to their inherent biocompatibility and resemblance to biological membranes [5]. Synthetic lipids, including lipids with modified hydrophobic chains or head groups, offer additional opportunities to tailor the properties of lipid-based nanocarriers for specific applications. The physicochemical properties of lipids, such as melting point, solubility, and phase behavior, play a crucial role in determining the stability and performance of lipid-based nanocarriers. Lipids with appropriate melting points and phase transition temperatures are selected to ensure the formation of stable lipid bilayers or colloidal structures under physiological conditions. The solubility of lipids in organic solvents and aqueous media influences the feasibility and efficiency of lipid-based nanocarrier fabrication processes, such as lipid film hydration or solvent evaporation. The ability of lipids to self-assemble into ordered structures, such as liposomes or lipid nanoparticles, is governed by their amphiphilic nature and intermolecular interactions, further emphasizing the importance of lipid selection in nanocarrier formulation [5, 6]. The lipid composition of nanocarriers dictates their biophysical properties, including membrane fluidity, permeability, and interactions with biological components. Lipids with varying chain lengths, saturation levels, and head group structures can modulate the fluidity and rigidity of lipid bilayers, influencing drug encapsulation efficiency and release kinetics. Specific lipid compositions may enhance the stability of nanocarriers in biological fluids, minimize premature drug leakage, and facilitate prolonged circulation times in vivo. Tailoring the lipid composition of nanocarriers allows for the optimization of drug delivery performance and the design of targeted delivery systems capable of navigating biological barriers and reaching specific tissues or cells.

Overall, lipid selection and properties play a pivotal role in the design and optimization of lipid-based nanocarriers for drug delivery applications [6]. By carefully considering the biocompatibility, physicochemical characteristics, and biophysical properties of lipids, researchers can develop nanocarriers with tailored functionalities and enhanced therapeutic efficacy.

#### 11.1.2.2 Drug Loading Techniques

Drug loading techniques are pivotal in the formulation of lipid-based nanocarriers, determining the efficiency and stability of drug encapsulation within these delivery systems. Various methods have been developed to incorporate therapeutic agents into lipid-based nanocarriers, each with their unique advantages and challenges. One commonly employed approach is passive encapsulation, where drugs are entrapped within the lipid matrix during the formation of nanocarriers through processes such as lipid hydration or nanoprecipitation. Passive encapsulation relies on the physicochemical properties of drugs and lipids to achieve drug loading, offering simplicity and versatility in formulation [7]. However, passive encapsulation may result in variable drug loading efficiency and limited control over drug release kinetics, necessitating optimization of formulation parameters to achieve desired drug loading levels and release profiles. Active loading techniques offer an alternative strategy to enhance drug encapsulation efficiency and control drug release from lipid-based nanocarriers. These techniques typically involve the use of gradients, such as pH or ion gradients, to drive the accumulation of drugs within the lipid core or bilayers of nanocarriers [8]. One example is the pH gradient loading method, where drugs are loaded into liposomes by exploiting differences in pH between the internal and external environments. By manipulating pH gradients, researchers can achieve high drug loading capacities and precise control over drug release kinetics, leading to improved therapeutic outcomes and reduced systemic toxicity. Active loading techniques enable the encapsulation of a broader range of drugs, including hydrophilic compounds and macromolecules, expanding the applicability of lipid-based nanocarriers in drug delivery.

Another approach to enhance drug loading in lipid-based nanocarriers is the use of lipid conjugates or prodrugs, where drugs are chemically modified to facilitate their incorporation into lipid matrices. Lipid conjugates can improve drug solubility, stability, and membrane permeability, enhancing drug loading efficiency and bioavailability. Moreover, lipid prodrugs offer the advantage of site-specific drug release, where drugs are released from lipid carriers upon enzymatic or chemical cleavage in target tissues or cells. This strategy enables spatial and temporal control over drug delivery, minimizing off-target effects and improving therapeutic efficacy [9]. However, the design and synthesis of lipid conjugates or prodrugs require careful consideration of chemical compatibility, stability, and pharmacokinetic properties to ensure successful integration into lipid-based nanocarriers. In addition to passive and active loading techniques, physical methods such as membrane extrusion or sonication can be employed to enhance drug encapsulation and homogenize lipid-based nanocarriers. These methods facilitate the formation of uniform lipid vesicles or nanoparticles with enhanced drug loading capacities and improved stability. Membrane extrusion, for instance, allows for the control of particle size and size distribution, optimizing drug encapsulation and delivery efficiency. Similarly, sonication can disrupt lipid aggregates and promote drug incorporation into lipid bilayers, leading to higher drug loading levels and more uniform distribution within nanocarriers.

#### 11.1.2.3 Stability Considerations

Stability considerations are paramount in the development of lipid-based nanocarriers to ensure their effectiveness and safety throughout the drug delivery process. One critical aspect of stability is the physical stability of lipid-based nanocarriers, which refers to their ability to maintain structural integrity and prevent aggregation or degradation over time. Lipid-based nanocarriers are susceptible to various destabilizing factors, including temperature fluctuations, pH changes, and mechanical stress, which can compromise their stability and drug delivery performance. Therefore, optimizing formulation parameters such as lipid composition [10], particle size, and surface properties is essential to enhance the physical stability of lipid-based nanocarriers and prolong their shelf-life under storage conditions. Chemical stability is another key consideration in the design of lipid-based nanocarriers, particularly concerning the degradation of lipids and encapsulated drugs during storage or in biological environments. Lipid oxidation, hydrolysis, and degradation can lead to changes in nanocarrier structure, loss of drug payload, and generation of toxic by-products, compromising therapeutic efficacy and patient safety. Strategies to improve chemical stability include the use of antioxidant additives, such as tocopherols or ascorbic acid, to mitigate lipid oxidation, and the selection of stable lipid materials with minimal susceptibility to hydrolysis or degradation. Encapsulation techniques that protect drugs from chemical degradation, such as active loading or covalent conjugation, can enhance the chemical stability of lipid-based nanocarriers and improve drug retention during storage and in vivo delivery.

Biological stability is critical for ensuring the compatibility of lipid-based nanocarriers with biological systems, including blood components, immune cells, and tissues. Upon administration, lipid-based nanocarriers interact with biological fluids and cells, which may trigger immune responses, clearance mechanisms, or adverse reactions. Surface modification strategies, such as PEGylation or surface coating with stealth polymers, can improve the biological stability of lipid-based nanocarriers by reducing protein adsorption, opsonization, and recognition by the reticuloendothelial system. Optimizing the physicochemical properties of lipid-based nanocarriers, such as size, charge, and surface chemistry, can minimize interactions with biological components and enhance their circulation time and biocompatibility in vivo. In addition to stability under storage and biological conditions [10], stability considerations extend to the stability of drug molecules encapsulated within lipid-based nanocarriers. Encapsulation within lipid matrices can protect drugs from degradation, metabolism, and premature release, enhancing their stability and bioavailability. However, the compatibility between drugs and lipid carriers must be carefully assessed to prevent chemical interactions or alterations in drug properties. Strategies to achieve controlled drug release from lipid-based nanocarriers, such as stimuli-responsive systems or sustained release formulations, can enhance drug stability and therapeutic efficacy by minimizing fluctuations in drug concentration and exposure.

Overall, stability considerations encompass a broad range of factors that influence the performance, safety, and effectiveness of lipid-based nanocarriers for drug delivery applications.

**Exercise 1**

**Fill ups**

**(Pharmacokinetic, Lipid-based nanocarriers, Hydrophobic, Cell types, Lipids)**

**1.** The selection of appropriate ..................is crucial in lipid-based nanocarrier formulation to achieve desired properties such as stability and drug loading capacity.

**2.** Various manufacturing techniques, including solvent evaporation and nanoprecipitation, are used in the preparation of ............................to ensure uniform particle size distribution.

**3.** Surface modification of lipid-based nanocarriers with targeting ligands can enhance their specificity for particular ...................or tissues, improving therapeutic efficacy.

**4.** The physicochemical characteristics of lipid-based nanocarriers, such as particle size and surface charge, influence their .........................behavior, including circulation time and cellular uptake.

**5.** Lipid-based nanocarriers offer advantages such as biocompatibility and the ability to encapsulate .................drugs, enhancing their solubility and bioavailability.

### *11.1.3 Characterization Techniques*

Characterization techniques play a crucial role in the development and evaluation of lipid-based nanocarriers, providing valuable insights into their physicochemical properties, stability, and performance as drug delivery systems. These techniques encompass a diverse array of analytical methods designed to assess various aspects of lipid-based nanocarriers, including particle size, morphology, surface charge, drug loading capacity, and drug release kinetics. By employing a combination of complementary characterization techniques, researchers can gain a comprehensive understanding of the structural and functional properties of lipid-based nanocarriers, guiding formulation optimization and facilitating quality control throughout the development process [11]. Moreover, characterization techniques enable researchers to elucidate the mechanisms underlying drug encapsulation, release, and interaction with biological systems, thereby informing rational design strategies to enhance the efficacy, safety, and translational potential of lipid-based nanocarriers for diverse therapeutic applications.

#### 11.1.3.1 Particle Size Analysis

Particle size analysis is a fundamental characterization technique used to assess the size distribution and dimensional characteristics of lipid-based nanocarriers, providing critical information regarding their physical properties and colloidal stability. Various methods are employed for particle size analysis, ranging from conventional techniques such as dynamic light scattering (DLS) and laser diffraction to more advanced approaches such as transmission electron microscopy (TEM) and atomic force microscopy (AFM). DLS is a widely utilized technique for measuring the hydrodynamic diameter of nanoparticles in solution, offering rapid and non-destructive analysis of particle size distribution. Laser diffraction, on the other hand, measures the intensity of light scattered by particles suspended in a dispersion medium, providing size distribution data based on the principle of Mie scattering [10, 11]. These techniques enable researchers to determine the average particle size, polydispersity index, and size distribution profile of lipid-based nanocarriers, facilitating the optimization of formulation parameters to achieve desired particle size and size homogeneity. Transmission electron microscopy (TEM) and atomic force microscopy (AFM) are imaging techniques commonly employed for visualizing the morphology and size of lipid-based nanocarriers at the nanoscale level. TEM offers high-resolution imaging of individual nanoparticles, providing detailed information regarding particle shape, structure, and surface morphology. AFM, on the other hand, utilizes a sharp probe to scan the surface of nanoparticles, allowing for precise measurement of particle height and size distribution. These imaging techniques complement conventional particle size analysis methods by providing direct visualization of lipid-based nanocarriers and facilitating qualitative assessment of their structural integrity and homogeneity.

Particle size analysis is essential for assessing the stability and colloidal behavior of lipid-based nanocarriers in various physiological and environmental conditions. Changes in particle size distribution can indicate alterations in nanocarrier structure, aggregation, or degradation, which may affect their drug delivery performance and biocompatibility. For instance, an increase in particle size or polydispersity may result in decreased colloidal stability, increased opsonization, and clearance by the reticuloendothelial system (RES), leading to reduced circulation time and diminished therapeutic efficacy [12]. Therefore, monitoring changes in particle size over time and under different storage or biological conditions is critical for ensuring the stability and functionality of lipid-based nanocarriers throughout their lifecycle.

Particle size analysis plays a crucial role in optimizing drug loading and release kinetics in lipid-based nanocarriers, as particle size can influence drug encapsulation efficiency, release rate, and tissue penetration. Smaller nanoparticles exhibit higher surface area-to-volume ratios, facilitating rapid drug release and diffusion, whereas larger nanoparticles may offer prolonged release profiles and enhanced stability. By modulating particle size and size distribution, researchers can tailor the pharmacokinetic and pharmacodynamic properties of lipid-based nanocarriers to achieve desired drug delivery outcomes, such as sustained release, targeted delivery, or controlled release in response to specific stimuli.

#### 11.1.3.2 Surface Charge Determination

Surface charge determination is a critical aspect of characterizing lipid-based nanocarriers, as it provides valuable information regarding their colloidal stability, interaction with biological systems, and potential for targeted drug delivery. The surface charge of nanoparticles is primarily determined by the presence of charged functional groups on their surface, such as ionizable groups from lipid head groups or surface-modifying agents [13]. Several techniques are commonly employed for surface charge determination, including zeta potential analysis, electrophoretic mobility measurements, and surface titration methods. Zeta potential analysis quantifies the electrokinetic potential at the nanoparticle surface, reflecting the magnitude and polarity of surface charge and providing insights into nanoparticle stability and dispersion behavior in solution. The surface charge of lipid-based nanocarriers plays a crucial role in governing their interaction with biological components, such as proteins, cell membranes, and extracellular matrix components. Nanoparticles with positively charged surfaces, known as cationic nanoparticles, may exhibit enhanced cellular uptake and intracellular delivery through electrostatic interactions with negatively charged cell membranes. Conversely, nanoparticles with negatively charged surfaces, termed anionic nanoparticles, may offer improved circulation stability and reduced nonspecific interactions with blood components, minimizing clearance by the reticuloendothelial system (RES) and enhancing their bioavailability. Therefore, understanding the surface charge properties of lipid-based nanocarriers is essential for designing delivery systems with optimized pharmacokinetic and pharmacodynamic profiles for specific therapeutic applications. Surface charge determination enables researchers to tailor the surface properties of lipid-based nanocarriers to achieve targeted drug delivery and controlled release. By modifying the surface charge through surface functionalization [14] or coating with charged polymers, researchers can enhance the specificity and selectivity of nanocarriers for target tissues or cells. For instance, cationic lipid-based nanocarriers can be engineered to facilitate electrostatic interactions with negatively charged cell surface receptors or extracellular matrix components, promoting targeted delivery to diseased tissues or tumor cells while minimizing off-target effects. Conversely, anionic nanocarriers can be designed to evade immune recognition and achieve prolonged circulation times by repelling negatively charged serum proteins and avoiding RES clearance. Surface charge determination is essential for assessing the stability and colloidal behavior of lipid-based nanocarriers in physiological environments. Nanoparticles with excessively high or low surface charges may exhibit aggregation, flocculation, or instability in biological fluids, compromising their drug delivery performance and therapeutic efficacy [13, 14]. Therefore, optimizing the surface charge of lipid-based nanocarriers to achieve a balance between colloidal stability and cellular interaction is crucial for ensuring their functionality and translational potential in clinical applications.

#### 11.1.3.3 Morphological Characterization

Morphological characterization is essential for assessing the structural integrity, shape, and surface morphology of lipid-based nanocarriers, providing valuable insights into their physical properties and colloidal behavior [10, 11]. Transmission electron microscopy (TEM) and scanning electron microscopy (SEM) are commonly used techniques for morphological characterization, offering high-resolution imaging capabilities to visualize individual nanoparticles and elucidate their structural features. TEM enables detailed examination of nanocarrier morphology at the nanoscale level, providing information regarding particle shape, size distribution, and internal structure. SEM, on the other hand, offers surface imaging of nanoparticles, allowing for the assessment of surface roughness, porosity, and morphology. These techniques play a crucial role in evaluating the structural integrity and homogeneity of lipid-based nanocarriers, guiding formulation optimization and quality control throughout the development process. In addition to TEM and SEM, atomic force microscopy (AFM) is employed for morphological characterization of lipid-based nanocarriers, offering high-resolution imaging of surface topography and mechanical properties at the nanoscale. AFM utilizes a sharp probe to scan the surface of nanoparticles, providing detailed information regarding particle height, surface roughness, and interaction forces. This technique complements electron microscopy methods by offering quantitative analysis of nanoparticle morphology and surface properties, facilitating the investigation of structural changes and interactions with biological systems. Moreover, AFM enables in situ imaging of lipid-based nanocarriers in solution, allowing for real-time monitoring of structural changes and dynamic processes such as drug release or cellular uptake.

Morphological characterization plays a crucial role in elucidating the mechanisms underlying drug encapsulation, release, and interaction with biological systems. By visualizing the morphology of lipid-based nanocarriers before and after drug loading, researchers can assess changes in particle size, shape, and structure, providing insights into drug encapsulation efficiency and drug-carrier interactions. Moreover, morphological characterization facilitates the investigation of drug release mechanisms from lipid-based nanocarriers, such as diffusion, degradation, or stimuli-responsive release [12–14]. Understanding the morphological changes of lipid-based nanocarriers under different physiological conditions or in the presence of biological components is essential for optimizing drug delivery performance and predicting in vivo behavior. Morphological characterization is instrumental in evaluating the stability and shelf-life of lipid-based nanocarriers under storage conditions. Changes in particle morphology, aggregation, or degradation over time can impact nanocarrier stability, drug release kinetics, and therapeutic efficacy. Therefore, monitoring morphological changes through techniques such as TEM, SEM, or AFM enables researchers to assess the physical stability of lipid-based nanocarriers and identify potential degradation mechanisms or formulation issues.

### 11.1.4 *Drug Release Mechanisms*

Understanding drug release mechanisms is crucial for designing effective lipid-based nanocarriers capable of delivering therapeutic agents with precise control over release kinetics and localization within the body. Drug release from lipid-based nanocarriers is governed by a complex interplay of factors, including carrier properties, drug characteristics, and environmental conditions [15]. Various mechanisms have been identified to regulate drug release from lipid-based nanocarriers, encompassing passive diffusion, carrier degradation, and stimuli-responsive release. By elucidating these mechanisms, researchers can optimize nanocarrier design and formulation strategies to achieve desired release profiles, enhance therapeutic efficacy, and minimize systemic side effects.

#### 11.1.4.1 Passive Diffusion

Passive diffusion is a fundamental mechanism governing drug release from lipid-based nanocarriers, whereby drugs are released from the carrier matrix through passive diffusion across biological membranes or extracellular matrices [16]. In this process, drugs move from regions of high concentration within the nanocarrier to regions of lower concentration in the surrounding environment, driven by concentration gradients and molecular diffusion. Passive diffusion is influenced by various factors, including drug physicochemical properties, carrier characteristics, and environmental conditions such as pH, temperature, and viscosity. Lipid-based nanocarriers provide an ideal platform for passive diffusion-based drug release due to their ability to encapsulate drugs within lipid bilayers or matrices, offering sustained release profiles and improved bioavailability compared to conventional dosage forms. The rate of passive diffusion-mediated drug release from lipid-based nanocarriers is influenced by the physicochemical properties of both the drug and the carrier system. Lipophilic drugs with high lipid solubility exhibit faster diffusion rates through lipid bilayers, leading to more rapid drug release from lipid-based nanocarriers. The lipid composition, structure, and fluidity of nanocarriers can impact passive diffusion kinetics by modulating the permeability and diffusivity of drugs across lipid membranes. The size, surface charge, and surface functionalization of lipid-based nanocarriers may influence passive diffusion by affecting carrier stability, interactions with biological membranes, and drug release kinetics.

Passive diffusion-mediated drug release from lipid-based nanocarriers plays a crucial role in various drug delivery applications, including sustained release formulations, transdermal patches, and targeted delivery systems. By leveraging passive diffusion mechanisms, lipid-based nanocarriers can achieve controlled release profiles, prolonged drug retention, and enhanced therapeutic efficacy [17]. Moreover, passive diffusion allows for the delivery of diverse therapeutic agents, including small molecules, peptides, and nucleic acids, by exploiting their inherent physicochemical properties and diffusivity across lipid membranes. This versatility makes passive diffusion an attractive strategy for formulating lipid-based nanocarriers tailored to specific therapeutic needs and disease indications.

While passive diffusion offers several advantages for drug release from lipid-based nanocarriers, it also presents challenges related to achieving precise control over release kinetics and localization. Factors such as drug loading efficiency, carrier stability, and membrane permeability can affect the reliability and consistency of passive diffusion-based drug release. Moreover, passive diffusion may not be suitable for delivering hydrophilic drugs or drugs with poor lipid solubility, limiting its applicability in certain therapeutic contexts.

#### 11.1.4.2 Controlled Release Strategies

Controlled release strategies are integral to the design of lipid-based nanocarriers, offering precise control over drug release kinetics, duration, and localization within the body. These strategies aim to optimize therapeutic outcomes by ensuring sustained drug delivery, minimizing systemic side effects, and enhancing patient compliance [18]. Various approaches have been developed to achieve controlled release from lipid-based nanocarriers, including formulation modifications, stimuli-responsive systems, and combination therapies. By tailoring these strategies to specific therapeutic needs and disease indications, researchers can harness the full potential of lipid-based nanocarriers for advanced drug delivery applications. Formulation modifications represent a straightforward approach to achieving controlled release from lipid-based nanocarriers, involving adjustments to carrier composition, structure, and surface properties. For instance, altering the lipid composition or ratio of lipids within nanocarriers can modulate their membrane fluidity, permeability, and drug release kinetics. Incorporating excipients or additives with sustained release properties, such as polymers, gels, or hydrogels, can provide sustained drug release profiles by controlling drug diffusion or dissolution rates. Engineering nanocarriers with specific particle sizes, shapes, or surface charges can influence drug release behavior and tissue distribution, enabling tailored delivery systems for optimized therapeutic efficacy. Stimuli-responsive systems offer a sophisticated approach to achieving controlled release from lipid-based nanocarriers, wherein drug release is triggered by external stimuli or physiological cues [16, 17]. These stimuli may include changes in pH, temperature, enzyme activity, or redox potential within target tissues or cellular environments. By incorporating responsive elements into lipid-based nanocarriers, such as pH-sensitive polymers, thermo-responsive lipids, or enzyme-cleavable linkers, researchers can design delivery systems capable of on-demand or site-specific drug release. Stimuli-responsive lipid-based nanocarriers offer potential advantages such as spatiotemporal control over drug delivery, enhanced targeting specificity, and reduced off-target effects, making them promising candidates for personalized medicine and precision therapeutics.

Combination therapies represent a synergistic approach to achieving controlled release from lipid-based nanocarriers, wherein multiple drugs or therapeutic agents are co-loaded within the same carrier system to achieve complementary or synergistic effects. By encapsulating drugs with different pharmacokinetic profiles or mechanisms of action, researchers can tailor drug release profiles and optimize therapeutic outcomes for complex diseases or multidrug-resistant conditions. Moreover, combining drugs with adjuvants, targeting ligands, or imaging agents within lipid-based nanocarriers can enhance treatment efficacy, reduce dosing frequency, and improve patient compliance. Combination therapies offer a versatile strategy for addressing diverse therapeutic challenges and maximizing the potential of lipid-based nanocarriers in personalized medicine and combination drug regimens [13]. Overall, controlled release strategies represent a cornerstone of lipid-based nanocarrier design, enabling precise modulation of drug release kinetics and therapeutic efficacy for a wide range of applications [18]. By leveraging formulation modifications, stimuli-responsive systems, and combination therapies, researchers can overcome challenges associated with conventional drug delivery approaches and develop innovative solutions for unmet clinical needs.

#### 11.1.4.3 Stimuli-Responsive Release

Stimuli-responsive release represents a cutting-edge approach in the field of drug delivery, where drug release from lipid-based nanocarriers is triggered by specific environmental cues or external stimuli (Fig. 11.2). These stimuli may include changes in pH, temperature, enzyme activity, redox potential, or light exposure within the target tissue or cellular microenvironment [19]. By incorporating responsive elements into the structure or composition of lipid-based nanocarriers, researchers can design delivery systems capable of precisely controlling drug release kinetics and spatial distribution, offering enhanced therapeutic efficacy, and reduced systemic side effects. One of the most widely studied stimuli for responsive release is pH, as many pathological conditions, such as tumors and inflammatory sites, exhibit distinct pH gradients compared to normal physiological environments [17, 18]. pH-responsive lipid-based nanocarriers can be designed to undergo structural changes or disassembly in response to acidic or basic pH conditions, triggering drug release at the desired site of action. For instance, pH-sensitive liposomes can exploit the acidic tumor microenvironment to enhance drug accumulation and release within tumor tissues while minimizing off-target effects in healthy tissues, thereby improving therapeutic outcomes for cancer therapy. Temperature-responsive lipid-based nanocarriers offer another promising approach for stimuli-responsive release, where drug release is triggered by changes in temperature within specific physiological or pathological regions [19]. Thermosensitive liposomes, for example, undergo phase transitions or structural changes in response to temperature fluctuations, leading to rapid drug release at elevated temperatures typical of fever, inflammation, or hyperthermic treatment modalities. By incorporating thermosensitive lipids or polymers into nanocarrier formulations, researchers can achieve controlled drug release profiles and spatially targeted delivery for various therapeutic applications.

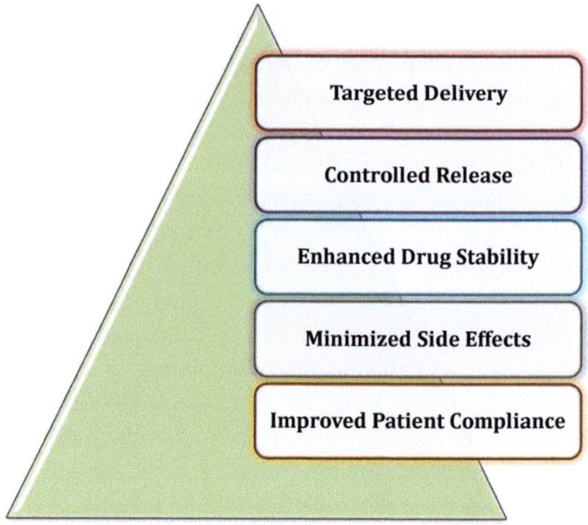


**Fig. 11.2** Advantages of stimuli-responsive drug delivery

**Exercise 2**

**Fill ups**

**(Swelling, Diffusion, Stimuli-responsive, Concentration gradient, Release)**

**1.** Drug release mechanisms include dissolution, ..............., and erosion.

**2.** Diffusion-controlled release involves the movement of drug molecules through a ..............................

**3.** Matrix systems rely on polymer ..................to control drug release.

**4.** Dissolution-controlled release depends on the drug's ............... from the dosage form.

**5.** ..........................systems use external triggers like pH or temperature to initiate controlled release.

Enzyme-responsive lipid-based nanocarriers have garnered significant interest for their ability to achieve triggered drug release in response to specific enzymatic activity within diseased tissues or cellular compartments. Enzyme-cleavable linkers or substrates can be incorporated into lipid-based nanocarriers to enable selective drug release upon enzymatic cleavage by disease-associated enzymes such as proteases, phospholipases, or hydrolases. This strategy allows for precise spatiotemporal control over drug delivery, minimizing systemic exposure and maximizing therapeutic efficacy while reducing off-target effects and toxicity [19]. Light-responsive lipid-based nanocarriers offer a versatile platform for stimuli-responsive release, where drug release can be triggered by exposure to specific wavelengths of light. Light-sensitive liposomes or lipid-based nanoparticles can be engineered to incorporate photo-responsive molecules, such as photosensitizers, photoacoustic agents, or photothermal agents, which undergo photochemical or photophysical reactions upon light irradiation, leading to controlled drug release or modulation of carrier properties. Light-responsive lipid-based nanocarriers hold promise for applications in photodynamic therapy, photothermal therapy, and optogenetics, enabling precise spatiotemporal control over drug delivery in response to external light stimuli.

### *11.1.5 Types of Lipid-Based Nanocarriers*

Lipid-based nanocarriers encompass a diverse range of delivery systems designed to encapsulate and deliver therapeutic agents with enhanced efficacy and safety. These nanocarriers leverage the unique properties of lipids to create versatile platforms for drug delivery, offering advantages such as biocompatibility, biodegradability, and tunable physicochemical properties. Several types of lipid-based nanocarriers have been developed, including liposomes, lipid nanoparticles, solid lipid nanoparticles, nanostructured lipid carriers, and lipid-based micelles [20], each with distinct structures, compositions, and functionalities tailored to specific therapeutic applications. By harnessing the inherent properties of lipids and employing innovative formulation strategies, lipid-based nanocarriers hold great promise for addressing unmet medical needs and advancing drug delivery technologies towards personalized and targeted therapies.

#### 11.1.5.1 Liposomes

Liposomes are versatile lipid-based nanocarriers composed of one or more concentric phospholipid bilayers surrounding an aqueous core (Fig. 11.3). These spherical vesicles offer a biocompatible and biodegradable platform for encapsulating hydrophilic, hydrophobic, or amphiphilic drugs, allowing for precise control over drug loading, release kinetics, and targeting specificity. Liposomes are characterized by their ability to self-assemble from amphiphilic lipid molecules, forming closed bilayer structures with a hydrophilic exterior and a hydrophobic interior, which mimics the structure of cell membranes [21]. This unique architecture enables liposomes to encapsulate drugs within their aqueous core or lipid bilayers, protecting them from degradation, metabolism, and premature release in biological fluids.

The size, surface charge, and lipid composition of liposomes can be tailored to achieve desired drug delivery properties and therapeutic outcomes. Liposomes can range in size from tens to hundreds of nanometers in diameter, offering versatility in drug encapsulation and tissue penetration. Surface modification techniques, such as PEGylation or ligand conjugation, can be employed to modify the surface charge and enhance the stability, circulation time, and targeting specificity of liposomes in vivo. Moreover, the lipid composition of liposomes can be optimized to control membrane fluidity, permeability, and drug release kinetics, allowing for tailored release profiles and optimized therapeutic efficacy. Liposomes offer several advantages as drug delivery vehicles, including the ability to encapsulate a wide range of

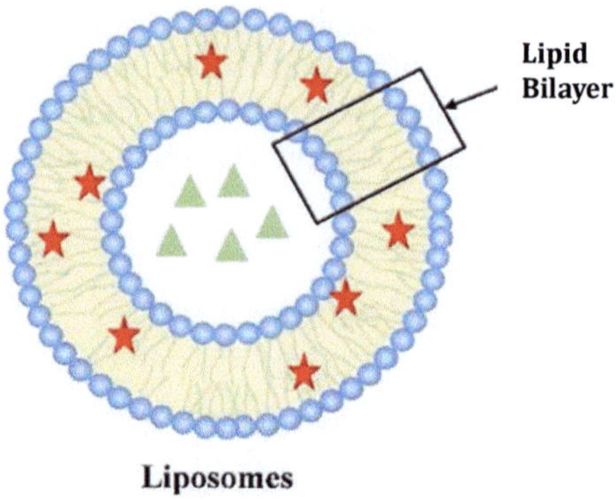


**Fig. 11.3** Illustrating liposomes: spherical vesicles composed of lipid bilayers

drugs with varying physicochemical properties, including small molecules, peptides, proteins, and nucleic acids. The lipid bilayer structure of liposomes provides a protective environment for encapsulated drugs, shielding them from enzymatic degradation and immune recognition, while facilitating their transport across biological barriers. Additionally, liposomes can be engineered to achieve controlled release profiles, targeted delivery, and triggered release in response to external stimuli or physiological cues, offering versatility in drug delivery applications (Fig. 11.4).

Liposomes have been extensively investigated for their potential in treating various diseases, including cancer, infectious diseases, and inflammatory disorders. In cancer therapy, liposomal formulations of chemotherapeutic agents offer advantages such as reduced systemic toxicity, enhanced tumor accumulation, and improved therapeutic

**Fig. 11.4** Advantages of liposomes as lipid-based nanocarriers

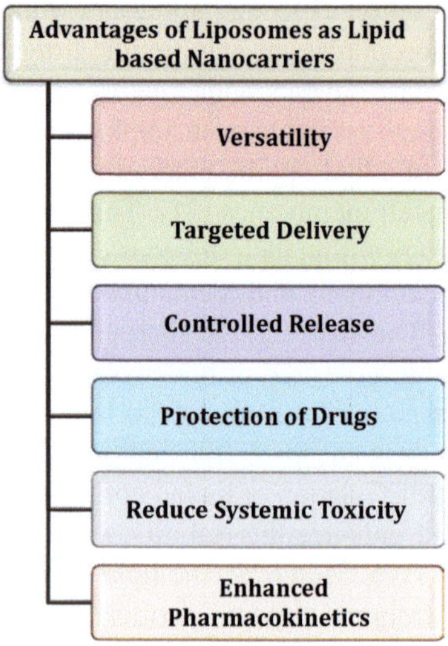

index compared to conventional chemotherapy [22]. Liposomes can also be functionalized by targeting ligands or antibodies to facilitate selective delivery to cancer cells or tumor microenvironments, enhancing treatment efficacy and minimizing off-target effects. Moreover, liposomal vaccines and gene delivery systems hold promise for immunotherapy and gene therapy applications, providing safe and efficient platforms for delivering antigens, adjuvants, or nucleic acids to target cells or tissues. Liposomes represent a versatile and promising class of lipid-based nanocarriers for drug delivery applications. Their unique structure, composition, and properties enable precise control over drug encapsulation, release, and targeting, offering advantages such as biocompatibility, biodegradability, and versatility in drug delivery.

#### 11.1.5.2 Lipid Nanoparticles

Lipid nanoparticles are a class of lipid-based nanocarriers that encompass various formulations, including solid lipid nanoparticles (SLNs) and nanostructured lipid carriers (NLCs), designed to encapsulate and deliver therapeutic agents with enhanced efficacy and stability. Unlike liposomes, which consist of lipid bilayers, lipid nanoparticles are composed of solid lipids or lipid blends dispersed in an aqueous phase or lipid matrix. Solid lipid nanoparticles are characterized by their crystalline lipid core, which provides structural integrity and stability, while nanostructured lipid carriers contain a mixture of solid and liquid lipids, offering additional flexibility in drug loading and release properties. The unique structure and composition of lipid nanoparticles confer several advantages for drug delivery applications. Solid lipid nanoparticles offer improved stability, reduced drug leakage, and enhanced biocompatibility compared to traditional liposomes, making them attractive carriers for poorly water-soluble drugs, peptides, proteins, and nucleic acids. Nanostructured lipid carriers, on the other hand, combine the benefits of solid lipid nanoparticles with increased drug loading capacity, improved drug entrapment efficiency, and controlled release profiles, making them versatile platforms for tailored drug delivery strategies [20, 21]. Lipid nanoparticles can be engineered to achieve controlled release profiles and targeted delivery through modulation of formulation parameters such as lipid composition, particle size, surface charge, and drug loading methods. By optimizing these parameters, researchers can tailor the release kinetics and tissue distribution of lipid nanoparticles to specific therapeutic applications, including sustained release formulations, localized delivery systems, and stimuli-responsive carriers. Surface modification techniques such as PEGylation or ligand conjugation can be employed to enhance the stability, circulation time, and targeting specificity of lipid nanoparticles in vivo, further expanding their utility in drug delivery.

Moreover, lipid nanoparticles have been investigated for their potential in various therapeutic areas, including cancer therapy, infectious diseases, and neurological disorders. In cancer therapy, lipid nanoparticles offer advantages such as improved drug solubility, prolonged circulation time, and enhanced tumor accumulation compared to conventional chemotherapy. Furthermore, lipid nanoparticles can be functionalized with targeting ligands or antibodies to facilitate selective delivery to tumor cells or tumor microenvironments, minimizing off-target effects and improving therapeutic outcomes.

Lipid nanoparticles represent a promising class of lipid-based nanocarriers for drug delivery applications, offering advantages such as biocompatibility, stability, and versatility in drug encapsulation and release.

#### 11.1.5.3 Comparison and Contrast

When comparing and contrasting liposomes and lipid nanoparticles as drug delivery systems, several key factors come into play, including their structure, composition, properties, and applications. Liposomes, composed of phospholipid bilayers surrounding an aqueous core, offer a versatile platform for encapsulating both hydrophilic and hydrophobic drugs. In contrast, lipid nanoparticles, such as solid lipid nanoparticles (SLNs) and nanostructured lipid carriers (NLCs), consist of solid or blended lipids dispersed in an aqueous phase or lipid matrix, providing enhanced stability and versatility in drug loading [23]. One of the primary distinctions between liposomes and lipid nanoparticles lies in their structural composition. Liposomes possess a lipid bilayer structure similar to cell membranes, which allows for encapsulation of drugs within the aqueous core or lipid bilayers. On the other hand, lipid nanoparticles have a solid lipid core or a combination of solid and liquid lipids, offering greater stability and reduced drug leakage compared to liposomes. This structural difference influences their drug loading capacity, release kinetics, and stability profiles, making liposomes suitable for a wider range of drug molecules, including hydrophilic and hydrophobic compounds. Liposomes and lipid nanoparticles exhibit differences in their drug release mechanisms and properties. Liposomes typically release drugs through passive diffusion or membrane fusion processes, resulting in rapid release kinetics and burst release profiles. In contrast, lipid nanoparticles offer controlled release profiles, prolonged drug retention, and reduced systemic toxicity due to their solid lipid matrix, which modulates drug diffusion and release rates [24]. Moreover, lipid nanoparticles can be engineered to achieve stimuli-responsive release, triggered by changes in pH, temperature, or enzymatic activity within target tissues or cellular microenvironments.

The surface properties of liposomes and lipid nanoparticles can be modified to enhance their stability, circulation time, and targeting specificity in vivo. Liposomes can be functionalized with targeting ligands or antibodies to facilitate selective delivery to specific cell types or tissues, whereas lipid nanoparticles can undergo surface modification with stealth polymers, such as polyethylene glycol (PEG), to evade immune recognition and prolong circulation time. These surface modifications play a crucial role in optimizing the pharmacokinetic and pharmacodynamic properties of liposomes and lipid nanoparticles, improving their efficacy and reducing off-target effects. Overall, while liposomes and lipid nanoparticles share similarities in their ability to encapsulate and deliver therapeutic agents, they exhibit distinct structural, compositional, and functional differences that influence their performance and applications in drug delivery. Understanding these differences is essential for selecting the most appropriate delivery system for specific therapeutic needs, whether it involves achieving controlled release kinetics, enhancing targeting specificity, or improving stability and biocompatibility.

### *11.1.6 Applications of Lipid-Based Nanocarriers*

Lipid-based nanocarriers have emerged as versatile and promising platforms for drug delivery, offering numerous applications across a wide range of therapeutic areas. These nanocarriers leverage the unique properties of lipids to encapsulate and deliver various types of therapeutic agents, including small molecules, peptides, proteins, nucleic acids, and imaging agents, with enhanced efficacy and safety. The applications of lipid-based nanocarriers span from cancer therapy and infectious diseases to neurological disorders and regenerative medicine [23, 24]. By harnessing innovative formulation strategies and surface engineering techniques, lipid-based nanocarriers hold great promise for addressing unmet medical needs, advancing personalized medicine, and improving patient outcomes in diverse clinical settings.

#### **11.1.6.1 Cancer Therapy**

Cancer therapy represents one of the most significant applications of lipid-based nanocarriers, owing to their ability to overcome many challenges associated with traditional chemotherapy, including poor solubility of anticancer drugs, nonspecific distribution, and systemic toxicity. Lipid-based nanocarriers, such as liposomes, solid lipid nanoparticles (SLNs), and nanostructured lipid carriers (NLCs), offer unique advantages for cancer therapy, including enhanced drug delivery to tumor sites, reduced off-target effects, and improved therapeutic index (Fig. 11.5). These nanocarriers can be engineered to encapsulate a wide range of anticancer drugs, including chemotherapeutic agents, targeted therapies, and nucleic acid-based therapeutics, allowing for precise modulation of drug release kinetics and pharmacokinetics [25]. One of the key advantages of lipid-based nanocarriers in cancer therapy is their ability to passively target tumor tissues through the enhanced permeability and retention (EPR) effect. Tumors often exhibit leaky vasculature and impaired lymphatic drainage, leading to preferential accumulation of nanocarriers in the tumor microenvironment. Lipid-based nanocarriers can exploit this phenomenon to achieve selective drug accumulation within tumors while minimizing exposure to healthy tissues, thereby enhancing therapeutic efficacy and reducing systemic side effects. Additionally, surface modification techniques, such as PEGylation or ligand conjugation, can further enhance tumor targeting and cellular uptake of lipid-based nanocarriers by facilitating interactions with tumor-specific receptors or antigens.

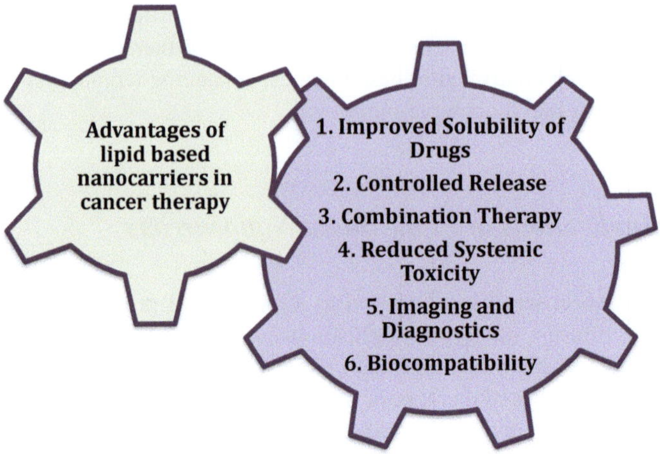


**Fig. 11.5** Advantages of lipid-based nanocarriers in cancer therapy

Lipid-based nanocarriers offer advantages for overcoming multidrug resistance (MDR) in cancer cells, a major obstacle to successful chemotherapy. MDR mechanisms, such as overexpression of drug efflux transporters and alterations in drug metabolism, often result in reduced intracellular drug accumulation and treatment failure. Lipid-based nanocarriers can bypass MDR mechanisms by encapsulating drugs within lipid bilayers or solid lipid matrices, protecting them from efflux transporters and facilitating intracellular drug delivery. Combination therapies involving multiple drugs co-loaded within lipid-based nanocarriers can overcome MDR and synergistically enhance antitumor activity, offering potential solutions for refractory cancers and relapsed disease.

In addition to traditional chemotherapy, lipid-based nanocarriers are being explored for targeted drug delivery and precision medicine approaches in cancer therapy. By functionalizing the surface of nanocarriers with targeting ligands, antibodies, or peptides, researchers can achieve selective delivery of anticancer drugs to specific cell types or tumor subtypes, improving treatment efficacy and minimizing off-target effects. Moreover, lipid-based nanocarriers can be engineered to respond to external stimuli or physiological cues, enabling triggered drug release within tumor tissues or cellular microenvironments, further enhancing therapeutic outcomes and reducing systemic toxicity. Moreover, lipid-based nanocarriers hold promise for combination therapies and personalized medicine approaches in cancer treatment. By co-encapsulating multiple drugs with different mechanisms of action or incorporating adjuvants, imaging agents, or therapeutic nucleic acids within lipid-based nanocarriers [25], researchers can tailor treatment regimens to individual patient needs and disease characteristics. Combination therapies involving chemotherapy, immunotherapy, and targeted therapies delivered via lipid-based nanocarriers offer potential synergistic effects, improved treatment outcomes, and reduced drug resistance, paving the way for more effective and personalized cancer treatment strategies.

#### 11.1.6.2 Treatment of Infectious Diseases

The treatment of infectious diseases represents another significant application of lipid-based nanocarriers, offering innovative solutions to combat microbial infections, including bacterial, viral, and fungal pathogens (Fig. 11.6). Lipid-based nanocarriers, such as liposomes, lipid nanoparticles, and lipid micelles, provide unique advantages for the delivery of antimicrobial agents, vaccines, and diagnostic agents, facilitating targeted delivery, controlled release, and enhanced therapeutic efficacy [26]. These nanocarriers can encapsulate a wide range of antimicrobial drugs, peptides, nucleic acids, or immunomodulators, enabling precise modulation of drug release kinetics and immune responses to combat infectious diseases effectively. One of the key advantages of lipid-based nanocarriers in the treatment of infectious diseases is their ability to enhance the therapeutic efficacy of antimicrobial drugs while minimizing systemic side effects. Lipid-based nanocarriers can improve the solubility, stability, and bioavailability of hydrophobic antimicrobial agents, enabling efficient drug delivery to target sites of infection. Moreover, lipid-based nanocarriers can protect antimicrobial drugs from enzymatic degradation or inactivation in biological fluids, facilitating prolonged drug retention and sustained release within infected tissues, thereby enhancing therapeutic outcomes, and reducing the development of drug resistance.

Lipid-based nanocarriers offer advantages for targeted drug delivery and site-specific release in the treatment of infectious diseases. By functionalizing the surface of nanocarriers with targeting ligands, antibodies, or peptides, researchers can achieve selective delivery of antimicrobial drugs to infected tissues, organs, or intracellular pathogens, improving treatment efficacy and minimizing off-target effects.

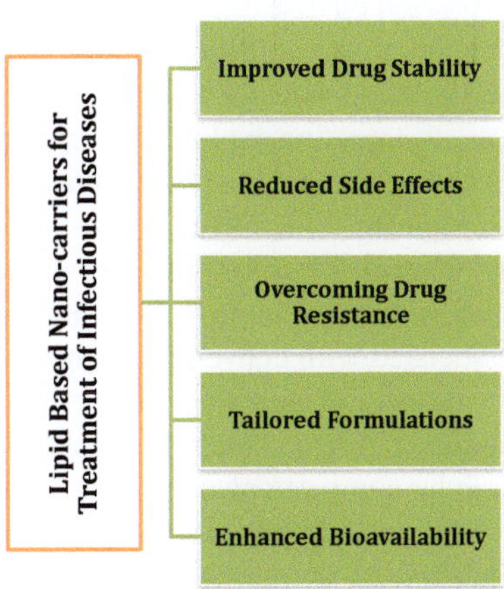


**Fig. 11.6** Advantages of lipid-based nanocarriers for treatment of infectious diseases

Moreover, lipid-based nanocarriers can be engineered to respond to specific physiological cues or environmental stimuli, enabling triggered drug release within infection sites or microbial microenvironments, further enhancing therapeutic outcomes and reducing systemic toxicity. In addition to traditional antimicrobial drugs, lipid-based nanocarriers are being explored for vaccine delivery and immunotherapy approaches in the treatment of infectious diseases [27]. Liposomes and lipid nanoparticles can encapsulate antigens, adjuvants, or immunomodulators, facilitating their delivery to antigen-presenting cells and enhancing immune responses against microbial pathogens. Moreover, lipid-based nanocarriers can be engineered to mimic the structure and composition of pathogens, such as viruses or bacteria, to enhance immune recognition and stimulate protective immune responses, offering potential solutions for vaccine development and infectious disease prevention [26, 27]. Lipid-based nanocarriers hold promise for combination therapies and synergistic approaches in the treatment of infectious diseases. By co-delivering multiple antimicrobial agents with different mechanisms of action or incorporating adjuvants, immunomodulators, or diagnostic agents within lipid-based nanocarriers, researchers can tailor treatment regimens to combat multidrug-resistant pathogens, improve treatment outcomes, and reduce the risk of treatment failure.

#### 11.1.6.3 Targeted Drug Delivery

Targeted drug delivery is a cornerstone of modern medicine, aiming to deliver therapeutic agents specifically to diseased tissues or cells while minimizing systemic exposure and off-target effects. Lipid-based nanocarriers have emerged as versatile platforms for targeted drug delivery, offering precise control over drug release kinetics, tissue distribution, and cellular uptake. These nanocarriers can be engineered to exploit various targeting mechanisms, including passive targeting through the enhanced permeability and retention (EPR) effect, as well as active targeting using ligands, antibodies, or peptides that bind to specific receptors or antigens overexpressed on diseased cells [14, 15]. One of the primary advantages of targeted drug delivery using lipid-based nanocarriers is the ability to enhance therapeutic efficacy and reduce systemic toxicity by concentrating drugs within target tissues or cells. Passive targeting exploits the abnormal vasculature and impaired lymphatic drainage of tumors, leading to preferential accumulation of nanocarriers within tumor tissues through the EPR effect. Lipid-based nanocarriers can exploit this phenomenon to achieve selective drug delivery to tumor sites while minimizing exposure to healthy tissues, thereby improving treatment outcomes, and reducing systemic side effects associated with conventional chemotherapy.

Targeted drug delivery using lipid-based nanocarriers offers advantages for overcoming biological barriers and achieving selective drug delivery to specific organs or cellular compartments. By functionalizing the surface of nanocarriers with targeting ligands or cell-penetrating peptides, researchers can facilitate receptor-mediated endocytosis or intracellular uptake of drugs by target cells, bypassing biological barriers such as the blood-brain barrier or the gastrointestinal epithelium. This targeted approach enables efficient drug delivery to sites of infection, inflammation, or neurological disorders, enhancing therapeutic efficacy and minimizing off-target effects. Targeted drug delivery using lipid-based nanocarriers enables personalized medicine approaches tailored to individual patient characteristics and disease profiles. By selecting targeting ligands or antibodies specific to disease biomarkers or genetic mutations, researchers can achieve precise targeting of therapeutic agents to patient-specific disease subtypes or molecular pathways, improving treatment outcomes and reducing the risk of treatment resistance [28]. Lipid-based nanocarriers can be engineered to respond to specific physiological cues or environmental stimuli, enabling triggered drug release within target tissues or cellular microenvironments, further enhancing therapeutic outcomes and reducing systemic toxicity. In addition to small molecule drugs, lipid-based nanocarriers are being explored for targeted delivery of nucleic acids, peptides, proteins, and imaging agents, offering potential applications in gene therapy, immunotherapy, and molecular imaging. Lipid-based nanocarriers can encapsulate nucleic acids such as siRNA, mRNA, or DNA, protecting them from degradation and facilitating their delivery to target cells or tissues for gene silencing, gene editing, or gene expression modulation.

Lipid-based nanocarriers can encapsulate therapeutic proteins or peptides for targeted delivery to specific cell types or tissues, offering potential applications in regenerative medicine, hormone replacement therapy, and enzyme replacement therapy. Also, lipid-based nanocarriers can be loaded with imaging agents such as contrast agents or fluorescent dyes, enabling non-invasive imaging and visualization of diseased tissues or cellular processes for diagnosis, monitoring, and treatment evaluation.

### 11.1.7 Challenges and Future Perspectives

As the field of lipid-based nanocarriers for drug delivery continues to advance, several challenges and future perspectives arise, shaping the trajectory of research and development in this area. Despite significant progress, challenges such as limited drug loading capacity, premature drug release [26, 27], and batch-to-batch variability remain, hindering the widespread clinical translation of lipid-based nanocarriers. Moreover, issues related to stability, scalability, and regulatory approval present additional hurdles that must be overcome to realize the full potential of lipid-based nanocarriers in clinical practice. However, with ongoing advancements in formulation techniques, surface engineering strategies, and manufacturing processes, these challenges are being addressed, paving the way for future innovations and breakthroughs in drug delivery technologies. Looking ahead, the future of lipid-based nanocarriers holds promise for personalized medicine, targeted therapies, and precision drug delivery, with applications ranging from cancer therapy and infectious diseases to neurological disorders and regenerative medicine. By addressing current challenges and capitalizing on emerging opportunities [27, 28], lipid-based nanocarriers are poised to revolutionize drug delivery and improve patient outcomes in diverse clinical settings.

#### 11.1.7.1 Biological Barriers

Biological barriers play a critical role in regulating the transport of drugs and therapeutic agents within the body, presenting challenges for effective drug delivery. These barriers encompass various physiological barriers, including the blood-brain barrier (BBB), gastrointestinal epithelium, skin barrier, and mucosal surfaces, which serve to protect tissues and organs from harmful substances while selectively allowing the passage of essential nutrients and molecules. The unique properties of these biological barriers, such as tight junctions, epithelial layers, and enzymatic barriers, can impede the penetration and absorption of drugs, limiting their therapeutic efficacy and bioavailability [14, 20, 21]. The blood-brain barrier (BBB) represents one of the most formidable biological barriers to drug delivery, restricting the passage of drugs from the bloodstream into the central nervous system (CNS). The BBB is composed of tightly packed endothelial cells lining the cerebral capillaries, which are interconnected by tight junctions that prevent the paracellular transport of hydrophilic molecules. The presence of efflux transporters, metabolic enzymes, and low pinocytic activity further limits drug penetration across the BBB. Overcoming the BBB presents a significant challenge in the treatment of neurological disorders, requiring innovative drug delivery strategies, such as nanoparticles, liposomes, or receptor-mediated transport systems, to enhance drug penetration and therapeutic efficacy in the CNS.

Similarly, the gastrointestinal epithelium serves as a major biological barrier to oral drug delivery, impacting the absorption and bioavailability of orally administered drugs. The intestinal epithelium is composed of a single layer of enterocytes with tight junctions, which regulate the paracellular transport of drugs across the intestinal barrier. Moreover, enzymatic barriers within the gastrointestinal tract, such as cytochrome P450 enzymes and gut microbiota, can metabolize drugs and reduce their systemic exposure. To overcome these barriers, drug delivery systems must be designed to enhance drug stability, solubility, and permeability across the intestinal epithelium, utilizing strategies such as nanoparticle-based formulations, mucoadhesive polymers, or intestinal permeation enhancers [26, 28]. Biological barriers at mucosal surfaces, such as the respiratory, ocular, and vaginal mucosa, pose challenges for drug delivery to these sites. The mucosal epithelium is characterized by a dense mucus layer, epithelial cells, and immune cells, which serve to protect underlying tissues from pathogens and foreign substances.

Nanoparticle-based formulations, mucoadhesive polymers, and surface modification techniques offer potential strategies for enhancing drug delivery to mucosal surfaces and improving therapeutic outcomes for various diseases and conditions.

#### 11.1.7.2 Toxicity Concerns

Toxicity concerns are a significant consideration in the development and application of drug delivery systems, including lipid-based nanocarriers [29]. While these nanocarriers offer numerous advantages for enhancing drug delivery and therapeutic efficacy, concerns regarding their potential toxicity to biological systems must be carefully evaluated. One primary concern is the biocompatibility of lipid-based nanocarriers, as they may interact with biological membranes, intracellular organelles, or immune cells, leading to adverse effects or immune responses. Understanding the biocompatibility profile of lipid-based nanocarriers is crucial for assessing their safety and minimizing potential toxic effects on cells, tissues, and organs. Concerns regarding the biodistribution and accumulation of lipid-based nanocarriers in vivo raise questions about their potential long-term effects and systemic toxicity [29]. Following administration, lipid-based nanocarriers may undergo systemic circulation, uptake by reticuloendothelial system (RES) organs such as the liver and spleen, or clearance via renal excretion, leading to the accumulation of lipid components or degradation products in tissues or organs. Accumulation of lipid-based nanocarriers in RES organs may elicit inflammatory responses, organ toxicity, or alterations in physiological functions [30], necessitating thorough investigation into their biodistribution, metabolism, and excretion pathways.

The potential for off-target effects and unintended interactions of lipid-based nanocarriers with biological molecules or pathways raises concerns about their safety and efficacy in clinical applications. Lipid-based nanocarriers may interact with plasma proteins, lipoproteins, or cell surface receptors, altering their pharmacokinetic and pharmacodynamic properties or triggering immune responses. The release of encapsulated drugs from lipid-based nanocarriers may lead to cytotoxic effects or tissue damage at off-target sites, necessitating strategies to enhance drug specificity and minimize off-target toxicity. Concerns regarding the stability and biodegradability of lipid-based nanocarriers raise questions about their potential accumulation or persistence in biological systems and the environment [31]. Poorly designed lipid-based nanocarriers may exhibit limited stability, leading to premature drug release, aggregation, or degradation in biological fluids, which can impact their safety and efficacy. Moreover, non-biodegradable lipid-based nanocarriers may accumulate in tissues or organs over time, posing risks of chronic toxicity or environmental pollution. Therefore, the development of biodegradable lipid-based nanocarriers with controlled degradation profiles is essential for minimizing long-term toxicity concerns and ensuring safe and sustainable drug delivery systems.

While lipid-based nanocarriers offer promising opportunities for improving drug delivery and therapeutic outcomes, concerns regarding their potential toxicity must be addressed through rigorous evaluation of their biocompatibility, biodistribution, off-target effects, and biodegradability.

### *11.1.8 Lipid Nanoparticles for RNA Delivery*

Lipid nanoparticles (LNPs) have emerged as promising vehicles for the delivery of RNA-based therapeutics, including messenger RNA (mRNA), small interfering RNA (siRNA), and microRNA (miRNA), owing to their ability to protect RNA molecules from degradation and facilitate their intracellular delivery. LNPs offer several advantages for RNA delivery, including high encapsulation efficiency, biocompatibility, and tunable physicochemical properties, making them attractive candidates for therapeutic applications [32]. LNPs can be engineered to overcome biological barriers, such as cellular membranes and endosomal compartments, enabling efficient cytoplasmic delivery of RNA payloads and achieving therapeutic effects. One of the primary advantages of LNPs for RNA delivery is their ability to encapsulate and protect RNA molecules from degradation by nucleases and other extracellular enzymes. LNPs can encapsulate RNA within their lipid bilayers or lipid cores, shielding them from enzymatic degradation in biological fluids and prolonging their circulation time in vivo. Moreover, LNPs can be designed to facilitate efficient cellular uptake and endosomal escape of encapsulated RNA payloads, enabling their delivery to the cytoplasm, where they can exert their therapeutic effects.

LNPs offer advantages for achieving targeted delivery of RNA therapeutics to specific cell types or tissues, enhancing therapeutic efficacy and minimizing off-target effects. By functionalizing the surface of LNPs with targeting ligands, antibodies, or peptides, researchers can achieve selective delivery of RNA payloads to diseased cells or tissues, such as cancer cells, immune cells, or neurons [33]. Targeted delivery of RNA therapeutics using LNPs offers potential applications in precision medicine, personalized therapies, and targeted cancer therapy, improving treatment outcomes and reducing systemic toxicity. LNPs can be engineered to achieve controlled release of RNA payloads, enabling sustained therapeutic effects and minimizing the need for frequent dosing. By modulating the physicochemical properties of LNPs, such as lipid composition, particle size, and surface charge, researchers can tailor their release kinetics and pharmacokinetic profiles to optimize therapeutic efficacy and patient compliance. Controlled release of RNA therapeutics using LNPs offers advantages for chronic diseases, genetic disorders, and long-term treatment regimens, providing sustained suppression or modulation of target gene expression.

Additionally, LNPs hold promise for co-delivery of multiple RNA payloads, enabling synergistic therapeutic effects and combination therapies for complex diseases [34]. By co-encapsulating different types of RNA molecules, such as mRNA, siRNA, and miRNA, within the same LNP formulation, researchers can target multiple disease pathways or genetic targets simultaneously, enhancing treatment efficacy and overcoming drug resistance. Co-delivery of RNA therapeutics using LNPs offers potential applications in cancer therapy, infectious diseases, and genetic disorders, providing innovative solutions for addressing unmet medical needs and improving patient outcomes.

### 11.1.9 Lipid-Based Nanocarriers in Clinical Trials

Lipid-based nanocarriers have garnered significant interest in clinical trials as promising platforms for drug delivery due to their unique properties and versatility. These nanocarriers, including liposomes, lipid nanoparticles [35], and lipid-based micelles, have been investigated for a wide range of therapeutic applications, including cancer therapy, infectious diseases, neurological disorders, and inflammatory conditions. Clinical trials involving lipid-based nanocarriers aim to evaluate their safety, efficacy, pharmacokinetics, and pharmacodynamics in humans, paving the way for their eventual translation into clinical practice. One area of extensive research in clinical trials involves the use of lipid-based nanocarriers for cancer therapy. Liposomes, in particular, have been extensively studied as drug delivery vehicles for chemotherapeutic agents, targeted therapies, and immunomodulators in various cancer types. Clinical trials have demonstrated the potential of liposomal formulations to improve drug solubility, prolong circulation time, and enhance tumor accumulation, leading to improved treatment outcomes and reduced systemic toxicity compared to conventional chemotherapy. Moreover, lipid-based nanocarriers are being explored for combination therapies, personalized medicine approaches, and precision oncology strategies, offering innovative solutions for addressing the heterogeneity and drug resistance observed in cancer patients [36]. In addition to cancer therapy, lipid-based nanocarriers have shown promise in clinical trials for the treatment of infectious diseases. Lipid nanoparticles, such as solid lipid nanoparticles (SLNs) and nanostructured lipid carriers (NLCs), have been investigated for the delivery of antimicrobial agents, vaccines, and antiviral drugs. Clinical trials involving lipid-based nanocarriers aim to assess their ability to enhance drug stability, bioavailability, and tissue penetration, thereby improving treatment outcomes for bacterial, viral, and fungal infections. Moreover, lipid-based nanocarriers offer advantages for targeted delivery and combination therapies, enabling synergistic effects and reducing the development of drug resistance.

Clinical trials involving lipid-based nanocarriers are exploring their potential applications in neurological disorders, such as Alzheimer's disease, Parkinson's disease, and brain tumors. Liposomal formulations of therapeutic agents, neuroprotective agents, and gene therapy vectors are being evaluated for their ability to cross the blood-brain barrier (BBB) and deliver drugs to the central nervous system (CNS) effectively [34–36]. Lipid-based nanocarriers are being investigated for their potential in regenerative medicine, drug delivery to the eye, and treatment of inflammatory conditions, providing novel approaches for addressing unmet medical needs and improving patient outcomes.

Clinical trials involving lipid-based nanocarriers are advancing our understanding of their safety, efficacy, and translational potential in diverse therapeutic areas.

**Exercise 3**

**Multiple Choice Questions**

1. Which of the following is a characteristic feature of lipid-based nanocarriers?
   a) High electrical conductivity
   b) Hydrophobicity
   c) Low drug loading capacity
   d) Large particle size

2. Lipid-based nanocarriers are used primarily for the delivery of:
   a) Hydrophilic drugs
   b) Large macromolecules
   c) Hydrophobic drugs
   d) Biologics

3. What is the most common type of lipid-based nanocarrier used in drug delivery?
   a) Polymeric micelles
   b) Liposomes
   c) Dendrimers
   d) Carbon nanotubes

4. Which of the following methods is commonly used in the preparation of lipid-based nanocarriers?
   a) Freeze-drying
   b) Spray drying
   c) Emulsification
   d) Centrifugation

5. Surface modification of lipid-based nanocarriers can be done to:
   a) Decrease drug stability
   b) Increase non-specific uptake
   c) Enhance targeting specificity
   d) Decrease circulation time

6. The primary advantage of lipid-based nanocarriers in drug delivery is:
   a) High cost
   b) Low biocompatibility
   c) Ability to encapsulate hydrophobic drugs
   d) Rapid clearance from the body

7. Lipid-based nanocarriers can enhance the bioavailability of poorly water-soluble drugs by:
   a) Increasing their hydrophilicity
   b) Decreasing their stability
   c) Enhancing their solubility
   d) Reducing their membrane permeability

8. Which of the following is NOT a potential application of lipid-based nanocarriers in drug delivery?
   a) Cancer therapy
   b) Gene therapy
   c) Cardiovascular disease treatment
   d) Insulin delivery via oral route

9. The size range of lipid-based nanocarriers typically falls within:
   a) 1-10 micrometers
   b) 100-500 nanometers
   c) 1-10 nanometers
   d) 100-500 micrometers

10. Which property of lipid-based nanocarriers is crucial for prolonged circulation time in the bloodstream?
   a) Large particle size
   b) Hydrophilic surface
   c) Positive surface charge
   d) Stealth coating (PEGylation)

## 11.2 Summary

Lipid-based nanocarriers have emerged as versatile and promising platforms for drug delivery across a wide range of therapeutic areas. From cancer therapy to the treatment of infectious diseases and neurological disorders, lipid-based nanocarriers offer innovative solutions to address the challenges of conventional drug delivery methods [1, 3, 4]. These nanocarriers, including liposomes, lipid nanoparticles, and lipid-based micelles, possess unique properties that enable them to encapsulate and deliver various types of therapeutic agents, such as small molecules, peptides, proteins, nucleic acids, and imaging agents, with enhanced efficacy and safety.

In the field of cancer therapy, lipid-based nanocarriers have shown significant potential for improving drug solubility, prolonging circulation time, and enhancing tumor accumulation, leading to improved treatment outcomes and reduced systemic toxicity compared to traditional chemotherapy. Moreover, lipid-based nanocarriers are being explored for targeted drug delivery and precision medicine approaches, allowing for personalized therapies tailored to individual patient needs and disease characteristics [18, 19]. In the realm of infectious diseases, lipid-based nanocarriers offer advantages for delivering antimicrobial agents, vaccines, and antiviral drugs, enhancing drug stability, bioavailability, and tissue penetration. Clinical trials involving lipid-based nanocarriers aim to assess their safety and efficacy in treating bacterial, viral, and fungal infections, with potential applications in precision medicine and combination therapies to combat multidrug-resistant pathogens.

Lipid-based nanocarriers are being investigated for their potential in neurological disorders, such as Alzheimer's disease, Parkinson's disease, and brain tumors, where they offer novel approaches for crossing the blood-brain barrier (BBB) and delivering therapeutic agents to the central nervous system (CNS). Lipid-based nanocarriers hold promise for regenerative medicine, drug delivery to the eye, and treatment of inflammatory conditions, providing innovative solutions for addressing unmet medical needs and improving patient outcomes. Overall, clinical trials involving lipid-based nanocarriers are advancing our understanding of their safety, efficacy, and translational potential in diverse therapeutic areas.

## Points to Remember

- Lipid-based nanocarriers can improve the solubility of poorly water-soluble drugs, enhancing their bioavailability and therapeutic efficacy.
- Lipids are naturally occurring molecules in the body and are generally well tolerated, making lipid-based nanocarriers biocompatible and suitable for drug delivery applications.
- Lipid-based nanocarriers can be engineered to achieve controlled release of drugs, allowing for sustained and prolonged therapeutic effects while minimizing side effects.
- By functionalizing the surface of lipid-based nanocarriers with targeting ligands, drugs can be delivered specifically to target tissues or cells, improving efficacy and reducing off-target effects.
- Lipid-based nanocarriers offer excellent stability, protecting encapsulated drugs from degradation and enzymatic metabolism, thereby extending their shelf life and improving storage conditions.
- Lipid-based nanocarriers can be tailored to accommodate various types of drugs, including small molecules, peptides, proteins, and nucleic acids, offering versatility in drug delivery applications.
- Surface modification of lipid-based nanocarriers with polyethylene glycol (PEGylation) can impart stealth properties, reducing recognition by the immune system and prolonging circulation time in the bloodstream.
- Lipid-based nanocarriers can facilitate intracellular drug delivery by overcoming biological barriers such as the cell membrane, enabling drugs to reach their intracellular targets more effectively.
- Lipid-based nanocarriers can be loaded with imaging agents such as fluorescent dyes or contrast agents, allowing for simultaneous drug delivery and real-time imaging of drug distribution in vivo.

## References

1. Plaza-Oliver M, Santander-Ortega MJ, Lozano MV (2021) Current approaches in lipid-based nanocarriers for oral drug delivery. Drug Deliv Transl Res 11(2):471–497. https://doi.org/11.1107/s13346-021-00908-7
2. Yang M, Gu Y, Tang X, Wang T, Liu J (2019) Advancement of lipid-based nanocarriers and combination application with physical penetration technique. Curr Drug Deliv 16(4):312–324. https://doi.org/11.2174/1567201816666190118125427
3. Mishra DK, Shandilya R, Mishra PK (2018) Lipid based nanocarriers: a translational perspective. Nanomed Nanotechnol Biol Med 14(7):2023–2050. https://doi.org/11.1116/j.nano.2018.05.021

4. Kothapalli P, Vasanthan M (2024) Lipid-based nanocarriers for enhanced delivery of plant-derived bioactive molecules: a comprehensive review. Ther Deliv 15(2):135–155. https://doi.org/11.4155/tde-2023-0116
5. Suhag D, Bhatia R, Das S, Shakeel A, Ghosh AB, Singh A et al (2015) Physically cross-linked pH-responsive hydrogels with tunable formulations for controlled drug delivery. RSC Adv 5(66):53963–53972. https://doi.org/11.1139/c5ra07424j
6. Shakeel A, Singh A, Das S et al (2017) Synthesis and morphological insight of new biocompatible smart hydrogels. J Polym Res 24:113. https://doi.org/11.1107/s11965-017-1267-7
7. Pardeshi C, Rajput P, Belgamwar V, Tekade A, Patil G, Chaudhary K, Sonje A (2012) Solid lipid based nanocarriers: an overview. Acta Pharm (Zagreb, Croatia) 62(4):433–472. https://doi.org/11.2478/v11007-012-0040-z
8. Ansari MT, Ramlan TA, Jamaluddin NN, Zamri N, Salfi R, Khan A, Sami F, Majeed S, Hasnain MS (2020) Lipid-based nanocarriers for cancer and tumor treatment. Curr Pharm Des 26(34):4272–4276. https://doi.org/11.2174/1381612826666200720235752
9. Kaur N, Sharma AK, Shakeel A, Kumar V, Singh A, Gupta A et al (2017) Therapeutic implications of superoxide dismutase and its importance in kinase drug discovery. Curr Top Med Chem 17(22). https://doi.org/11.2174/1568026617666170307112837
10. Carbone C, Cupri S, Leonardi A, Puglisi G, Pignatello R (2013) Lipid-based nanocarriers for drug delivery and targeting: a patent survey of methods of production and characterization. Pharm Patent Anal 2(5):665–677. https://doi.org/11.4155/ppa.13.43
11. Chen C, Chen C, Li Y, Gu R, Yan X (2023) Characterization of lipid-based nanomedicines at the single-particle level. Fundam Res 3(4):488–504. https://doi.org/10.1016/j.fmre.2022.09.011
12. Danaei M, Dehghankhold M, Ataei S, Davarani FH, Javanmard R, Dokhani A, Khorasani S, Mozafari M (2018) Impact of particle size and polydispersity index on the clinical applications of lipidic nanocarrier systems. Pharmaceutics 10(2):57. https://doi.org/10.3390/pharmaceutics10020057
13. Priya S, Desai VM, Singhvi G (2022) Surface modification of lipid-based nanocarriers: a potential approach to enhance targeted drug delivery. ACS Omega 8(1):74–86. https://doi.org/10.1021/acsomega.2c05976
14. Rathore R, Suhag D, Wan F, Verma R, Thakur A, Thakur P (2023) Functionalization of biomaterials. In: Suhag D, Thakur A, Thakur P (eds) Integrated nanomaterials and their applications. Springer, Singapore. https://doi.org/10.1007/978-981-99-6105-4_11
15. Ding C, Li Z (2017) A review of drug release mechanisms from nanocarrier systems. Mater Sci Eng C Biomimetic Mater Sens Syst 76:1440–1453. https://doi.org/10.1016/j.msec.2017.03.130
16. Son G, Lee B, Cho C (2017) Mechanisms of drug release from advanced drug formulations such as polymeric-based drug-delivery systems and lipid nanoparticles. J Pharm Invest 47(4):287–296. https://doi.org/10.1007/s40005-017-0320-1
17. Kokate A, Chang JH, Park MS (2023) Absorption of drugs via passive diffusion and carrier-mediated pathways. Wiley, 61–78. https://doi.org/10.1002/9781119660699.ch5
18. Saeed Jan M, Alam W, Shabnam M (2023) Fundamentals applications of controlled release drug delivery. IntechOpen. https://doi.org/10.5772/intechopen.113283
19. Trehan K, Saini M, Thakur S (2023) Stimuli-responsive material in controlled release of drug. In: Engineering materials (pp. 535–561). https://doi.org/10.1007/978-981-99-6698-1_18
20. Noore S, Pathania S, Fuciños P, O'Donnell CP, Tiwari BK (2024) Lipid-based nanocarriers. In: Nanocarriers for controlled release and target delivery of bioactive compounds. SpringerBriefs in food, health, and nutrition. Springer, Cham. https://doi.org/10.1007/978-3-031-57488-7_2
21. Nsairat H, Khater D, Sayed U, Odeh F, Al Bawab A, Alshaer W (2022) Liposomes: structure, composition, types, and clinical applications. Heliyon 8(5):e09394. https://doi.org/10.1016/j.heliyon.2022.e09394
22. Rommasi F Esfandiari N (2021) Liposomal nanomedicine: applications for drug delivery in cancer therapy. Nanoscale Res Lett 16(1). https://doi.org/10.1186/s11671-021-03553-8
23. Xu L, Wang X, Liu Y, Yang G, Falconer RJ, Zhao C (2021) Lipid nanoparticles for drug delivery. Adv NanoBiomed Res 2(2). https://doi.org/10.1002/anbr.202100109

24. Scioli Montoto S, Muraca G, Ruiz ME (2020) Solid lipid nanoparticles for drug delivery: pharmacological and biopharmaceutical aspects. Front Mol Biosci 7:587997. https://doi.org/10.3389/fmolb.2020.587997
25. Liu D, Zhang N (2010) Cancer chemotherapy with lipid-based nanocarriers. Crit Rev in Ther Drug Carrier Syst 27(5):371–417. https://doi.org/10.1615/critrevtherdrugcarriersyst.v27.i5.10
26. Simnani FZ, Singh D, Patel P, Choudhury A, Sinha A, Nandi A, Samal SK, Verma SK, Panda PK (2023) Nanocarrier vaccine therapeutics for global infectious and chronic diseases. Mater Today 66:371–408. https://doi.org/10.1016/j.mattod.2023.04.008
27. Bezbaruah R, Chavda VP, Nongrang L, Alom S, Deka K, Kalita T, Ali F, Bhattacharjee B, Vora L (2022) Nanoparticle-based delivery systems for vaccines. Vaccines 10(11):1946. https://doi.org/10.3390/vaccines10111946
28. Xu Y, Fourniols T, Labrak Y, Préat V, Beloqui A, des Rieux A (2022) Surface modification of lipid-based nanoparticles. ACS Nano 16(5):7168–7196. https://doi.org/11.1121/acsnano.2c02347
29. Rathore R, Suhag D, Wan F, Thakur A, Thakur P (2023) Toxicology and toxicity studies of nano-biomaterials. In: Suhag D, Thakur A, Thakur P (eds) Integrated nanomaterials and their applications. Springer, Singapore. https://doi.org/11.1107/978-981-99-6115-4_20
30. Subramanian P (2021) Lipid-based nanocarrier system for the effective delivery of nutraceuticals. Molecules (Basel, Switzerland) 26(18):5511. https://doi.org/11.3390/molecules26185511
31. Niu Z, Conejos-Sánchez I, Griffin BT, O'Driscoll CM, Alonso MJ (2016) Lipid-based nanocarriers for oral peptide delivery. Adv Drug Deliv Rev 116(Pt B):337–354. https://doi.org/11.1116/j.addr.2016.04.001
32. Hou X, Zaks T, Langer R, Dong Y (2021) Lipid nanoparticles for mRNA delivery. Nat Rev Mater 6(12):1078–1094. https://doi.org/10.1038/s41578-021-00358-0
33. Puri A, Loomis K, Smith B, Lee JH, Yavlovich A, Heldman E, Blumenthal R (2009) Lipid-based nanoparticles as pharmaceutical drug carriers: from concepts to clinic. Crit Rev Ther Drug Carrier Syst 26(6):523–580. https://doi.org/11.1615/critrevtherdrugcarriersyst.v26.i6.11
34. Aziz Hazari S, Kaur H, Karwasra R, Abourehab MAS, Ali Khan A, Kesharwani P (2023) An overview of topical lipid-based and polymer-based nanocarriers for treatment of psoriasis. Int J Pharm 638:122938. https://doi.org/11.1116/j.ijpharm.2023.122938
35. Patel D, Patel B, Thakkar H (2021) Lipid based nanocarriers: promising drug delivery system for topical application. Eur J Lipid Sci Technol 123(5). https://doi.org/11.1102/ejlt.202000264
36. Gupta P, Neupane YR, Aqil M, Kohli K, Sultana Y (2023) Lipid-based nanoparticle-mediated combination therapy for breast cancer management: a comprehensive review. Drug Deliv Transl Res 13(11):2739–2766. https://doi.org/10.1007/s13346-023-01366-z

# Chapter 12
# Challenges and Future Directions

## 12.1 Introduction

Biomaterials have emerged as indispensable tools in modern therapeutic applications, offering unique opportunities to address a myriad of medical challenges. These materials, both natural and synthetic, interact with biological systems to promote healing, provide mechanical support, and deliver therapeutic agents. The utilization of biomaterials spans a wide range of medical fields, including tissue engineering, regenerative medicine, drug delivery, and medical device manufacturing. As our understanding of biological processes deepens and technological advancements accelerate, the landscape of biomaterials in therapeutic applications continues to evolve rapidly. The exploration of biomaterials in therapeutic contexts is driven by the urgent need to develop innovative solutions for pressing medical issues [1]. Chronic diseases, traumatic injuries, and age-related degenerative conditions pose significant burdens on healthcare systems worldwide. Biomaterials offer promising avenues to address these challenges by providing scaffolds for tissue regeneration, carriers for targeted drug delivery, and platforms for diagnostic and therapeutic interventions. Moreover, the interdisciplinary nature of biomaterials research fosters collaborations between materials scientists, biologists, engineers, and clinicians, catalyzing groundbreaking discoveries and novel applications. Despite the remarkable progress achieved in the field of biomaterials, several challenges persist that hinder their widespread adoption and efficacy [2]. Issues such as material biocompatibility, immunogenicity, mechanical properties, and degradation rates pose significant hurdles that must be overcome to ensure the safety and effectiveness of biomaterial-based therapies. The complexity of biological systems presents intricate hurdles, including the need for precise control over material-cell interactions, the design of multifunctional materials, and the integration of biomaterials with host tissues.

 
D. Suhag et al., *Handbook of Biomaterials for Medical Applications, Volume 1*,
Biomedical Materials for Multi-functional Applications,
https://doi.org/10.1007/978-981-97-4818-1_12

Looking ahead, the future of biomaterials in therapeutic applications holds immense promise, driven by ongoing technological innovations and scientific discoveries. Emerging trends such as the development of nanomaterials, advanced fabrication techniques, and the advent of personalized medicine are poised to revolutionize the field, offering tailored solutions that cater to individual patient needs. The integration of biomaterials with cutting-edge technologies such as artificial intelligence, bioprinting, and bioinformatics holds the potential to unlock new frontiers in healthcare delivery and patient outcomes. We aim to explore the current landscape of biomaterials in therapeutic applications, highlighting both the challenges and opportunities that lie ahead [3]. We will examine the existing limitations of biomaterials, explore emerging technologies and trends, discuss regulatory considerations and ethical implications, and envision future directions for the field. By fostering a deeper understanding of the complexities and potentials of biomaterials, we endeavor to contribute to the advancement of biomedical science and the improvement of patient care.

### 12.1.1 Overview of Biomaterials in Therapeutic Applications

Biomaterials play a pivotal role in therapeutic applications by providing a platform for tissue regeneration, drug delivery, and medical device manufacturing. These materials, which can be derived from natural sources or synthetically engineered, interact with biological systems to promote healing, and improve patient outcomes. In the context of tissue engineering and regenerative medicine, biomaterial scaffolds serve as templates for cell growth and tissue formation, offering a means to repair or replace damaged organs and tissues. Moreover, biomaterials are integral components of medical devices such as implants, prosthetics, and surgical instruments, where they contribute to the functionality and biocompatibility of these devices. One of the key advantages of biomaterials is their ability to mimic the extracellular matrix (ECM) of native tissues, providing a supportive microenvironment for cellular activities. By modulating the composition, structure, and mechanical properties of biomaterials, researchers can tailor their properties to specific therapeutic applications, enhancing biocompatibility and tissue integration [1, 2]. Biomaterials can be engineered to exhibit controlled release properties, enabling precise delivery of therapeutic agents such as drugs, growth factors, and genes to target tissues or cells. The versatility of biomaterials extends beyond traditional therapeutic modalities, encompassing emerging fields such as regenerative medicine, drug delivery, and personalized medicine. In regenerative medicine, biomaterials hold the potential to revolutionize the treatment of degenerative diseases and traumatic injuries by promoting tissue repair and regeneration. Through the development of innovative biomaterial scaffolds, researchers aim to overcome the limitations of conventional treatments and provide more effective solutions for patients suffering from conditions such as osteoarthritis, cardiovascular disease, and spinal cord injuries. In the realm of drug delivery, biomaterials offer a means to improve the efficacy and safety of therapeutic agents by enhancing their pharmacokinetics and targeting specific sites of action. By encapsulating drugs within biocompatible carriers or conjugating them to biomaterial surfaces, researchers can achieve sustained release, minimize systemic toxicity [3, 4], and enhance therapeutic outcomes. The advent of personalized medicine has spurred interest in the development of biomaterial-based platforms for patient-specific diagnostics and therapies, enabling tailored treatment strategies based on individual genetic profiles and disease characteristics.

As the field of biomaterials continues to evolve, researchers are exploring novel materials and fabrication techniques to address current challenges and expand the scope of therapeutic applications [5]. From the development of bioactive ceramics and biodegradable polymers to the integration of nanotechnology and advanced manufacturing methods, the landscape of biomaterials is constantly evolving to meet the demands of modern medicine.

### *12.1.2 Current Challenges in Biomaterials Utilization*

Despite the tremendous potential of biomaterials in therapeutic applications, several challenges persist that impede their widespread adoption and effectiveness (Fig. 12.1). One of the primary hurdles is ensuring the biocompatibility of biomaterials, as interactions between the material and the biological environment can trigger immune responses or adverse reactions. Additionally, achieving optimal mechanical properties while maintaining biocompatibility remains a significant challenge, particularly in load-bearing applications such as orthopedic implants [6]. The degradation rates of biomaterials must be carefully controlled to ensure adequate tissue regeneration and minimize the risk of inflammation or foreign body reactions. Addressing these challenges requires a multidisciplinary approach, combining expertise in materials science, biology, engineering, and medicine to develop innovative solutions that meet the complex demands of therapeutic applications.

#### 12.1.2.1 Material Biocompatibility

Material biocompatibility is a critical aspect of biomaterials design and utilization, ensuring that the materials interact harmoniously with biological systems without eliciting adverse reactions. Achieving biocompatibility involves considering various factors, including the chemical composition, surface properties, and degradation behavior of the biomaterial. The ideal biomaterial should be non-toxic, non-immunogenic, and capable of integrating seamlessly with surrounding tissues to promote healing and functional restoration [7]. However, achieving these characteristics poses significant challenges, as the biological response to biomaterials can vary widely depending on factors such as the site of implantation, duration of exposure, and patient-specific factors. One of the primary concerns in material biocompatibility is the potential for immune reactions or inflammatory responses to biomaterial

**Fig. 12.1** Challenges in biomaterials utilization

implants. When introduced into the body, biomaterials can elicit immune responses mediated by inflammatory cells and cytokines, leading to tissue damage and implant failure. Strategies to mitigate immunogenicity include modifying the surface chemistry of biomaterials to reduce protein adsorption and cell adhesion, thereby minimizing the activation of immune cells. The development of bioinert materials such as titanium and certain polymers aims to minimize immune recognition and promote tissue integration [8]. Another aspect of material biocompatibility is the cytotoxicity of biomaterials, which refers to their potential to cause harm to cells or tissues. Certain biomaterials may release toxic byproducts during degradation or leach harmful chemicals into the surrounding environment, compromising cell viability and tissue function. To assess cytotoxicity, researchers employ a range of in vitro and in vivo assays to evaluate the effects of biomaterial extracts on cell morphology, proliferation, and viability. Strategies to improve cytocompatibility include incorporating bioactive molecules or surface modifications that enhance cell adhesion and promote tissue regeneration while minimizing adverse effects on cellular function.

The surface properties of biomaterials play a crucial role in determining their biocompatibility, as interactions between the material surface and surrounding tissues influence cellular adhesion, proliferation, and differentiation [9]. Surface modifications such as coatings, functionalization, and patterning can be employed to modulate these interactions and promote favorable tissue responses. For example, biomaterials can be functionalized with cell-adhesive peptides or growth factors to enhance cell attachment and stimulate tissue regeneration. Surface patterning techniques such as micro- and nano structuring can create topographical cues that guide cell behavior and promote tissue integration. In addition to considering the biocompatibility of individual biomaterials, the design of biomaterial-based implants must account for the complex interplay between the material and the host immune system. The foreign body response (FBR) is a natural immune reaction triggered by the presence of biomaterials in the body, characterized by the formation of a fibrous capsule around the implant. While the FBR is an essential protective mechanism, excessive fibrotic encapsulation can compromise the functionality of implants and limit their long-term performance [8, 9]. Strategies to modulate the FBR include incorporating anti-inflammatory agents, designing implant surfaces that promote M2 macrophage polarization, and developing materials that elicit minimal foreign body reactions. Despite significant advances in biomaterials research, achieving optimal biocompatibility remains an ongoing challenge due to the complex and dynamic nature of biological systems. The development of predictive models and screening assays for assessing biocompatibility early in the design process holds promise for accelerating the translation of biomaterials from the bench to the bedside.

#### 12.1.2.2 Immunogenicity

Immunogenicity is a critical consideration in the development and utilization of biomaterials for therapeutic applications, as the immune system plays a pivotal role in the response to foreign materials introduced into the body. When biomaterials are implanted or administered, they can elicit immune responses mediated by various components of the innate and adaptive immune systems, leading to inflammation, tissue damage, and implant failure. Understanding and mitigating immunogenicity are essential for ensuring the safety and efficacy of biomaterial-based therapies [10, 11]. One of the primary challenges in managing immunogenicity is minimizing the activation of immune cells such as macrophages, dendritic cells, and lymphocytes in response to biomaterials. Upon encountering foreign materials, these cells can initiate inflammatory cascades, release pro-inflammatory cytokines, and trigger adaptive immune responses, which can compromise the biocompatibility and functionality of biomaterial implants. Strategies to reduce immunogenicity include modifying the surface properties of biomaterials to minimize protein adsorption and cell adhesion, thereby attenuating immune recognition and activation. In addition to innate immune responses, biomaterials can also elicit adaptive immune responses characterized by the production of antibodies and activation of T lymphocytes. The development of anti-material antibodies and T cell-mediated responses can lead to hypersensitivity reactions, allergic responses, and the formation of immune complexes, which may exacerbate inflammation and contribute to implant rejection. To mitigate adaptive immune responses, researchers explore strategies such as immune modulation, antigen-specific tolerance induction, and the design of biomaterials that promote regulatory T cell activation.

The choice of biomaterial composition and structure can significantly influence its immunogenic potential. Certain materials, such as metallic alloys and synthetic polymers, may inherently trigger immune responses due to their chemical properties or degradation products [12]. In contrast, biomaterials derived from natural sources, such as decellularized tissues and extracellular matrices, may exhibit reduced immunogenicity and enhanced biocompatibility. By selecting materials with favorable immunological profiles and engineering biomaterials to mimic the native tissue microenvironment, researchers can minimize immune recognition and improve tissue integration. The route of biomaterial administration and the site of implantation can impact the magnitude and nature of the immune response. Biomaterials delivered via systemic routes, such as intravenous injection, may encounter immune cells in circulation and trigger systemic inflammatory responses. In contrast, biomaterials implanted directly into tissues or organs may elicit localized immune reactions influenced by the tissue microenvironment and resident immune cells [10, 12]. Understanding the immunological dynamics at different implantation sites is crucial for optimizing biomaterial design and therapeutic outcomes. Despite significant progress in understanding and mitigating immunogenicity, challenges remain in predicting and managing immune responses to biomaterials in diverse clinical settings. Factors such as patient variability, pre-existing immune conditions, and the presence of concomitant therapies can influence the immune response to biomaterial implants. Thus, personalized approaches that account for individual immune profiles and disease characteristics are essential for optimizing the safety and efficacy of biomaterial-based therapies. Immunogenicity is a complex and multifaceted aspect of biomaterials utilization that requires careful consideration throughout the design, development, and clinical translation of biomaterial-based therapies.

#### 12.1.2.3 Mechanical Properties

The mechanical properties of biomaterials are crucial considerations in their design and application, as they determine the material's ability to withstand physiological forces and maintain structural integrity in vivo. Biomaterials used in therapeutic applications must exhibit mechanical properties that match or closely resemble those of the native tissues they are intended to replace or support. Key mechanical properties include stiffness, strength, toughness, and elasticity, each of which plays a distinct role in determining the performance and longevity of biomaterial implants. Stiffness, often quantified by Young's modulus, reflects the resistance of a material to deformation under an applied load. Biomaterials with stiffness values that closely match those of native tissues can distribute mechanical forces more evenly, reducing stress concentrations and minimizing the risk of implant failure or tissue damage [12]. However, achieving the optimal balance of stiffness while maintaining biocompatibility and tissue integration poses a significant challenge, particularly in load-bearing applications such as orthopedic implants and cardiovascular stents. Strength is another critical mechanical property that influences the load-bearing capacity and structural integrity of biomaterial implants. Biomaterials must withstand physiological forces without undergoing plastic deformation or fracture to ensure long-term performance and patient safety. Strategies to enhance material strength include selecting materials with high tensile and compressive strength, optimizing processing techniques to minimize defects and imperfections, and incorporating reinforcing agents or structural modifications to improve load-bearing capacity. Toughness, defined as the ability of a material to absorb energy and deform plastically before fracturing, is essential for biomaterials exposed to dynamic loading conditions or impact forces. Biomaterials with high toughness can withstand sudden changes in mechanical stress without catastrophic failure, reducing the risk of implant rupture or fatigue [13]. Enhancing material toughness often involves modifying the material composition, microstructure, or processing methods to promote energy dissipation and plastic deformation mechanisms. Elasticity, represented by parameters such as the modulus of elasticity and resilience, describes the ability of a material to return to its original shape after deformation. Biomaterials with high elasticity can undergo reversible deformation in response to mechanical loading, enabling them to accommodate physiological movements and deformations without permanent damage. Achieving optimal elasticity is critical for biomaterials used in soft tissue engineering, where flexibility and compliance are essential for mimicking native tissue function and promoting tissue integration.

In addition to these fundamental mechanical properties, biomaterials must also exhibit durability and fatigue resistance to withstand long-term exposure to physiological conditions and repetitive loading cycles. Material degradation, wear, and fatigue can compromise the mechanical integrity of implants over time, leading to mechanical failure, tissue damage, and adverse clinical outcomes. Therefore, comprehensive testing and evaluation of biomaterials under physiological conditions are essential for assessing their mechanical performance and predicting long-term durability in vivo [14]. Advances in materials science, biomechanics, and computational modeling have facilitated the design and optimization of biomaterials with tailored mechanical properties for specific therapeutic applications.

#### 12.1.2.4 Degradation Rates

The degradation rate of biomaterials is a critical parameter that influences their performance and biocompatibility in therapeutic applications. Biomaterial degradation refers to the process by which the material undergoes chemical breakdown or physical deterioration over time in response to environmental factors such as pH, temperature, and enzymatic activity. Controlled degradation is desirable in many biomedical applications, particularly in tissue engineering and drug delivery, where biomaterial scaffolds or carriers are designed to degrade gradually and be replaced by native tissue or to release therapeutic agents at a controlled rate. The degradation kinetics of biomaterials can vary widely depending on factors such as material composition, structure, morphology, and environmental conditions [11]. For example, biodegradable polymers such as poly (lactic-co-glycolic acid) (PLGA) and polycaprolactone (PCL) degrade through hydrolytic cleavage of their ester bonds, with degradation rates influenced by the polymer molecular weight, crystallinity, and hydrophilicity. Similarly, natural biomaterials such as collagen and chitosan degrade via enzymatic degradation by proteases or lysozyme, with degradation rates modulated by the presence of specific enzymes and the cross-linking density of the material. Controlling the degradation rate of biomaterials is essential for ensuring optimal tissue regeneration, drug release kinetics, and long-term biocompatibility. Rapid degradation can lead to premature loss of mechanical integrity, inflammatory responses, and insufficient tissue ingrowth, whereas slow degradation may hinder tissue remodeling and prolong the presence of foreign materials in the body, increasing the risk of chronic inflammation or fibrotic encapsulation. Therefore, understanding the mechanisms and kinetics of biomaterial degradation is crucial for designing biomaterials with tailored degradation profiles for specific therapeutic applications. Strategies to modulate the degradation rate of biomaterials include adjusting material composition, incorporating degradation-resistant moieties or cross-linking agents, and controlling environmental factors such as pH, temperature, and enzymatic activity. For example, modifying the composition of biodegradable polymers by blending or copolymerization can alter their degradation kinetics and mechanical properties [15]. Similarly, surface modifications such as coatings or functionalization can enhance the stability and degradation resistance of biomaterials while preserving their biocompatibility and bioactivity.

In tissue engineering applications, biomaterial degradation must be synchronized with tissue regeneration processes to facilitate cell infiltration, extracellular matrix deposition, and vascularization. Scaffold degradation can be tailored to match the rate of tissue ingrowth and remodeling, ensuring seamless integration of the scaffold with host tissues. Additionally, in drug delivery systems, the degradation rate of biomaterial carriers dictates the release kinetics of encapsulated therapeutic agents, influencing drug bioavailability, efficacy, and safety. By optimizing the degradation rate of biomaterial carriers, researchers can achieve controlled drug release profiles that meet the specific requirements of therapeutic interventions.

Despite significant progress in understanding and controlling biomaterial degradation, challenges remain in predicting and optimizing degradation kinetics in complex biological environments. The heterogeneity of tissue microenvironments, variability in enzymatic activity, and patient-specific factors can influence the degradation behavior of biomaterials in vivo, complicating the design and optimization process [12, 13]. Therefore, interdisciplinary approaches combining materials science, biology, engineering, and computational modeling are essential for advancing our understanding of biomaterial degradation and developing innovative strategies to modulate degradation rates for enhanced therapeutic outcomes.

#### 12.1.2.5 Sterilization Methods

Sterilization of biomaterials is a critical step in ensuring the safety and efficacy of medical devices, implants, and tissue-engineered constructs used in therapeutic applications. Sterilization methods aim to eliminate or inactivate microbial contaminants, including bacteria, viruses, fungi, and spores, to prevent infections and adverse reactions in patients. The choice of sterilization method depends on various factors such as material compatibility, device complexity, regulatory requirements, and the desired level of sterility assurance. One of the most commonly used sterilization methods is steam sterilization, also known as autoclaving, which utilizes high-pressure steam to penetrate biomaterials and achieve microbial inactivation. Autoclaving is highly effective, efficient, and widely applicable for heat-stable materials such as metals, glass, and certain polymers [14]. However, heat-sensitive biomaterials may undergo degradation or structural changes during autoclaving, limiting its suitability for certain applications. Ethylene oxide (EO) sterilization is another widely used method for heat-sensitive biomaterials and medical devices. EO gas penetrates deeply into porous materials and kills microorganisms by alkylating their cellular components. EO sterilization is effective against a broad spectrum of microorganisms and compatible with a wide range of materials, including plastics, elastomers, and biological tissues. However, EO is highly toxic and flammable, requiring careful handling, ventilation, and off-gassing procedures to ensure safety.

Radiation-based sterilization methods, such as gamma irradiation and electron beam (e-beam) irradiation, utilize ionizing radiation to disrupt the DNA and cellular structures of microorganisms, rendering them non-viable. Gamma irradiation is commonly used for sterilizing medical devices, biomaterials, and tissue allografts due to its ability to penetrate dense materials and achieve uniform sterilization. E-beam irradiation, on the other hand, is well-suited for sterilizing heat-sensitive materials and complex geometries [11], as it delivers controlled doses of radiation without generating heat. Another sterilization method gaining popularity is hydrogen peroxide gas plasma sterilization, which utilizes a combination of hydrogen peroxide vapor and low-temperature plasma to achieve microbial inactivation. Gas plasma sterilization is effective for heat- and moisture-sensitive materials and does not leave toxic residues or byproducts, making it suitable for a wide range of medical devices and biomaterials. However, gas plasma sterilization systems can be costly and require specialized equipment and expertise for operation.

Sterilization methods such as filtration, chemical sterilants, and aseptic processing are also employed for specific applications and materials. Filtration methods utilize membranes or filters to physically remove microorganisms from liquids or gases, making them suitable for sterilizing cell culture media, drug solutions, and biological fluids. Chemical sterilants, such as hydrogen peroxide vapor, peracetic acid, and chlorine dioxide, are used for disinfecting surfaces, equipment, and environmental surfaces in healthcare facilities. Aseptic processing involves maintaining sterile conditions throughout the manufacturing process to prevent microbial contamination of biomaterials and medical devices. Selecting the appropriate sterilization method is crucial for ensuring the safety, efficacy, and regulatory compliance of biomaterials used in therapeutic applications. Each sterilization method has its advantages, limitations, and compatibility requirements, which must be carefully evaluated based on the specific characteristics of the biomaterial, device, and intended use.

**Exercise 1**

**Fill ups**

**(Mechanical, Degradation rate, Scalability, Biomaterials, Biocompatibility)**

**1.** One current challenge in biomaterials utilization is .........................assessment.

**2.** Addressing the issue of ...............................remains a hurdle in biomaterials applications.

**3.** Achieving optimal ......................properties is a key concern in the utilization of biomaterials.

**4.** Overcoming regulatory barriers is essential for advancing ....................in clinical settings.

**5.** Enhancing the .........................of biomaterials is critical for their widespread adoption in various industries.

### *12.1.3 Emerging Biomaterials Technologies*

Emerging biomaterials technologies represent a frontier of innovation in biomedical engineering, offering novel approaches to address complex medical challenges and improve patient outcomes. These technologies encompass a diverse array of materials, fabrication techniques, and functionalization strategies that enable precise control over material properties and interactions with biological systems [11]. From advanced nanomaterials and bioinspired polymers to cutting-edge 3D printing and tissue engineering platforms, emerging biomaterials technologies hold the potential to revolutionize diagnostics, therapeutics, and regenerative medicine.

#### 12.1.3.1 Nanomaterials

Nanomaterials have emerged as a transformative platform in the field of biomaterials, offering unique opportunities to engineer materials with tailored properties and functionalities at the nanoscale. These materials, typically ranging in size from 1 to 100 nm, exhibit distinct physical, chemical, and biological properties compared to their bulk counterparts, making them highly versatile for a wide range of biomedical applications. One of the key advantages of nanomaterials is their large surface area-to-volume ratio, which enables enhanced interactions with biological molecules and cells, facilitating targeted drug delivery, biosensing, and tissue engineering. Nanoparticles, nanofibers, and nanocomposites are among the most commonly explored nanomaterials in biomedical research, owing to their unique physicochemical properties and tunable characteristics [15]. Nanoparticles, such as liposomes, polymeric nanoparticles, and quantum dots, can encapsulate therapeutic agents, imaging agents, or nucleic acids for targeted delivery to specific tissues or cells. Their small size allows them to penetrate biological barriers and accumulate at target sites, minimizing systemic toxicity and maximizing therapeutic efficacy. Nanofibers, fabricated through techniques such as electrospinning or self-assembly, offer a biomimetic scaffold for tissue engineering and regenerative medicine. These nanofibrous scaffolds mimic the architecture and mechanical properties of the extracellular matrix, providing a supportive microenvironment for cell adhesion, proliferation, and differentiation. By incorporating bioactive molecules or growth factors into nanofibers, researchers can promote tissue regeneration and repair in applications such as wound healing, bone regeneration, and nerve regeneration.

Nanocomposites, comprised of nanoscale fillers dispersed within a matrix material, exhibit synergistic properties that combine the advantages of both components. For example, nanocomposite hydrogels can exhibit enhanced mechanical strength, stability, and bioactivity compared to traditional hydrogels, making them promising candidates for tissue engineering, drug delivery, and biomedical imaging. By precisely controlling the composition, morphology, and processing parameters of nanocomposites, researchers can tailor their properties to specific biomedical applications [16]. Despite the tremendous potential of nanomaterials in biomedical research, several challenges must be addressed to realize their full clinical translation. Concerns such as biocompatibility, toxicity, and long-term safety profiles remain critical considerations in the design and development of nanomaterial-based therapies. Additionally, issues related to scale-up production, regulatory approval, and clinical validation pose hurdles that must be overcome to bring nanomaterials from the laboratory to the clinic. The ethical, societal, and environmental implications of nanomaterials in healthcare require careful consideration to ensure responsible innovation and equitable access to emerging technologies. Addressing these challenges necessitates interdisciplinary collaboration between scientists, engineers, clinicians, regulators, and policymakers to foster a holistic approach to nanomaterial development and implementation in healthcare.

Nanomaterials represent a paradigm shift in biomaterials research, offering unprecedented opportunities to revolutionize diagnostics, therapeutics, and regenerative medicine.

#### 12.1.3.2 Hydrogels

Hydrogels are three-dimensional networks of crosslinked hydrophilic polymers that have the ability to absorb and retain large amounts of water. These materials exhibit unique properties, such as high-water content, biocompatibility, and tunable mechanical properties, making them highly versatile for a wide range of biomedical applications (Fig. 12.2). Hydrogels can be synthesized from natural polymers, such as alginate, collagen, and hyaluronic acid, or synthetic polymers, such as polyethylene glycol (PEG) and polyacrylamide, or a combination of both. One of the key advantages of hydrogels is their ability to mimic the extracellular matrix (ECM) of native tissues, providing a supportive microenvironment for cell adhesion, proliferation [17], and differentiation. This makes hydrogels promising candidates for tissue engineering and regenerative medicine applications, where they can be used as scaffolds to promote tissue repair and regeneration in damaged or diseased tissues. Additionally, the porous structure of hydrogels allows for the diffusion of nutrients, oxygen, and signaling molecules, facilitating cellular activities and tissue remodeling. Hydrogels can be tailored to exhibit a wide range of mechanical properties, from soft and compliant to stiff and resilient, by adjusting parameters such as polymer composition, cross-linking density, and network architecture. This tunability enables researchers to design hydrogels with mechanical properties that match those of specific tissues or organs, making them suitable for applications such as cartilage repair, wound healing, and drug delivery.

In addition to their role as scaffolds for tissue engineering, hydrogels are also utilized as drug delivery vehicles due to their ability to encapsulate and release therapeutic agents in a controlled manner. Drug-loaded hydrogels can be administered locally to target tissues or implanted directly into the body, where they gradually release the encapsulated drugs over time. This localized drug delivery approach minimizes systemic side effects and improves therapeutic outcomes by maintaining therapeutic concentrations of drugs at the target site. Hydrogels can be engineered to respond to external stimuli, such as pH, temperature, light, and electric fields, enabling triggered drug release and on-demand modulation of mechanical properties. These stimuli-responsive hydrogels, also known as smart hydrogels, hold

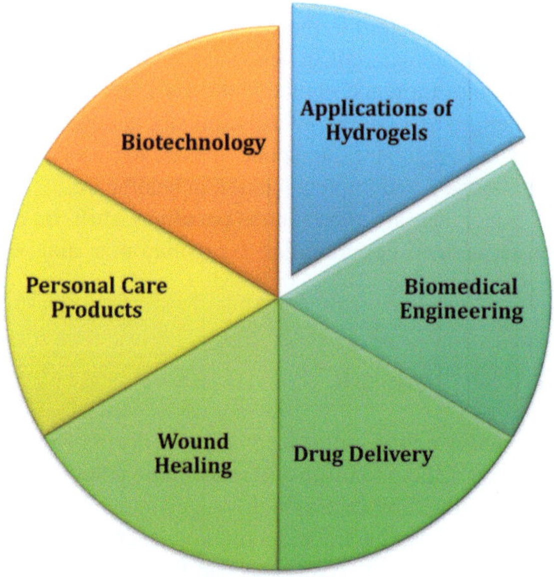


**Fig. 12.2** Various applications of hydrogels

promise for applications such as triggered drug delivery, biosensing, and tissue engineering, where precise control over material properties and drug release kinetics is desired. Despite their numerous advantages, hydrogels also face challenges such as limited mechanical strength, poor long-term stability, and potential immunogenicity [17]. Researchers are actively investigating strategies to overcome these challenges by developing reinforced hydrogels, incorporating bioactive molecules or growth factors, and optimizing fabrication techniques to improve the mechanical properties and stability of hydrogel-based constructs.

Hydrogels are versatile biomaterials with a wide range of applications in tissue engineering, drug delivery, and biomedical research. Their unique properties, including high-water content, tunable mechanical properties, and biocompatibility, make them attractive candidates for addressing complex medical challenges and improving patient outcomes.

#### 12.1.3.3 3D Printing

3D printing, also known as additive manufacturing, has revolutionized the field of biomedical engineering by enabling the fabrication of complex three-dimensional structures with unprecedented precision and customization. This transformative technology allows researchers and clinicians to create patient-specific implants, tissue scaffolds, and medical devices tailored to individual anatomical characteristics and therapeutic needs (Fig. 12.3). 3D printing works by depositing successive layers of material, such as polymers, ceramics, or metals, according to computer-aided design (CAD) models, allowing for the precise control of shape, size, and internal architecture. One of the key advantages of 3D printing in biomedical applications is its ability to produce patient-specific implants and prosthetics that match the unique anatomical features of individual patients. By utilizing medical imaging data, such as computed tomography (CT) or magnetic resonance imaging (MRI) scans, clinicians can design and fabricate implants that fit precisely to the patient's anatomy, reducing the risk of implant rejection, discomfort, and complications. This personalized approach improves patient outcomes and enhances the functionality and longevity of medical implants. In addition to patient-specific customization, 3D printing enables the rapid prototyping and iterative design of biomedical devices, accelerating the development process and reducing time-to-market. Researchers can quickly iterate through design variations, test prototypes in vitro and in vivo, and refine designs based on feedback, leading to the development of more effective and innovative medical devices. This iterative design process fosters innovation and collaboration across disciplines [18], driving advancements in biomedical engineering and healthcare delivery.

3D printing offers unparalleled versatility in fabricating complex tissue scaffolds for regenerative medicine and tissue engineering applications. By precisely controlling the spatial distribution of biomaterials and incorporating bioactive molecules or cells, researchers can engineer scaffolds that mimic the structure and function of native tissues, promoting cell adhesion, proliferation, and differentiation. These bio-fabricated scaffolds hold promise for repairing or replacing damaged tissues

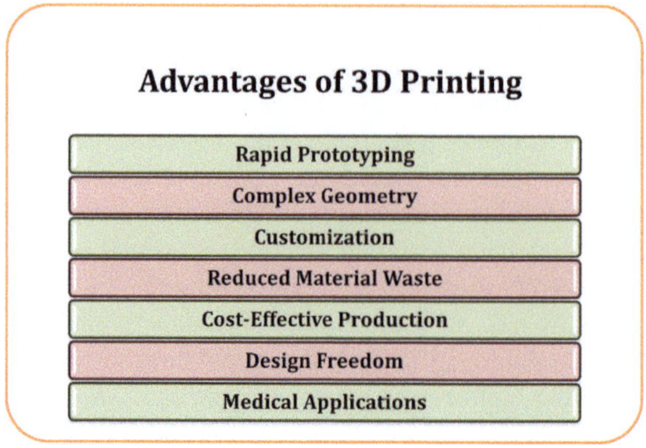


**Fig. 12.3** Advantages of 3D printing

and organs, offering potential solutions for conditions such as bone defects, cartilage injuries, and organ failure. 3D printing enables the fabrication of multi-material constructs with heterogeneous properties, allowing for the integration of functional features such as drug delivery systems, sensors, and microfluidic channels into biomedical devices. This capability opens new opportunities for developing smart implants and devices that can monitor physiological parameters, deliver therapeutics on-demand, and facilitate personalized medicine approaches [16, 17]. By integrating sensing and actuation capabilities into 3D-printed devices, researchers can create closed-loop systems that respond dynamically to changes in the patient's condition, improving treatment outcomes and patient care. Despite its numerous advantages, challenges such as material selection, process optimization, and regulatory approval remain obstacles to the widespread adoption of 3D printing in biomedical applications. Advancements in imaging technologies, computational modeling, and bio fabrication techniques are driving further innovation in the field, unlocking new possibilities for 3D printing in personalized medicine, regenerative therapies, and point-of-care diagnostics.

3D printing holds immense promise for revolutionizing biomedical engineering and healthcare delivery by enabling the fabrication of patient-specific implants, tissue scaffolds, and medical devices with unprecedented precision and customization. As technology continues to evolve and interdisciplinary collaborations flourish, the impact of 3D printing on personalized medicine, regenerative therapies, and medical device innovation is poised to grow, ushering in a new era of patient-centric healthcare and improved quality of life.

#### 12.1.3.4 Smart Biomaterials

Smart biomaterials represent a class of advanced materials that exhibit dynamic, responsive behavior in response to external stimuli, enabling precise control over their properties and functions (Fig. 12.4). These materials integrate sensing, actuation, and feedback mechanisms to adapt to changes in their environment, making them highly versatile for a wide range of biomedical applications. Smart biomaterials can respond to various stimuli, including pH, temperature, light, electric fields, and biological signals, allowing for on-demand modulation of material properties and functionalities. One of the key advantages of smart biomaterials is their ability to mimic the dynamic and adaptive nature of biological systems, enabling biomaterials to interact intelligently with living tissues and cells. By incorporating responsive elements such as stimuli-responsive polymers, nanomaterials, or biological molecules into biomaterial matrices, researchers can design materials that actively respond to physiological cues and regulate cellular activities [18]. This biomimetic approach holds promise for applications such as drug delivery, tissue engineering, and regenerative medicine, where precise control over material-cell interactions is essential for therapeutic efficacy. In drug delivery applications, smart biomaterials can be engineered to release therapeutic agents in response to specific stimuli present at the target site, such as changes in pH or enzyme activity. Stimuli-responsive drug delivery systems offer advantages such as site-specific drug delivery, controlled release kinetics, and reduced systemic toxicity, improving therapeutic outcomes and patient compliance.

Smart biomaterials can be utilized for biosensing and diagnostic applications, where they can detect and respond to biomarkers or disease-related signals in real-time. Biosensing platforms based on smart biomaterials offer advantages such as high sensitivity, specificity, and rapid response times, enabling early detection of diseases and monitoring of physiological parameters. In tissue engineering and regenerative medicine, smart biomaterials can play a crucial role in promoting tissue repair and regeneration by modulating cellular behavior and tissue microenvironments. Responsive scaffolds can dynamically adjust their properties in response to changes in the local environment, facilitating cell adhesion, proliferation, and differentiation. Smart biomaterials can mimic the spatiotemporal presentation of biochemical cues

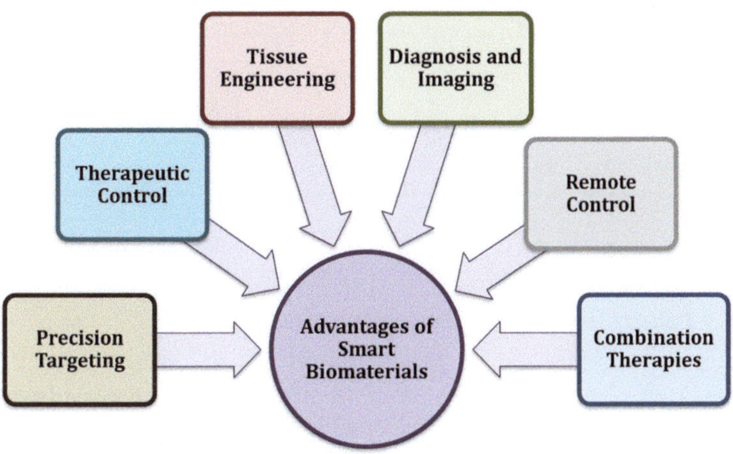


**Fig. 12.4** Advantages of smart biomaterials

in the native tissue microenvironment, guiding tissue morphogenesis and promoting functional tissue regeneration.

Despite their tremendous potential, challenges such as biocompatibility, stability, and scalability remain obstacles to the widespread adoption of smart biomaterials in clinical applications. Researchers are actively addressing these challenges by developing biocompatible responsive materials, optimizing fabrication techniques, and conducting rigorous preclinical testing to evaluate safety and efficacy [19]. Interdisciplinary collaborations between materials scientists, biologists, engineers, and clinicians are essential for advancing the field and translating smart biomaterials from the laboratory to the clinic. Smart biomaterials represent a paradigm shift in biomaterials research, offering unprecedented opportunities to develop dynamic, responsive materials with tailored properties and functionalities for biomedical applications.

**Exercise 2**

**Fill ups**

**(3D bioprinting, Healthcare, Integration, Biomaterials, Tissue engineering)**

**1.** A promising emerging biomaterials technology is ..........................

**2.** Novel approaches in nanotechnology are shaping the future of.......................

**3.** The ........................of smart materials is revolutionizing biomaterials applications.

**4.** Advancements in ......................hold great potential for next-generation biomaterials.

**5.** Emerging biomaterials technologies are driving innovation in medical and ...........................fields.

### *12.1.4 Future Directions and Opportunities*

In the realm of biomaterials and biomedical engineering, the horizon brims with promising avenues and transformative opportunities that hold the potential to redefine healthcare paradigms. As technology continues to advance at a rapid pace, the future of biomaterials unfolds with boundless possibilities, offering innovative solutions to address pressing medical challenges and improve patient outcomes.

Emerging trends such as personalized medicine, regenerative therapies, and point-of-care diagnostics pave the way for the development of biomaterials with enhanced functionality, biocompatibility, and therapeutic efficacy. Moreover, interdisciplinary collaborations between materials scientists, biologists, clinicians, and engineers are poised to catalyze innovation and accelerate the translation of bench side discoveries into clinical applications. As we embark on this journey towards the future, the landscape of biomaterials stands as a beacon of hope, offering new horizons for transforming healthcare delivery and advancing human health and well-being.

#### 12.1.4.1 Personalized Medicine

Personalized medicine, also known as precision medicine, represents a paradigm shift in healthcare that aims to tailor medical treatments and interventions to the individual characteristics of each patient. This approach recognizes that individuals vary in their genetic makeup, lifestyle factors, environmental exposures, and disease susceptibility, leading to differences in treatment response and outcomes. By leveraging advances in genomics, molecular diagnostics, and data analytics, personalized medicine seeks to identify biomarkers and molecular signatures that can predict disease risk, prognosis, and therapeutic response, enabling clinicians to make informed decisions and optimize treatment strategies for each patient. One of the key principles of personalized medicine is the integration of patient-specific data, such as genetic information, biomarker profiles, and clinical parameters, into the decision-making process [20]. By analyzing genetic variants, gene expression patterns, and other molecular markers, clinicians can identify subpopulations of patients who are likely to respond favorably to specific treatments or are at increased risk of adverse drug reactions. This precision-guided approach enables targeted therapies that are tailored to the unique genetic and biological characteristics of each patient, maximizing therapeutic efficacy while minimizing side effects. In oncology, personalized medicine has revolutionized cancer treatment by enabling the identification of molecular alterations driving tumor growth and metastasis. Through genomic profiling and molecular diagnostics, oncologists can identify actionable mutations, such as EGFR mutations in lung cancer or BRAF mutations in melanoma and prescribe targeted therapies that specifically inhibit the molecular pathways driving tumor progression. This precision oncology approach has led to significant improvements in treatment outcomes, with higher response rates, prolonged progression-free survival, and improved quality of life for cancer patients. Beyond cancer, personalized medicine holds promise for a wide range of medical conditions, including cardiovascular disease, neurological disorders, autoimmune diseases, and rare genetic disorders. By identifying genetic risk factors, disease biomarkers, and therapeutic targets, personalized medicine approaches can inform preventive strategies, early detection methods, and targeted interventions tailored to the individual needs of patients. For example, in cardiovascular medicine, genetic testing can identify individuals at increased risk of inherited cardiac conditions, such as hypertrophic cardiomyopathy or familial hypercholesterolemia, enabling early intervention and personalized management plans to prevent adverse events.

Personalized medicine extends beyond the realm of therapeutics to include personalized diagnostics, prognostics, and monitoring tools that enable real-time assessment of disease progression and treatment response. Advances in wearable sensors, mobile health technologies, and digital health platforms allow patients to actively participate in their healthcare by monitoring physiological parameters, tracking medication adherence, and communicating with healthcare providers remotely [21]. These personalized health monitoring solutions empower patients to take proactive steps to manage their health and make informed decisions in collaboration with their healthcare team. Despite its tremendous potential, personalized medicine faces challenges such as data privacy, regulatory oversight, and healthcare disparities that must be addressed to realize its full impact on healthcare delivery. The integration of personalized medicine into routine clinical practice requires investments in infrastructure, workforce training, and patient education to ensure equitable access and adoption across diverse populations. Nevertheless, with continued advancements in genomic technologies, computational biology, and translational research, personalized medicine holds the promise of transforming healthcare delivery from a one-size-fits-all approach to a tailored, patient-centric model that delivers better outcomes and improved quality of life for individuals worldwide.

#### 12.1.4.2 Targeted Drug Delivery

Targeted drug delivery represents a sophisticated approach to drug administration that aims to deliver therapeutic agents specifically to the site of action within the body while minimizing systemic exposure and off-target effects. This precision-guided approach holds promise for enhancing the therapeutic efficacy of drugs, reducing side effects, and improving patient outcomes across a wide range of medical conditions. By leveraging targeted drug delivery systems, clinicians can achieve higher drug concentrations at the desired site of action, thereby maximizing therapeutic benefit while minimizing toxicity to healthy tissues. Key strategies in targeted drug delivery involves the use of drug carriers or delivery vehicles that can transport therapeutic agents to the target site. These carriers, which can be nanoparticles, liposomes, micelles, or polymer-based constructs, are designed to encapsulate drugs and protect them from degradation, metabolism, and premature clearance in the body [22]. By functionalizing these carriers with targeting ligands, such as antibodies, peptides, or aptamers, researchers can enhance their specificity for diseased tissues or cells, enabling precise delivery of therapeutic agents to the intended site of action. Nanoparticle-based drug delivery systems, in particular, have attracted significant attention due to their unique properties, such as small size, large surface area, and tunable surface chemistry. These nanoparticles can passively accumulate in tumors through the enhanced permeability and retention (EPR) effect, which exploits the leaky vasculature and impaired lymphatic drainage characteristic of solid tumors.

Additionally, nanoparticles can be functionalized with targeting ligands that recognize overexpressed receptors or antigens on the surface of cancer cells, further enhancing their specificity and accumulation at the tumor site.

Furthermore, targeted drug delivery systems can be engineered to respond to specific stimuli present at the target site, such as changes in pH, temperature, or enzyme activity. Stimuli-responsive drug carriers can release therapeutic agents in a controlled manner in response to these environmental cues, thereby improving drug bioavailability and reducing systemic toxicity [23]. For example, pH-sensitive nanoparticles can release drugs in response to the acidic tumor microenvironment, while thermo-responsive hydrogels can release drugs in response to local hyperthermia induced by external sources such as laser or ultrasound. In addition to cancer therapy, targeted drug delivery holds promise for treating a wide range of diseases and conditions, including inflammatory disorders, infectious diseases, neurological disorders, and cardiovascular diseases. By delivering drugs specifically to inflamed tissues, infected cells, or diseased organs, targeted drug delivery systems can achieve therapeutic concentrations while minimizing exposure to healthy tissues, reducing the risk of adverse effects, and improving patient compliance.

Despite its tremendous potential, targeted drug delivery faces challenges such as scalability, reproducibility, and regulatory approval that must be addressed to facilitate clinical translation and widespread adoption. The complexity of disease biology and the heterogeneity of patient populations present obstacles that require innovative solutions and personalized approaches to achieve optimal therapeutic outcomes.

#### 12.1.4.3 Biomimicry and Bioinspiration

Biomimicry and bioinspiration have emerged as powerful strategies in the field of biomaterials, offering innovative approaches to design, fabrication, and functionalization that draw inspiration from nature's own materials and structures. By studying the remarkable diversity and sophistication of biological materials found in organisms ranging from plants and animals to microorganisms and marine organisms, researchers seek to emulate their unique properties and functionalities to create biomaterials with tailored properties and enhanced performance. These biomimetic and bioinspired biomaterials hold promise for a wide range of applications in biomedicine, biotechnology, and beyond [24]. One area where biomimicry and bioinspiration have made significant strides is in the development of structural biomaterials with superior mechanical properties and structural integrity. By mimicking the hierarchical organization and composition of natural materials such as bone, collagen, and nacre, researchers have been able to design biomimetic composites that exhibit exceptional strength, toughness, and flexibility. These biomimetic materials hold promise for applications such as bone grafts, dental implants, and load-bearing implants, where mechanical stability and biocompatibility are critical requirements.

Biomimetic and bioinspired biomaterials can also replicate the bioactive properties of natural tissues and extracellular matrices, promoting cell adhesion, proliferation, and tissue regeneration. By incorporating bioactive molecules, such as growth factors, peptides, or extracellular matrix proteins, into biomaterial matrices, researchers can create scaffolds that mimic the biochemical cues present in native tissues, guiding cell behavior and promoting tissue regeneration. These bioinspired scaffolds hold promise for applications such as wound healing, tissue engineering, and organ transplantation, where promoting tissue regeneration and repair is paramount. In addition to structural and bioactive properties, biomimetic and bioinspired biomaterials can also replicate the dynamic and responsive behavior of biological systems. By incorporating stimuli-responsive polymers, nanomaterials, or molecular switches into biomaterial matrices, researchers can create smart biomaterials that respond to changes in their environment, such as pH, temperature, or mechanical stress. These smart biomaterials can be programmed to release drugs, deliver therapeutic agents, or modulate their mechanical properties in response to specific cues, enabling precise control over drug delivery, tissue regeneration, and other biomedical applications. Furthermore, bioimetic and bioinspired biomaterials can also replicate the self-healing and self-repair mechanisms observed in biological systems, enabling materials with enhanced durability and longevity. By incorporating self-healing polymers, microcapsules filled with healing agents, or dynamic covalent bonds into biomaterial matrices, researchers can create materials that autonomously repair damage and restore functionality when subjected to mechanical or chemical stress [22, 24]. These self-healing biomaterials hold promise for applications such as medical implants, prosthetics, and wearable devices, where durability and reliability are critical requirements.

Despite their potential, biomimetic and bioinspired biomaterials also present challenges, such as scalability, reproducibility, and regulatory approval, that must be addressed to facilitate clinical translation and widespread adoption. Additionally, the complexity of biological systems and the need for interdisciplinary collaboration require innovative engineering solutions and rigorous testing to ensure safety, reliability, and performance.

#### 12.1.4.4 Bio Fabrication Techniques

Fabrication techniques play a crucial role in the development of biomaterials, providing the means to create structures with tailored properties and functionalities for various biomedical applications. These techniques encompass a diverse array of methods, ranging from conventional manufacturing processes to advanced nanofabrication technologies, each offering unique advantages and capabilities. One of the most widely used fabrication techniques in biomaterials is solvent casting and particulate leaching, which involves dissolving a polymer in a solvent, mixing it with particles or porogens, casting the mixture into a mold, and then leaching out the porogens to create a porous structure. This technique is commonly used to fabricate porous scaffolds for tissue engineering and regenerative medicine, where the interconnected pore structure provides a supportive environment for cell growth and tissue ingrowth. Another commonly employed fabrication technique is electrospinning [25], which utilizes an electric field to draw polymer solutions or melts into ultrafine fibers that are collected on a grounded substrate. Electrospun fibers can be engineered to mimic the structure and mechanical properties of native extracellular matrix, making them suitable for applications such as wound dressings, drug delivery systems, and tissue scaffolds. The high surface area-to-volume ratio of electrospun fibers enables efficient drug loading and release, while their porous structure facilitates cell infiltration and nutrient transport. Additive manufacturing, also known as 3D printing, has emerged as a versatile fabrication technique for biomaterials, enabling precise control over the geometry, architecture, and composition of complex structures. 3D printing allows researchers to create patient-specific implants, tissue scaffolds, and medical devices with unprecedented customization and resolution. By layering biomaterials or bioinks according to computer-aided designs, 3D printing can create intricate structures with tailored mechanical, biological, and chemical properties, making it ideal for personalized medicine and tissue engineering applications.

In addition to these techniques, microfluidics has gained prominence as a fabrication method for biomaterials, particularly in the development of lab-on-a-chip devices and microscale tissue models. Microfluidic devices utilize tiny channels and chambers to manipulate fluids and particles at the microscale, enabling precise control over biochemical and biophysical cues. These devices can be used for applications such as cell culture, drug screening, and disease modeling, offering advantages such as high throughput, low sample volume, and precise spatial and temporal control [26]. Self-assembly techniques, such as molecular self-assembly and colloidal self-assembly, have emerged as powerful methods for fabricating biomaterials with hierarchical structures and complex functionalities. These techniques rely on the spontaneous organization of molecules or particles into ordered structures driven by non-covalent interactions, such as hydrogen bonding, van der Waals forces, and hydrophobic interactions. Self-assembled biomaterials can exhibit unique properties such as programmable assembly, dynamic responsiveness, and stimuli-responsiveness, making them promising candidates for applications such as drug delivery, biosensing, and tissue engineering. Despite their diverse capabilities, each fabrication technique has its strengths and limitations, and the choice of technique depends on factors such as desired material properties, geometric complexity, scalability, and cost-effectiveness.

### 12.1.5 Regulatory Considerations and Ethical Implications

- **Regulatory Approval**: Compliance with regulatory requirements is essential for the development and commercialization of biomaterials-based products. Researchers must navigate complex regulatory frameworks, such as those established by the Food and Drug Administration (FDA) in the United States or the European Medicines Agency (EMA) in Europe, to ensure that biomaterials meet safety, efficacy, and quality standards before they can be introduced into clinical practice.
- **Biocompatibility Testing**: Biomaterials must undergo rigorous biocompatibility testing to assess their compatibility with biological systems and ensure they do not elicit adverse reactions or immune responses. Testing may include in vitro assays, animal studies, and clinical trials to evaluate cytotoxicity, immunogenicity, and long-term biocompatibility.
- **Material Characterization**: Accurate characterization of biomaterials' physical, chemical, and mechanical properties is critical for assessing their safety, performance, and stability.
- **Ethical Considerations**: Ethical considerations play a significant role in the development and use of biomaterials, particularly in research involving human subjects or animals.
- **Intellectual Property Protection**: Intellectual property protection is essential for incentivizing innovation and investment in biomaterials research and development.
- **Environmental Impact**: The environmental impact of biomaterials production, use, and disposal is an important consideration, particularly as the field moves towards sustainable and eco-friendly materials.
- **Social Equity**: Access to biomaterials-based therapies and technologies must be equitable and inclusive to ensure that all individuals, regardless of socioeconomic status or geographical location, can benefit from advances in healthcare.
- **Public Perception and Acceptance**: Public perception and acceptance of biomaterials-based products can influence their adoption and use in clinical practice.

**Exercise 3**

**Multiple Choice Questions**

1. Which of the following is a current challenge in the therapeutic application of biomaterials?
   a) Biocompatibility assessment
   b) Regulatory compliance
   c) Scalability
   d) All of the above

2. What is a key hurdle in the widespread adoption of biomaterials in clinical settings?
   a) Degradation rate
   b) Mechanical properties
   c) Regulatory barriers
   d) Manufacturing costs

3. Which emerging technology holds great potential for next-generation biomaterials?
   a) 3D bioprinting
   b) Conventional molding
   c) Casting techniques
   d) None of the above

4. What area of biomaterials is currently driving innovation in medical and healthcare fields?
   a) Smart materials
   b) Tissue engineering
   c) Nanotechnology
   d) All of the above

5. Achieving optimal _________ is a key concern in the utilization of biomaterials.
   a) Chemical composition
   b) Mechanical properties
   c) Degradation rate
   d) All of the above

6. What aspect of biomaterials is critical for their successful integration into the body?
   a) Biocompatibility
   b) Mechanical strength
   c) Surface roughness
   d) Color

7. Which regulatory aspect must biomaterials meet for safe utilization in clinical applications?
   a) ISO 9001 certification
   b) CE marking
   c) FDA approval
   d) None of the above

8. What technology allows for the controlled release of drugs from biomaterials?
   a) Nanotechnology
   b) 3D printing
   c) Smart materials
   d) Conventional molding

9. Overcoming __________ barriers is essential for advancing biomaterials in clinical settings.
   a) Technical
   b) Manufacturing
   c) Regulatory
   d) All of the above

10. What emerging field is contributing to the development of novel biomaterials with enhanced properties?
   a) Biocompatible polymers
   b) Bioinformatics
   c) Regenerative medicine
   d) None of the above

## 12.2 Summary

This chapter delves into the dynamic landscape of biomaterials in therapeutic applications, exploring the challenges encountered and future directions that have shaped its evolution. There are multifaceted challenges faced by researchers and practitioners in harnessing biomaterials for therapeutic purposes, ranging from biocompatibility to functional integration. These challenges have spurred innovative solutions and propelled the field forward. Achieving optimal biocompatibility stands as a paramount challenge in biomaterials therapeutics, necessitating materials that seamlessly interact with biological systems while minimizing adverse reactions. Overcoming this challenge has required a nuanced understanding of material properties and their impact on biological responses, leading to advancements in surface modifications, biofunctionalization, and biomimetic design strategies. Controlling the degradation kinetics of biomaterials poses another significant hurdle, particularly in the realm of drug delivery and tissue regeneration. Striking a balance between degradation rates to meet therapeutic needs while avoiding premature degradation or persistence has driven the development of stimuli-responsive materials, nanotechnology-enabled platforms, and innovative biomimetic designs. Navigating the immune response elicited by implanted biomaterials represents a critical challenge, as immune reactions can significantly influence device functionality and therapeutic efficacy. Strategies to modulate the immune response, including surface engineering and the incorporation of immunomodulatory agents, have advanced alongside our understanding of immune-biomaterial interactions. This chapter provides insights into the challenges and future directions in biomaterials for therapeutic applications, illustrating how the field has evolved through innovation and collaboration. By fostering interdisciplinary collaboration, embracing ethical principles, and embracing technological innovation, researchers can harness the power of biomaterials to pave the way for a brighter future in healthcare and beyond. Ultimately, the journey towards realizing the full potential of biomaterials in therapeutic applications is marked by collaboration, innovation, and a steadfast commitment to improving human health and well-being.

## Points to Remember

- Future biomaterials will likely incorporate bioactive properties to promote tissue integration and regeneration, reducing the risk of rejection and improving long-term implant success rates.
- Advances in 3D printing technology will enable the precise fabrication of complex biomaterial structures, allowing for patient-specific implants and scaffolds tailored to individual needs.
- Biomaterials equipped with sensing and responsive capabilities will enable real-time monitoring of physiological parameters and controlled drug release, enhancing therapeutic outcomes and patient care.
- Integration of nanotechnology into biomaterials will enable precise control over material properties at the nanoscale, leading to enhanced mechanical strength, bioactivity, and drug delivery capabilities.
- Future biomaterials may combine biological components with synthetic materials to create biohybrid constructs with enhanced functionality and compatibility for various biomedical applications.
- Bioresorbable biomaterials will play a significant role in temporary implants and drug delivery systems, gradually degrading and being replaced by native tissue as healing progresses, thereby eliminating the need for surgical removal.
- Biomaterials will be tailored to individual patient characteristics, including genetic makeup, lifestyle factors, and disease state, enabling personalized treatment approaches and optimizing therapeutic outcomes.
- Advances in biomaterials and tissue engineering will pave the way for the development of functional artificial organs and tissues, offering new solutions for organ transplantation and regenerative medicine.
- Biomaterials inspired by natural structures and processes will continue to be developed, mimicking the hierarchical organization and properties of biological tissues for enhanced biocompatibility and functionality.

## References


1. Kantak MN, Bharate SS (2022) Analysis of clinical trials on biomaterial and therapeutic applications of chitosan: a review. Carbohyd Polym 278:128999. https://doi.org/10.1016/j.carbpol.2021.128999
2. Han X, Alu A, Liu H, Shi Y, Wei X, Cai L, Wei Y (2022) Biomaterial-assisted biotherapy: a brief review of biomaterials used in drug delivery, vaccine development, gene therapy, and stem cell therapy. Bioact Mater 17:29–48. https://doi.org/10.1016/j.bioactmat.2022.01.012
3. Troy E, Tilbury MA, Power AM, Wall JG (2021) Nature-based biomaterials and their application in biomedicine. Polymers 13(19):3321. https://doi.org/10.3390/polym13193321
4. Karp JM, Langer R (2007) Development and therapeutic applications of advanced biomaterials. Curr Opin Biotechnol 18(5):454–459. https://doi.org/10.1016/j.copbio.2007.09.008
5. Trucillo P (2024) Biomaterials for drug delivery and human applications. Materials (Basel, Switzerland) 17(2):456. https://doi.org/10.3390/ma17020456
6. Chen X, Wu G, Feng Z, Dong Y, Zhou W, Li B, Bai S, Zhao Y (2016) Advanced biomaterials and their potential applications in the treatment of periodontal disease. Crit Rev Biotechnol 36(4):760–775. https://doi.org/10.3109/07388551.2015.1035693
7. Suhag D, Bhatia R, Das S, Shakeel A, Ghosh AB, Singh A et al (2015) Physically cross-linked pH-responsive hydrogels with tunable formulations for controlled drug delivery. RSC Adv 5(66):53963–53972. https://doi.org/10.1039/c5ra07424j
8. Kaur N, Sharma AK, Shakeel A, Kumar V, Singh A, Gupta A et al (2017) Therapeutic implications of superoxide dismutase and its importance in kinase drug discovery. Curr Top Med Chem 17(22). https://doi.org/10.2174/1568026617666170307122837
9. Cecen B, Hassan S, Li X, Zhang YS (2024) Smart biomaterials in biomedical applications: current advances and possible future directions. Macromol Biosci 24(3):e2200550. https://doi.org/10.1002/mabi.202200550
10. Uludağ H (2014) Grand challenges in biomaterials. Front Bioeng Biotechnol 2:43. https://doi.org/10.3389/fbioe.2014.00043
11. Williams DF (2019) Challenges with the development of biomaterials for sustainable tissue engineering. Front Bioeng Biotechnol 7:127. https://doi.org/10.3389/fbioe.2019.00127
12. Suhag D, Chauhan M, Shakeel A, Das S (2020) Emerging trends in nanotheranostics. In: Saxena S, Khurana S (eds) NanoBioMedicine. Springer, Singapore. https://doi.org/10.1007/978-981-32-9898-9_14
13. Suhag D, Thakur P, Thakur A (2023) Introduction to nanotechnology. In: Suhag D, Thakur A, Thakur P (eds) Integrated nanomaterials and their applications. Springer, Singapore. https://doi.org/10.1007/978-981-99-6105-4_1
14. Prasad K, Bazaka O, Chua M, Rochford M, Fedrick L, Spoor J, Symes R, Tieppo M, Collins C, Cao A, Markwell D, Ostrikov KK, Bazaka K (2017) Metallic biomaterials: current challenges and opportunities. Materials (Basel, Switzerland) 10(8):884. https://doi.org/10.3390/ma10080884
15. Zakhary KE, Thakker JS (2017) Emerging biomaterials in trauma. Oral Maxillofac Surg Clin North Am 29(1):51–62. https://doi.org/10.1016/j.coms.2016.08.010
16. Wang Y, Chen G, Zhang H, Zhao C, Sun L, Zhao Y (2021) Emerging functional biomaterials as medical patches. ACS Nano 15(4):5977–6007. https://doi.org/10.1021/acsnano.0c10724
17. Victor SP, Selvam S, Sharma CP (2019) Recent advances in biomaterials science and engineering research in India: a minireview. ACS Biomater Sci Eng 5(1):3–18. https://doi.org/10.1021/acsbiomaterials.8b00233
18. Bian N, Chu C, Rung S, Huangphattarakul V, Man Y, Lin J, Hu C (2023) Immunomodulatory biomaterials and emerging analytical techniques for probing the immune micro-environment. Tissue eng Regen Med 20(1):12–24. https://doi.org/10.1007/s13770-022-00491-z
19. Kumari S, Chatterjee K (2021) Biomaterials-based formulations and surfaces to combat viral infectious diseases. APL Bioeng 5(1):012503. https://doi.org/10.1063/5.0029486

20. Chouhan D, Dey N, Bhardwaj N, Mandal BB (2019) Emerging and innovative approaches for wound healing and skin regeneration: current status and advances. Biomaterials 216:129267. https://doi.org/10.1016/j.biomaterials.2019.129267
21. Ashique S, Sandhu NK, Chawla V, Chawla PA (2021) Targeted drug delivery: trends and perspectives. Curr Drug Deliv 18(10):1435–1455. https://doi.org/10.2174/1567201818666210609161301
22. Jain KK (2020) An overview of drug delivery systems. Methods Mol Biol (Clifton, N.J.) 2059:1–54. https://doi.org/10.1007/978-1-4939-9798-5_1
23. Rathore R, Suhag D, Wan F, Thakur A, Thakur P (2023) Toxicology and toxicity studies of nano-biomaterials. In: Suhag D, Thakur A, Thakur P (eds) Integrated nanomaterials and their applications. Springer, Singapore. https://doi.org/10.1007/978-981-99-6105-4_20
24. Rathore R, Suhag D, Wan F, Verma R, Thakur A, Thakur P (2023) Functionalization of biomaterials. In: Suhag D, Thakur A, Thakur P (eds) Integrated nanomaterials and their applications. Springer, Singapore. https://doi.org/10.1007/978-981-99-6105-4_12
25. Iqbal J, Anwar F, Afridi S (2017) Targeted drug delivery systems and their therapeutic applications in cancer and immune pathological conditions. Infect Disord Drug Targets 17(3):149–159. https://doi.org/10.2174/1871526517666170606102623
26. Bahrami B, Hojjat-Farsangi M, Mohammadi H, Anvari E, Ghalamfarsa G, Yousefi M, Jadidi-Niaragh F (2017) Nanoparticles and targeted drug delivery in cancer therapy. Immunol Lett 190:64–83. https://doi.org/10.1016/j.imlet.2017.07.015

# Intext Exercises Answers

## Chapter 1

### Exercise 1

1. Healing
2. Bedside
3. Material Science
4. Human Body.

### Exercise 2

1. Organic Sources
2. Cellulose
3. Biocompatibility
4. Tissue Regeneration
5. Drug Delivery.

### Exercise 3

1. (d) Derived from natural sources
2. (c) Hybrid biomaterials
3. (c) Polyethylene glycol (PEG)
4. (c) Enhanced mechanical properties
5. (d) Cross-linking
6. (c) Biocompatibility issues
7. (a) Hydroxyapatite
8. (a) Porosity
9. (d) Optical fiber manufacturing
10. (d) Better control over properties.



D. Suhag et al., *Handbook of Biomaterials for Medical Applications, Volume 1*, Biomedical Materials for Multi-functional Applications,
https://doi.org/10.1007/978-981-97-4818-1

## Chapter 2

### Exercise 1

1. Strength, Elastic modulus, Toughness
2. Stress
3. Biocompatibility
4. Cellular response
5. Structural.

### Exercise 2

1. Composition
2. Origin
3. Functionality
4. Degradation
5. Applications.

### Exercise 3

1. (c) Polysaccharides
2. (a) Synthetic and natural
3. (b) Mimic natural tissues and organs
4. (b) Are slowly broken down and absorbed by the body
5. (c) Repair and replacement of bones and joints
6. (c) Scanning electron microscopy (SEM)
7. (b) Chemical composition
8. (c) Surface energy
9. (c) Toughness
10. (d) Bioactivity.

## Chapter 3

### Exercise 1

1. Biodegradation
2. Degradation
3. Bulk
4. Rate
5. Release.

### Exercise 2

1. Tissue engineering
2. Restenosis
3. Cell growth
4. Efficacy
5. Biomedical implants.

### Exercise 3

1. (d) Degradation into non-toxic byproducts
2. (c) Enzymatic hydrolysis
3. (c) Polyglycolic acid
4. (d) All of the above
5. (b) They promote tissue regeneration
6. (b) Implant location
7. (a) Surgical removal
8. (d) All of the above
9. (c) Corrosion
10. (b) Temporary medical interventions.

## Chapter 4

### Exercise 1

1. Drugs
2. Small size
3. Nanoparticles
4. Biological barriers
5. Potential.

### Exercise 2

1. Nano biomaterials
2. Therapeutic
3. Cancer cells
4. Drug penetration
5. Cancer treatments.

### Exercise 3

1. (c) Solvent evaporation
2. (c) High surface area to volume ratio
3. (d) Intravenous
4. (b) Gold nanoparticles
5. (d) Near-infrared fluorescence imaging
6. (c) Passive targeting via enhanced permeability and retention (EPR) effect
7. (a) Liposomes
8. (c) Gold nanoparticles
9. (b) Surface plasmon resonance (SPR)
10. (c) By increasing surface area for biomolecule binding.

## Chapter 5

### Exercise 1

1. Polymerize
2. Polymerization
3. Biocompatibility
4. Topography
5. Interactions.

### Exercise 2

1. Adhesion
2. Functionality
3. Regenerative medicine
4. Efficacy
5. Biomaterials.

### Exercise 3

1. (a) They consist of monomers linked by covalent bonds
2. (b) Polymerization
3. (a) Monomer
4. (b) Bone
5. (c) To enhance therapeutic effects
6. (c) Biomedical implants
7. (c) Scanning electron microscopy (SEM)
8. (c) Energy-dispersive X-ray spectroscopy (EDS)
9. (c) Thermogravimetric analysis (TGA)
10. (c) Tensile testing.

## Chapter 6

### Exercise1

1. 3D structure
2. Scaffold
3. Adhesion
4. Tissue regeneration
5. Extracellular matrix.

### Exercise 2

1. Surface modification
2. Scaffold biocompatibility
3. Functional
4. Surface roughness
5. Tissue engineering.

### Exercise 3

1. (c) Mimicking the natural extracellular matrix
2. (d) Tailoring scaffold architecture to guide tissue regeneration
3. (c) To provide structural support and guide cell growth
4. (d) Biocompatibility and porosity
5. (d) Biodegradability and biocompatibility
6. (a) Uniform pore size
7. (c) Enhancing cell attachment and proliferation
8. (c) Coating with bioactive molecules
9. (c) By providing guidance for cell growth and tissue regeneration
10. (c) Enhancing tissue regeneration outcomes.

## Chapter 7

### Exercise 1

1. 3D bioprinting
2. Biomaterials
3. Electron Beam Melting
4. Binder Jetting
5. Photopolymer material.

### Exercise2

1. Medical devices
2. Clinical translation, Commercialization
3. Adoption
4. Efficacy
5. Bioprinted tissue.

### Exercise 3

1. (b) Creating three-dimensional objects using biological materials
2. (d) Customization and personalization
3. (c) Space exploration
4. (c) Biological materials, such as living cells and biomaterials
5. (a) Selective Laser Sintering (SLS)
6. (c) It designs digital models of the desired object
7. (d) Orthopedics
8. (d) Challenges in vascularization and tissue integration
9. (d) Medical segmentation
10. (c) Continued growth with advancements in technology and expanded applications.

## Chapter 8

### Exercise 1

1. Tissue regeneration
2. Retention
3. Systemic
4. Hydrogels
5. Blood.

### Exercise 2

1. Hydrogels
2. Polymer concentration
3. Drug encapsulation
4. Biomacromolecules
5. Degradation.

### Exercise 3

1. (b) Polymers
2. (c) High water content
3. (c) Rapid drug degradation
4. (d) Through physical erosion
5. (d) All of the above
6. (c) Both hydrophobic and hydrophilic drugs
7. (b) Hydrogels have a porous structure
8. (d) All of the above
9. (c) Reduced side effects
10. (c) Continued research and development for enhanced drug delivery systems.

## Chapter 9

### Exercise 1

1. Biomineralization
2. Calcium carbonate
3. Hydroxypatite
4. Silica ahells
5. Bone graft.

### Exercise 2

1. Bio ceramics
2. Biocompatibility
3. Osseointegration
4. Zirconia
5. Cell adhesion.

### Exercise 3

1. (b) Materials that mimic the structure and function of natural tissues
2. (b) To repair damaged tissues by promoting regeneration
3. (b) Hydroxyapatite
4. (c) To improve the integration of implants with surrounding tissues
5. (a) Hydrogels
6. (c) By mimicking the mechanical properties of natural tissues
7. (c) It promotes direct bonding of the implant with the bone
8. (c) Resistance to corrosion and degradation
9. (c) They promote tissue regeneration and reduce inflammation
10. (d) High toxicity.

## Chapter 10

### Exercise 1

1. Diagnostic, Therapeutic
2. Gold
3. Ligands
4. Monitoring
5. Tailored treatments.

### Exercise 2

1. Therapeutic agents
2. Small size
3. Therapeutic molecules
4. Off-target effects
5. Drug release.

### Exercise 3

1. (c) To combine diagnosis and therapy in a single approach
2. (b) MRI (Magnetic Resonance Imaging)
3. (d) All of the above
4. (c) It allows for real-time monitoring of treatment response.
5. (c) It allows for personalized treatment strategies.
6. (b) Positron Emission Tomography (PET)
7. (b) To improve patient outcomes
8. (d) All of the above
9. (b) Personalized medicine
10. (c) Nanoparticle conjugated with a fluorescent dye and a drug.

## Chapter 11

### Exercise 1

1. Lipids
2. Lipid- Based Nanocarriers
3. Cell Types
4. Pharmacokinetics
5. Hydrophobic.

### Exercise 2

1. Diffusion
2. Concentration Gradient
3. Swelling
4. Release
5. Stimuli-responsive.

### Exercise 3

1. (b) Hydrophobicity
2. (c) Hydrophobic drugs
3. (b) Liposomes
4. (c) Emulsification
5. (c) Enhance targeting specificity
6. (c) Ability to encapsulate hydrophobic drugs
7. (c) Enhancing their solubility
8. (d) Insulin delivery via oral route
9. (b) 100–500 nm
10. (d) Stealth coating (PEGylation).

## Chapter 12

### Exercise 1

1. Biocompatibility
2. Degradation rate
3. Mechanical
4. Biomaterials
5. Scalability.

### Exercise 2

1. 3D bioprinting
2. Biomaterials
3. Integration
4. Tissue engineering
5. Healthcare.

## Exercise 3

1. (d) All of the above
2. (c) Regulatory barriers
3. (a) 3D bioprinting
4. (d) All of the above
5. (d) All of the above
6. (a) Biocompatibility
7. (c) FDA approval
8. (a) Nanotechnology
9. (c) Regulatory
10. (c) Regenerative medicine.

# Glossary

| **Angiogenic** |
|---|
| Formation of new blood vessels from existing ones. This physiological process is essential for various biological functions, including tissue growth and repair, wound healing, and the development of organs |
| **Augmentation** |
| Enhancement or reinforcement of biological functions or structures using synthetic or engineered materials that mimic natural tissues. For example, augmentation using biomimetic materials may involve strengthening bones with implants made of hydroxyapatite to restore or improve skeletal integrity |
| **Amphiphilic molecules** |
| Molecules that have both hydrophilic (water-attracting) and hydrophobic (water-repelling) regions |
| **Bioactive coatings** |
| Specialized layers applied to biomaterial surfaces to enhance their interaction with biological systems. These coatings are designed to promote specific biological responses, such as cell adhesion, proliferation, differentiation, or tissue integration. They are commonly used in medical devices, implants, and tissue engineering scaffolds to improve their biocompatibility and performance |
| **Biomimetic** |
| Design and development of materials and structures that mimic or are inspired by biological systems found in nature. The goal of biomimetic biomaterials is to replicate the functionalities, properties, and structures of natural biological materials for various applications in medicine, engineering, and materials science |
| **Bioresorbable** |
| Materials that have the ability to degrade and be absorbed by the body over time. These materials are designed to gradually break down and be metabolized or eliminated through natural biological processes, without leaving permanent residues or foreign bodies within the body |

(continued)





D. Suhag et al., *Handbook of Biomaterials for Medical Applications, Volume 1*, Biomedical Materials for Multi-functional Applications,
https://doi.org/10.1007/978-981-97-4818-1

(continued)

| |
|---|
| **Bioluminescent** |
| Emission of light produced by living organisms, typically as a result of a biochemical reaction involving luciferase enzymes and their substrates. In theranostics, bioluminescent techniques can be utilized for both diagnostic and therapeutic purposes |
| **Brachytherapy** |
| It could potentially refer to a therapeutic approach that branches out or diversifies based on individual patient characteristics, disease progression, or treatment response |
| **Bioinspiration** |
| Process of drawing inspiration from biological systems, structures, and processes to design and develop new materials or improve existing ones |
| **Biomimicry** |
| Innovation that seeks sustainable solutions to human challenges by emulating nature's time-tested patterns and strategies. It involves studying biological systems, processes, and structures to gain inspiration and insights for designing products, technologies, and systems that are more efficient, resilient, and environmentally friendly |
| **Biomolecule grafting** |
| Attaching or linking biomolecules (such as proteins, peptides, DNA, or carbohydrates) onto a substrate or surface. This technique is commonly used in various fields, including biomedicine, biotechnology, and materials science, to functionalize surfaces for specific applications |
| **Contrast Agents** |
| Substances used to enhance the visibility of specific tissues, organs, or structures in medical imaging techniques such as X-ray, computed tomography (CT), magnetic resonance imaging (MRI), and ultrasound. These agents are introduced into the body to improve the contrast between different tissues or to highlight areas of interest, aiding in the diagnosis and monitoring of diseases and medical conditions |
| **Cell adhesion** |
| Cells attach to and interact with surfaces, other cells, or extracellular matrices. It plays a fundamental role in various biological processes, including tissue formation, wound healing, immune response, embryonic development, and maintenance of tissue integrity. Cell adhesion involves complex molecular interactions mediated by specific cell adhesion molecules (CAMs) and receptors |
| **Chondrogenic** |
| Cells that are involved in or related to the formation and development of cartilage tissue. Cartilage is a specialized connective tissue that provides structural support and flexibility to various parts of the body, including joints, ears, nose, and the trachea |
| **Chemotherapy** |
| Treatment of diseases, particularly cancer, using drugs or medications. In the context of cancer treatment, chemotherapy typically involves the administration of cytotoxic drugs that target rapidly dividing cells, including cancer cells. These drugs work by interfering with the cell's ability to grow and divide, ultimately leading to cell death |
| **Colloidal stability** |
| Ability of colloidal systems, such as nanoparticles, liposomes, or micelles, to maintain their physical and chemical integrity over time when dispersed in a liquid medium |

(continued)

(continued)

| |
|---|
| **Coacervation** |
| Phase separation process in which a solution containing two or more components forms separate liquid phases. This occurs when the solubility of one or more components decreases significantly, leading to the formation of a dense phase (coacervate) and a dilute phase |
| **Click chemistry** |
| Set of highly selective and efficient chemical reactions that are used to join biomolecules or bioactive compounds with specific functionalities to form desired products. These reactions are characterized by their high yields, fast reaction rates, and compatibility with biological systems |
| **Computational modeling** |
| Involves simulating the interactions between biomolecules and materials at the molecular or atomic level. This can include studying how biomolecules (such as proteins, enzymes, or DNA) interact with surfaces of materials, how drugs or therapeutic agents bind to specific targets, or how biomaterials influence cellular responses |
| **3D Printing** |
| Also known as additive manufacturing, refers to the process of creating three-dimensional objects layer by layer using computer-aided design (CAD) models. In biomaterials, 3D printing is utilized to fabricate customized implants, scaffolds, prosthetics, and tissue constructs for various medical and biological applications |
| **Drug-eluting** |
| Ability of a material or device to release therapeutic drugs or substances over time. This controlled drug release is often used in medical applications to deliver medications directly to target tissues or organs, thereby improving treatment efficacy while minimizing systemic side effects |
| **Dental implants** |
| Artificial tooth roots made from biomimetic materials like titanium or hydroxyapatite, which are surgically positioned into the jawbone beneath the gum line. These implants provide a stable base for artificial teeth (crowns, bridges, or dentures) to be mounted onto, mimicking the natural function and appearance of teeth |
| **Dystrophies** |
| A group of disorders characterized by the progressive degeneration or weakening of tissues, especially in muscles or other bodily structures |
| **Electrospinning** |
| Versatile and precise technique used to produce nanofibers from a wide range of materials, including polymers, ceramics, and composites. It involves the application of an electric field to a polymer solution or melt, which leads to the formation of ultrafine fibers with diameters typically ranging from tens to hundreds of nanometers |
| **Emulsification** |
| Process of dispersing one liquid into another immiscible liquid to form an emulsion. This is typically achieved by vigorously mixing the two liquids together along with an emulsifying agent or surfactant, which helps stabilize the resulting mixture |
| **Extracellular Enzymes** |
| Enzymes that are produced by cells but are released outside of the cell into the surrounding environment |

(continued)

(continued)

| Extrusion-based bioprinting |
|---|
| 3D bioprinting technique that uses a syringe-based extrusion system to deposit bioinks containing living cells and biomaterials in a controlled manner. It is one of the most widely used bioprinting methods due to its versatility, ease of use, and ability to print complex structures with high cell densities |
| **Fabrication** |
| Creating or manufacturing materials and devices that are used in medical and biological applications. This process involves the selection of appropriate biomaterials, such as polymers, ceramics, metals, or composites, and the utilization of various techniques to shape, assemble, and modify these materials into desired forms |
| **Fibronectin** |
| High-molecular-weight glycoprotein found in the extracellular matrix (ECM) of connective tissues, including blood vessels, skin, tendons, and bones. It plays a crucial role in cell adhesion, migration, growth, and differentiation during tissue development, wound healing, and various physiological processes. Fibronectin is synthesized and secreted by various cell types, including fibroblasts, endothelial cells, and macrophages |
| **Feasibility** |
| Assessment of whether a proposed method or approach for delivering a drug is practical, achievable, and likely to succeed in achieving its intended goals. This assessment considers factors such as the drug's physicochemical properties, the intended target site within the body, potential barriers to delivery, available technology, safety concerns, regulatory considerations, and economic viability |
| **Fibrosis** |
| Fibrosis refers to the formation of excess fibrous connective tissue in an organ or tissue in response to injury or chronic inflammation. This fibrous tissue can lead to scarring and can impair the normal function of the affected organ or tissue |
| **Gamma irradiation** |
| Sterilize biomaterials by exposing them to gamma rays emitted by a radioactive source, typically cobalt-60 or cesium-137. This radiation effectively kills microorganisms such as bacteria, viruses, and fungi, making the biomaterials safe for use in medical applications |
| **Hydrogels** |
| Type of biomaterial composed of a three-dimensional network of hydrophilic polymers that can absorb and retain a significant amount of water or biological fluids. They are characterized by their high-water content, soft and flexible texture, and biocompatibility, making them useful in various biomedical and tissue engineering applications |
| **Homogeneous** |
| A uniform or consistent composition or nature of a substance or system throughout. This term is often used in various scientific and everyday contexts to describe the uniformity or lack of variability within a given entity |
| **Hydroxyapatite** |
| Mimics the structure and composition of natural bone mineral. It's often used in medical and dental applications, such as bone grafts and dental implants, due to its biocompatibility and ability to integrate with surrounding tissues |

(continued)

(continued)

| Homogeneity |
|---|
| Uniform distribution of a drug or therapeutic agent within a formulation or delivery system. It ensures that each dose administered to the patient contains the same amount of active ingredient, leading to consistent therapeutic outcomes |
| **Hinder** |
| Create difficulties or obstacles that slow down or impede progress, movement, or development. It implies making something more challenging or causing a delay in completing a task |
| **Hyperthermia** |
| Elevated body temperature that exceeds the normal range. In medical terminology, hyperthermia specifically refers to a condition where the body's core temperature rises above the normal physiological level due to external factors or underlying health conditions. It is important to distinguish hyperthermia from fever, which is a regulated increase in body temperature usually in response to infection or illness |
| **Immunogenicity** |
| Ability of a biomaterial to induce an immune response in the body. This immune response can be either beneficial or detrimental, depending on the specific characteristics of the biomaterial and the nature of the immune reaction |
| **Implantology** |
| Branch of dentistry or medicine that focuses on the design, development, and use of implants made from biomaterials to replace missing teeth, support prosthetic devices, or restore biological function |
| **Inkjet-based bioprinting** |
| 3D bioprinting technique that utilizes inkjet technology to deposit bioinks containing living cells and biomaterials in a precise and controlled manner. This method is inspired by traditional inkjet printing used in graphic arts but adapted to print biological materials and cells for tissue engineering, regenerative medicine, and biomedical research applications |
| **Jurisdiction** |
| Athority or scope of control that regulatory bodies, research institutions, healthcare providers, and government agencies have over matters related to biomedical research, healthcare practices, medical ethics, and public health within a defined area or domain. It encompasses legal, ethical, and regulatory aspects that govern activities and decisions in the field of biomedicine |
| **Liposomes** |
| Microscopic vesicles composed of lipid bilayers, resembling cell membranes, that enclose an aqueous core. They are commonly used as delivery systems for various substances, including drugs, nutrients, and cosmetic ingredients |
| **Metalloproteinases** |
| Family of enzymes that play a crucial role in the degradation and remodeling of extracellular matrix (ECM) proteins in tissues. They are zinc-dependent endopeptidases, and they catalyze the cleavage of peptide bonds within protein molecules |
| **Osseointegration** |
| Direct structural and functional connection between an artificial implant, such as a dental implant or orthopedic implant, and the surrounding bone tissue. It is a critical process in biomaterials and implantology because it ensures the stability and long-term success of the implant within the body |

(continued)

(continued)

**Osteoconduction**

Field of biomaterials and bone tissue engineering. It refers to the ability of a biomaterial to provide a physical framework or structure that promotes the ingrowth of new bone tissue into its porous structure. Essentially, osteoconduction facilitates the formation of new bone by providing a scaffold for bone cells to attach, proliferate, and differentiate

**Organoids**

Three-dimensional (3D) cell cultures that mimic the structure and function of organs or tissues in vitro (outside the body). They are typically derived from stem cells, induced pluripotent stem cells (iPSCs), or organ-specific progenitor cells and can self-organize into structures that recapitulate the architecture and cellular diversity of the organ they represent

**Ophthalmology**

Branch of medicine that deals with anatomy, physiology, and diseases of the eye. Ophthalmologists are medical doctors who specialize in diagnosing and treating eye disorders, performing surgeries, prescribing medications, and providing vision care

**Organ-on-a-chip models**

Microfluidic devices that mimic the structure, function, and physiological environment of human organs or tissues in a controlled in vitro system. Organ-on-a-chip technology has emerged as a powerful tool in biomedicine for studying organ physiology, disease mechanisms, drug responses, and personalized medicine

**Porcelain**

Dental material used in restorative dentistry. Porcelain dental materials are ceramic compounds that mimic the appearance of natural teeth. They are often used for dental crowns, bridges, and veneers due to their ability to closely resemble the color, translucency, and texture of natural tooth enamel. Porcelain is biocompatible, meaning it is well-tolerated by the body, and it offers excellent aesthetic results

**Prosthetics**

Branch of medicine and engineering involved in the design, manufacture, and application of artificial body parts or devices to replace or enhance missing or impaired body parts. Prosthetic devices, commonly known as prostheses, can range from simple devices like artificial limbs or braces to more complex implants such as artificial organs or sensory aids

**Pathophysiology**

The study of the abnormal physiological processes that occur within the body, leading to the development of disease or abnormal conditions

**Proteolytic**

Breaking down proteins into smaller peptides or amino acids by enzymes called proteases. These enzymes catalyze the hydrolysis of peptide bonds, leading to the cleavage of proteins into fragments

**PEGylation**

Drug delivery where polyethylene glycol (PEG) molecules are attached to a drug or therapeutic agent. This modification can improve the pharmacokinetic properties of the drug by increasing its solubility, stability, and circulation time in the body

(continued)

(continued)

| Propelled |
|---|
| Systems used to move or deliver drugs within the body. This could involve various methods such as diffusion, convection, active transport, or engineered delivery vehicles like nanoparticles, microparticles, liposomes, or even biological vectors |
| **Pharmacokinetics** |
| Study of how a drug moves through the body. It involves processes such as absorption, distribution, metabolism, and excretion (ADME) of a drug, as well as the factors that influence these processes. Understanding pharmacokinetics helps in determining the optimal dosage regimen for a drug and predicting its effects on the body over time |
| **Photoacoustic** |
| Non-invasive imaging technique that combines the principles of laser-induced photoacoustic effect and ultrasound detection to visualize biological tissues and structures with high resolution and contrast |
| **Perfusable** |
| Ability of a structure or system to allow the passage or flow of fluids, such as blood, fluids used in experimentation, or other liquid substances |
| **Pluripotent stem cells** |
| Ability to differentiate into cells from all three germ layers: ectoderm, mesoderm, and endoderm. These cells are characterized by their capacity for self-renewal and their potential to differentiate into various cell types found in the human body |
| **Preoperative** |
| Period of time before a surgical procedure takes place. It encompasses the activities, assessments, preparations, and interventions that occur leading up to the surgery |
| **Renaissance** |
| The Renaissance was a period in European history that spanned roughly from the fourteenth to the seventeenth century. It marked a significant cultural, intellectual, and artistic revival after the Middle Ages. The term "Renaissance" is derived from the French word meaning "rebirth," and during this time, there was a renewed interest in classical learning, literature, philosophy, art, and science |
| **Radiopacity** |
| Ability of a material to block or attenuate X-rays, making it visible on radiographic images such as X-rays, computed tomography (CT) scans, or fluoroscopy. In the context of biomaterials, radiopacity is an important characteristic for medical devices, implants, and materials used in surgical procedures |
| **Restenosis** |
| Re-narrowing or re-occlusion of a blood vessel or a tubular structure in the body, typically occurring after a medical intervention such as angioplasty or stenting and can impact the long-term effectiveness of treatments and patient outcomes |
| **Radiotracers** |
| Specialized compounds labeled with a radioactive isotope that can be used for both diagnostic imaging and therapy. These radiotracers are designed to selectively target specific molecules or receptors associated with diseases, such as cancer |
| **Scaffolds** |
| Structures or frameworks used to support tissue regeneration or growth in biological systems |

(continued)

(continued)

| |
|---|
| **Spatiotemporal** |
| Precise control and understanding of both space (location) and time (duration) in the administration and distribution of drugs within the body. It involves strategies to target specific tissues or cells at particular times, optimizing therapeutic efficacy while minimizing side effects |
| **Sterilants** |
| Processes used to eliminate or deactivate all forms of microbial life, including bacteria, viruses, fungi, and spores, from the biomaterial. Essential for ensuring that biomaterials intended for medical use are free from harmful pathogens that could cause infection or other adverse effects when implanted or used in the body |
| **Stiffness** |
| Resistance of a material to deformation when a force is applied. It indicates how much a material resists being bent, stretched, or compressed. Stiffness is an important consideration in biomaterials because it can influence the performance, durability, and compatibility of implants, scaffolds, and other biomedical devices within the body |
| **Spatiotemporal** |
| Interplay or interaction between space and time, often used in various fields such as physics, biology, geography, and computer science to describe phenomena that occur in both space and time dimensions simultaneously |
| **Stochastic** |
| Phenomena that involve randomness, chance, or probability rather than deterministic cause-and-effect relationships. Stochastic processes are characterized by variability and uncertainty, and they play a significant role in various aspects of biomedical research, data analysis, modeling, and understanding biological systems |
| **Therapeutics** |
| Branch of medicine concerned with the treatment of diseases and the management of medical conditions. It involves the use of various interventions, such as medications, physical therapies, surgical procedures, and lifestyle modifications, to alleviate symptoms, cure diseases, or improve patients' quality of life |
| **Tread** |
| To proceed with caution or to be cautious and vigilant in one's actions or decisions, especially in a situation that is delicate, risky, or potentially dangerous. It suggests being mindful of potential consequences, taking calculated steps, and avoiding hasty or reckless behavior. The phrase often implies the need to consider various factors and to act prudently to avoid making mistakes or causing harm |
| **Tomography** |
| Imaging techniques that reconstruct cross-sectional images or slices of an object or body using data obtained from multiple angles or perspectives. Tomography is widely used in various fields, including medical imaging, geology, archaeology, and materials science, to visualize internal structures in a non-destructive manner |
| **Thermoresponsive** |
| The ability of a material to undergo reversible changes in its physical or chemical properties in response to changes in temperature |

(continued)

(continued)

| **Thermosensitive** |
|---|
| Material or substance that undergoes a change in its physical properties in response to changes in temperature |
| **Transdermal** |
| Method of administering medication through the skin, typically via a patch or gel. This allows the medication to be absorbed directly into the bloodstream, bypassing the digestive system |
| **Tunable Porosity** |
| Ability to adjust or control the size or density of the pores within a material or substance. This adjustment allows for specific properties or functions to be tailored to meet particular needs or applications. Tunable porosity is often desirable in various fields such as materials science, chemistry, and engineering for applications like filtration, drug delivery, and catalysis |
| **Theranostic** |
| Theranostics combines “therapy” and “diagnostics,” emphasizing the integration of diagnostic and therapeutic capabilities within a single approach or platform. This approach allows for more precise and effective diagnosis, treatment planning, and monitoring of treatment responses |
| **Vascularization** |
| Forming new blood vessels, including capillaries, arteries, and veins, within tissues or engineered constructs. This process is crucial for ensuring adequate blood supply, oxygenation, nutrient delivery, waste removal, and overall tissue function |
| **Viscoelasticity** |
| Material’s combined characteristics of viscosity (resistance to flow) and elasticity (ability to return to its original shape after deformation). This property is exhibited by a wide range of materials, including polymers, biological tissues, and some fluids, and it describes how these materials behave under different types of stress or deformation |
| **$\pi$–$\pi$ stacking** |
| A non-covalent interaction between aromatic rings in molecules where the electron clouds of adjacent aromatic systems overlap, leading to attractive forces between them. In bioactive polymers, $\pi$–$\pi$ stacking interactions can play a significant role in shaping the structure, properties, and functions of the polymers |

# Resources and Further Reading

## Chapter 2: Fundamentals of Biomaterials

### Key Textbooks

1. "Biomaterials Science: An Introduction to Materials in Medicine" by Buddy D. Ratner et al. This textbook is essential for anyone entering the field, providing a comprehensive overview of the fundamentals.
2. "Introduction to Biomaterials: Basic Theory with Engineering Applications" by Joon Park and R.S. Lakes, offering detailed technical insights into material properties and their biological interactions.

### Primary Journals

Journal of Biomedical Materials Research: Covers a broad spectrum of topics in biomaterials science, including material synthesis and character evaluations.

Biomaterials: Focuses on the development and clinical application of new biomaterials.

1. Acta Biomaterialia: Publishes research that emphasizes the structure–property relationships of biomaterials.

Review Article: "Biomaterials: A Forecast for the Future" by L. Hench and J. Jones, Biomaterials, 1999. This review discusses the future of biomaterials from a historical perspective.

Research Article: "Synthesis and Characterization of Hydroxyapatite-based Biomaterials" by Smith et al., Journal of Biomedical Materials Research, 2018. This article details innovative methods in creating bioactive ceramics for bone replacement.



D. Suhag et al., *Handbook of Biomaterials for Medical Applications, Volume 1*, Biomedical Materials for Multi-functional Applications,
https://doi.org/10.1007/978-981-97-4818-1

## Chapter 3: Biodegradable Biomaterials in Biomedicine

### Key Textbooks

1. "Biomaterials Science: An Introduction to Materials in Medicine" by Buddy D. Ratner, Allan S. Hoffman, Frederick J. Schoen, and Jack E. Lemons—Provides a comprehensive overview of biomaterials, including biodegradable polymers and their applications in medicine.
2. "Biodegradable Polymers in Clinical Use and Clinical Development" edited by Abraham J. Domb, Neeraj Kumar, and Aviva Ezra—Focuses on the clinical applications and the development process of biodegradable polymers in various therapeutic areas.
3. "Handbook of Biodegradable Polymeric Materials and Their Applications" by Surya Mallapragada and Balaji Narasimhan—Covers the synthesis, properties, and medical applications of biodegradable polymers.

### Primary Journals

1. "Enhanced Performance of the Bioartificial Pancreas Using Biodegradable Polymer Scaffolds" by J. H. Lee et al.—This article, published in Tissue Engineering, discusses the use of biodegradable polymer scaffolds to improve the function of bioartificial pancreases.
2. "Biodegradable Nanocomposites with Enhanced Cell Proliferation and Differentiation for Bone Tissue Engineering" by Q. Chen et al.—Published in Biomaterials, this research focuses on the development of biodegradable nanocomposites aimed at enhancing bone tissue regeneration.
3. "A Biodegradable Polymer as a Cytokine Delivery System for Inducing Bone Formation" by S. M. Richardson et al.—Found in Nature Biotechnology, this paper explores the use of biodegradable polymers for the controlled delivery of cytokines to stimulate bone formation.

Review Articles

1. "Biodegradable Polymers as Biomaterials" by Johnna S. Temenoff and Antonios G. Mikos—This review, published in Progress in Polymer Science, provides a comprehensive overview of biodegradable polymers used as biomaterials, covering their synthesis, degradation mechanisms, and clinical applications.
2. "Biodegradable Synthetic Polymers: Preparation, Functionalization and Biomedical Application" by Dirk Schmaljohann—Published in Advanced Drug Delivery Reviews, this article reviews the preparation methods, functionalization techniques, and the range of biomedical applications of biodegradable synthetic polymers.
3. "Recent Advances in Biodegradable Nanomaterials for Biomedical Applications" by X. Zhang et al.—Found in Biomaterials Science, this review discusses recent developments in biodegradable nanomaterials, focusing on their applications in drug delivery, tissue engineering, and as diagnostic tools.

## Chapter 4: Nanomaterials in Biomedicine

### Key Textbooks

1. "Nanotechnology for Biomedical Imaging and Diagnostics: From Nanoparticle Design to Clinical Applications" by Mikhail Y. Berezin, which explores the interface of nanotechnology and biomedical applications.

### Primary Journals

1. Nanomedicine: Nanotechnology, Biology, and Medicine: Offers a mix of research articles focused on clinical applications of nanotechnology.
2. International Journal of Nanomedicine: Provides research on nanoscale materials and their biomedical applications.

Review Article: "Nanomaterials in Medicine: New Trends and Prospects" by Li et al., Nano Today, 2020. A comprehensive review of the latest trends in nanomaterials for medical applications.

Research Article: "Targeted Nanoparticles for Cancer Therapy: A Review" by Pan et al., Biomaterials, 2019. This article explores the design and application of targeted nanoparticles in oncology.

## Chapter 5: Bioactive Polymers

### Key Textbooks

1. "Bioactive Polymers" by Vijay Kumar Thakur, focusing on polymer science for biomedical applications.

### Primary Journals

1. Biomacromolecules: Covers all aspects of macromolecular bio related materials, including bioactive polymers.
2. Progress in Polymer Science: Features reviews on the latest advancements in polymer chemistry and its various applications in biomedicine.

Review Article: "Bioactive Polymers and Their Applications in Drug Delivery" by Gupta and Kumar, Progress in Polymer Science, 2021. This review provides an overview of bioactive polymers used for controlled drug delivery.

Research Article: "Engineering Bioactive Self-Healing Antibacterial Exopolymers for Biomedical Applications" by Zhang et al., Science Advances, 2020. A novel study on bioactive polymers with self-healing and antibacterial properties.

## Chapter 6: Scaffold Design in Tissue Engineering

### Key Textbooks

1. "Principles of Tissue Engineering" by Lanza, Langer, and Vacanti, the definitive guide to tissue engineering, covering everything from basic principles to detailed engineering of tissues.

**Primary Journals**

1. Tissue Engineering Part A, B, and C: Series covering different aspects of tissue engineering, from fundamental research to clinical applications.

Review Article: "Scaffolds in Tissue Engineering Bone and Cartilage" by O'Brien, BMC Medicine, 2011. The article reviews the design, fabrication, and clinical application of scaffolds.

Research Article: "The Use of 3D Printing in the Fabrication of Tissue Engineered Bone Scaffolds" by Turner et al., Biomaterials, 2020. Discusses advancements in scaffold design using 3D printing technologies.

**Chapter 7: 3D Printing in Biomedicine**

**Key Textbooks**

1. "3D Printing in Medicine" by Deepak M. Kalaskar, which offers a practical guide to the medical applications of 3D printing technology.

**Primary Journals**

1. 3D Printing and Additive Manufacturing: Discusses the innovative approaches in 3D printing for the medical field.
2. Bio fabrication: Focuses on using additive manufacturing, bioprinting, and related techniques.

Review Article: "3D Printing Technologies for Drug Delivery: A Review" by Goyanes et al., Drug Discovery Today, 2016. Reviews the application of 3D printing in developing personalized medicine solutions.

Research Article: "Application of 3D Printed Porous Titanium Alloy in Orthopedic Implants" by Liu et al., Journal of Orthopaedic Surgery and Research, 2019. Explores the biomechanical properties and applications of 3D-printed materials in orthopedics.

**Chapter 8: Hydrogels in Drug Delivery**

**Key Textbooks**

1. "Hydrogel Micro and Nanoparticles" by L. Andrew Lyon and Michael J. Serpe, offering insight into the synthesis and application of hydrogel particles in medicine.

**Primary Journals**

1. Journal of Controlled Release: Focuses on the controlled release and delivery of drugs and other agents.
2. European Journal of Pharmaceutics and Biopharmaceutics: Discusses issues relevant to the pharmaceutical and biopharmaceutical aspects of technology.

Review Article: "Hydrogels for Hydrophobic Drug Delivery: Classification and Current Trends" by Agrawal and Gupta, European Journal of Pharmaceutics and Biopharmaceutics, 2017.

Research Article: "Synthesis and Characterization of pH-Sensitive Hydrogels for Colon Targeted Drug Delivery" by Singh et al., Journal of Controlled Release, 2018.

**Chapter 9: Biomimetic Materials**

**Key Textbooks**

1. "Biomimetics: Nature-Based Innovation" by Yoseph Bar-Cohen, which covers the fundamentals of mimicking nature to solve human problems, with a focus on materials.

**Primary Journals**

1. Bioinspiration & Biomimetics: Publishes research that looks to natural systems to solve technological problems.
2. Biomimetics: Focuses specifically on synthetic methods that replicate biological processes.

Review Article: "Biomimetic Materials and Design: Biointerfacial Strategies, Tissue Engineering, and Targeted Drug Delivery" by Sartor et al., Materials Science and Engineering: R: Reports, 2019.

Research Article: "Biomimetic Hydrogels for Controlled Biomolecule Delivery to Augment Bone Regeneration" by Lee et al., Advanced Healthcare Materials, 2019.

**Chapter 10: Theranostics: Combining Diagnosis and Therapy**

**Key Textbooks**

1. "Theranostics: An Emerging Strategy for Imaging and Therapy Applications" by Darryl Bornhop, exploring the combined use of diagnostic and therapeutic strategies.

**Primary Journals**

1. Theranostics: Offers articles that combine therapies with diagnostic tests to enhance treatment monitoring and patient care.
2. Advanced Drug Delivery Reviews: Reviews the latest trends in drug delivery technologies that can be applied to theranostics.

Review Article: "Theranostics: The Future of Personalized Medicine" by Kang et al., Journal of Clinical Medicine, 2017.

Research Article: "Nanoparticle-based Theranostic Agents" by Chen et al., Advanced Drug Delivery Reviews, 2020.

**Chapter 11 : Drug Delivery via Lipid-Based Nanocarriers**

**Key Textbooks**

1. "Lipid Nanocarriers in Cancer Diagnosis and Therapy" by Eliana B. Souto, detailing the use of lipid-based systems in the targeted delivery of therapeutic agents.

### Primary Journals

1. International Journal of Pharmaceutics: Features research on all aspects of pharmaceutics, including lipid nanocarriers.
2. Lipids in Health and Disease: Focuses on the role of lipids in health and their potential therapeutic applications.

Review Article: "Lipid-Based Nanocarriers for Drug Delivery and Diagnosis" by Mu et al., Advanced Drug Delivery Reviews, 2017.

Research Article: "Development and Characterization of Lipid Nanoparticles for Oral Delivery of Paclitaxel" by Sharma et al., Journal of Controlled Release, 2018.

Review Article: "Lipid-Based Nanocarriers for Drug Delivery and Diagnosis" by Mu et al., Advanced Drug Delivery Reviews, 2017.

Research Article: "Development and Characterization of Lipid Nanoparticles for Oral Delivery of Paclitaxel" by Sharma et al., Journal of Controlled Release, 2018.

## Chapter 12: Challenges and Future Directions

### Key Textbooks

1. "The Biomaterials: Silver Jubilee Compendium" by David Williams, which reflects on the past, present, and future of biomaterials science.

### Primary Journals

1. Advanced Healthcare Materials: Covers cutting-edge research in the development of new materials for healthcare applications.
2. Materials Today: Provides authoritative information on the latest advancements in materials science, including biomaterials.

Review Article: "The Future of Biomaterials: A Nexus of Compounding Challenges and Evolving Technology" by Hall et al., Acta Biomaterialia, 2021.

Research Article: "New Frontiers in Biomaterials for Overcoming Bacterial Resistance" by Kumar et al., Journal of Biomedical Materials Research Part A, 2019.

MIX
Papier aus verantwortungsvollen Quellen
Paper from responsible sources
FSC® C105338

If you have any concerns about our products,
you can contact us on
**ProductSafety@springernature.com**

In case Publisher is established outside the EU,
the EU authorized representative is:
**Springer Nature Customer Service Center GmbH**
**Europaplatz 3, 69115 Heidelberg, Germany**

Printed by Libri Plureos GmbH
in Hamburg, Germany